Smart Innovation, Systems and Technologies

Volume 365

The Smart Innovation, Systems and Technologies book series encompasses the topics of knowledge, intelligence, innovation and sustainability. The aim of the series is to make available a platform for the publication of books on all aspects of single and multi-disciplinary research on these themes in order to make the latest results available in a readily-accessible form. Volumes on interdisciplinary research combining two or more of these areas is particularly sought.

The series covers systems and paradigms that employ knowledge and intelligence in a broad sense. Its scope is systems having embedded knowledge and intelligence, which may be applied to the solution of world problems in industry, the environment and the community. It also focusses on the knowledge-transfer methodologies and innovation strategies employed to make this happen effectively. The combination of intelligent systems tools and a broad range of applications introduces a need for a synergy of disciplines from science, technology, business and the humanities. The series will include conference proceedings, edited collections, monographs, handbooks, reference books, and other relevant types of book in areas of science and technology where smart systems and technologies can offer innovative solutions.

High quality content is an essential feature for all book proposals accepted for the series. It is expected that editors of all accepted volumes will ensure that contributions are subjected to an appropriate level of reviewing process and adhere to KES quality principles.

Indexed by SCOPUS, EI Compendex, INSPEC, WTI Frankfurt eG, zbMATH, Japanese Science and Technology Agency (JST), SCImago, DBLP.

All books published in the series are submitted for consideration in Web of Science.

Kazumi Nakamatsu · Roumen Kountchev ·
Srikanta Patnaik · Jair M. Abe
Editors

Advanced Intelligent Technologies for Information and Communication

Proceedings of 3rd International Conference on Advanced Intelligent Technologies (ICAIT 2022)

Springer

Editors
Kazumi Nakamatsu
University of Hyogo
Kobe, Japan

Roumen Kountchev
Technical University of Sofia
Sofia, Bulgaria

Srikanta Patnaik
Interscience Institute of Management
and Technology
Interscience Research Network
Bhubaneswar, Odisha, India

Jair M. Abe
Paulista University
São Paulo, Brazil

ISSN 2190-3018 ISSN 2190-3026 (electronic)
Smart Innovation, Systems and Technologies
ISBN 978-981-99-5202-1 ISBN 978-981-99-5203-8 (eBook)
https://doi.org/10.1007/978-981-99-5203-8

This Springer imprint is published by the registered company Springer Nature Singapore Pte Ltd.
The registered company address is: 152 Beach Road, #21-01/04 Gateway East, Singapore 189721, Singapore

ICAIT-2022 Organization

Honorary Chair

Prof. Lakhmi C. Jain, KES International, UK

General Chair

Prof. Kazumi Nakamatsu, University of Hyogo, Japan

Conference Co-chairs

Prof. Jair M. Abe, Pauliata University, Brazil
Assoc. Prof. Ari Aharari, Sojo University, Japan

International Advisory Board Chair

Prof. Srikanta Patnaik, Interscience Institute of Management & Technology
Interscience Research Network, India

Program Chair

Prof. Roumen Kountchev, Technical University of Sofia, Bulgaria

Organizing Co-chairs

Prof. Xiaogang Qi, Xidian University, China
Prof. Xu Yuelin, The Sci.&Tech. Near-Surface Detection Laboratory, China
Dr. Shoulin Yin, Shenyang Normal University, China
Prof. Hang Li, Shenyang Normal University, China

ICAIT-2022 International Advisory Board and Program Committee

Srikanta Patnaik, India
Roumen Kountchev, Bulgaria
Jair M. Abe, Brazil
Ari Aharari, Japan
Nashwa El-Bendary, Egypt
Aboul Ela Hassanian, Egypt
Tulay Yildirim, Turkey
Altas Ismail, Turkey
Valentina E. Balas, Romania
Ajith Abraham, USA
Vincenzo Piuri, Italy
Morgado Dias, Portugal
Nicolae Paraschiv, Romania
Sachio Hirokawa, Japan
Siddhartha Bhattacharyya, India
Ladjel Bellatreche, France
Juan D. Velasquez Silva, Chile
Hossam A. Gaber, Canada
Ngoc Thanh Nguyen, Poland
Mario Divan, USA
Petia Koprinkova, Bulgaria
Jawad K. Ali, Iraq
Sufyan T. Faraj Al-Janabi, Iraq
Hercules A. Prado, Brazil
Lorna Uden, UK
Sunil Kumar Khatri, India
Mabrouk Omrani, USA
Ding Li Ya, Japan
Won Seok Yang, Japan
Marius Olteanu, Romania
Veska Georgieva, Bulgaria
Rouminia Kountchev, Bulgaria

Preface

The purpose of the international conference series AIT (Advanced Intelligent Technologies) is to provide an interdisciplinary platform for exchanging ideas and discussion on wider topics in terms of the most advanced intelligent technologies by academic researchers, graduate students who are studying theories and applications of intelligent technologies, and engineers and practitioners concerned with intelligent technologies applied to various fields.

The first international conference on AIT was launched as the 1st International Conference on Agriculture and IT/IoT/ICT (ICAIT-2019) in Xi'an, China, during November 22–24, 2019, and the topics of ICAIT-2019 were mainly on agriculture and intelligent network technologies. In order to cover much more interdisciplinary topics in terms of intelligent technologies, the second conference on AIT has been held as the 2nd International Conference on Advanced Intelligent Technologies (ICAIT-2021) subtitled Intelligent Technologies for Industries. ICAIT-2021 was originally supposed to be held in Xi'an, China during October 23–25, 2021, but the world was fighting against COVID-19 pandemic, and there was no doubt that the safety and well-being of our participants are most important. Considering the health and safety of everyone, we had to make a tough decision and convert ICAIT-2021 into fully online conference via the Internet. Both the proceedings of ICAIT-2019 and ICAIT-2021 have been published in the same book series, Smart Innovation, Systems and Technologies by Springer. As ICAIT-2019 and ICAIT-2021, the 3rd International Conference on Advanced Intelligent Technologies (ICAIT-2022) was held as a fully online conference via the Internet during December 29–31, 2022.

The topics of ICAIT-2022 include all aspects of intelligent technologies applicable to various fields as follows:

Artificial Intelligence, Artificial Life, Automated Reasoning
Automated Manufacturing, Automatic Car Driving, Autonomous Robot,
Bayesian Networks, Bee-colony Optimization, Bio-inspired Systems,
Business Process Modelling, Cloud Computing, Clustering,
Communication Network Systems, Complex Systems, Computational
Intelligence,

Data Analysis, Data Mining, Data Processing Systems, Deep Learning,
Distributed Systems, E-business, E-commerce, E-learning,
Embedded Systems, Emotion Recognition, Energy Technologies,
Enterprise Network Management, Expert Systems, Evolutionary Computing,
Factory Automation, Farm Management, Forecasting Systems, Fuzzy Computing,
Fuzzy Controller, Fuzzy Logic, Fuzzy Set, Fuzzy Systems,
Genetic Algorithm, Hardware Design, Hardware Emulation, Heuristics,
High Performance Computing, Hybrid Intelligent Systems,
Image Processing, Information Security, Intelligent Agent Technologies,
Intelligent Control Systems, Intelligent Information Systems,
Intelligent Logistics for Industry, Intelligent Monitoring Systems,
Intelligent Network Routing, Intelligent Numerical Control,
Ontology Techniques, Process Control, Intelligent Production Systems,
Intelligent Safety Verification, Intelligent Teaching Systems,
Intelligent Transportation Systems, Internet of Things, Internet Security,
Machine Learning, Mathematical Foundation of Intelligence,
Multi-Agent Systems, Natural Language Processing, Neural Networks,
Neuro Computing, Optimization Techniques, Pattern Recognition,
Quantum Computing, Risk Management Systems, Signal Processing,
Virtual Reality, etc.

One invited and 37 regular papers among submitted 58 papers from China and Brazil were accepted at ICAIT-2022. This volume is devoted to presenting all those accepted papers of ICAIT 2022. The contents of the volume include so many interdisciplinary variations in the field of Theory and Application of Intelligent Technology. Some chapters present applications of intelligent techniques to civil and mechanical engineering, some other chapters deal with big data techniques in the field of tourism.

Lastly, we wish to express our sincere appreciation to all individuals and the program committee for their review of all the submissions, which is vital to the success of ICAIT-2022, and also to the members of the organizing committee who had dedicated their time and efforts in planning, promoting, organizing, and helping the conference. Special appreciation is extended to ICAIT-2022 Honorary Chair: Prof. Dr. Lakhmi C. Jain, KES International, UK, who gave us great support to publish this volume, and keynote speakers, Prof. Jair M. Abe, Paulista University, Sao Paulo, Brazil and Prof. Adi Baykasoglu, Dokuz Eylul University, Turkey who made very beneficial speeches for the conference audience titled Towards Paraconsistent Engineering, Data-Driven Dynamic Multi-Criteria Decision Making via Learning of Fuzzy Cognitive Maps, respectively. Prof. Jair M. Abe also contributed one invited paper together with his student titled: VSM 4.0 (Automated Value Stream Mapping): A Digital Transformation from Lean in the Context of I4.0.

Kobe, Japan
Sofia, Bulgaria
Bhubaneswar, India
Sao Paulo, Brazil
December 2022

Kazumi Nakamatsu
Roumen Kountchev
Srikanta Patnaik
Jair M. Abe

Contents

About the Editors

Dr. Kazumi Nakamatsu received the Ms. Eng. and Dr. Sci. from Shizuoka University and Kyushu University, Japan, respectively. His research interests encompass various kinds of logic and their applications to Computer Science, especially paraconsistent annotated logic programs and their applications. He has developed some paraconsistent annotated logic programs called Annotated Logic Program with Strong Negation (ALPSN), Vector ALPSN (VALPSN), Extended VALPSN (EVALPSN) and before-after EVALPSN (bf-EVALPSN) recently and applied them to various intelligent systems such as a safety verification-based railway interlocking control system and process order control. He is an author of over 180 papers and 30 book chapters and 15 edited books published by prominent publishers. Kazumi Nakamatsu has chaired various international conferences, workshops and invited sessions, and he has been a member of numerous international program committees of workshops and conferences in the area of Computer Science. He has served as the editor-in-chief of the International Journal of Reasoning-based Intelligent Systems (IJRIS), and he is now the founding editor of IJRIS, and an editorial board member of many international journals. He has contributed numerous invited lectures at international workshops, conferences and academic organizations. He also is a recipient of numerous research paper awards.

Prof. Roumen Kountchev Ph.D., D.Sc. is a professor at the Faculty of Telecommunications, Department of Radio Communications and Video Technologies, Technical University of Sofia, Bulgaria. Areas of interest: digital signal and image processing, image compression, multimedia watermarking, video communications, pattern recognition and neural networks. Prof. Kountchev has 350 papers published in magazines and proceedings of conferences; 20 books; 47 book chapters; and 21 patents. He had been the principal investigator of 38 research projects. At present, he is a member of Euro Mediterranean Academy of Arts and Sciences and the president of Bulgarian Association for Pattern Recognition (a member of Intern. Association for Pattern Recognition). He is the editorial board member of: Intern. J. of Reasoning-based Intelligent Systems; Intern. J. Broad Research in Artificial Intelligence and Neuroscience; KES Focus Group on Intelligent Decision

Technologies; Egyptian Computer Science J.; Intern. J. of Bio-Medical Informatics and e-Health; and Intern. J. Intelligent Decision Technologies.

Dr. Srikanta Patnaik is presently working as the director of International Relation and Publication of SOA University. He is a full professor at the Department of Computer Science and Engineering, SOA University, Bhubaneswar, India. He has received his Ph.D. (Engineering) in Computational Intelligence from Jadavpur University, India, in 1999. He has supervised more than 25 Ph.D. theses and 60 master's theses in the area of Computational Intelligence, Machine Learning, Soft Computing Applications and Re-Engineering. Dr. Patnaik has published around 100 research papers in international journals and conference proceedings. He is the author of 2 text books and 52 edited volumes and few invited book chapters, published by leading international publisher like Springer-Verlag, Kluwer Academic, etc.

He is the editor-in-chief of International Journal of Information and Communication Technology and International Journal of Computational Vision and Robotics published from Inderscience Publishing House, England, and International Journal of Computational Intelligence in Control, published by MUK Publication, the editor of Journal of Information and Communication Convergence Engineering and the associate editor of Journal of Intelligent and Fuzzy Systems (JIFS), which are all Scopus Index journals. He is also the editor-in-chief of Book Series on "Modeling and Optimization in Science and Technology" published from Springer, Germany, and Advances in Computer and Electrical Engineering (ACEE) and Advances in Medical Technologies and Clinical Practice (AMTCP), published by IGI Global, USA. Dr. Patnaik has traveled more than 20 countries across the globe to deliver invited talks and keynote address at various places. He is also a visiting professor at some of the universities in China, South Korea and Malaysia.

Dr. Jair M. Abe received B.A. and M.Sc. in Pure Mathematics from the University of Sao Paulo, Brazil, and also received the doctorate degree and Livre-Docente title from the same University. He was the coordinator of Logic Area of Institute of Advanced Studies at the University of Sao Paulo, Brazil, in 1987–2019 and the full professor at Paulista University, Brazil. His research interest topics include Paraconsistent Annotated Logics and AI, ANN in Biomedicine and Automation, among others. He is the senior member of IEEE. Professor Abe is a studious of a family of Paraconsistent Annotated Logic which is used to solve many complex problems in engineering. He has authored/edited books on paraconsistent and related logic published by Springer, Germany, and other reputed publishers. He is the recipient of many awards including medals for his academic performance and also received many best papers awards. Professor Abe is the editor-in-chief of International Journal of Reasoning-based Intelligent Systems. Presently, Professor Abe serves as an associate editor and a member of the Editorial Board of some journals related to the intelligent systems and applications. Professor Abe has supervised a number of Ph.D. candidates successfully and presented a number of keynote addresses. He has authored/co-authored around 300+ publications including books, research papers, research reports, etc. Professor Abe's research interests include System Design using Conventional and Artificial Intelligence Techniques, Paraconsistent Annotated Logic, Human factors in Aviation, Intelligent Decision Making, Teaching &Learning practices and Cognitive Studies.

Invited Paper

VSM 4.0 (Automated Value Stream Mapping): A Digital Transformation from Lean in the Context of I4.0

Nilton Cesar França Teles and Jair M. Abe

Abstract Many companies widely adopt value stream mapping (VSM) as a lean manufacturing method. VSM reduces production process waste by mapping, analysing and shaping value streams. Despite its potential, VSM does not cover the promising new opportunities of Industry 4.0 (I4.0) digitisation to consolidate information flows, reduce processing times and increase flexibility and productivity in complex environments, among other related topics. As a result, companies began to adopt VSM 4.0, which combines the traditional hand-structured VSM with the scanning technology of I4.0. This article aims to identify the benefits obtained by companies that adopt VSM 4.0.

1 Introduction

To survive in the market, companies seek to be more efficient, productive and with quality in their processes. Toyota's production system, Lean Manufacturing, has been used for years to reduce waste and increase customer value [1, 2]. Among the methods preferred by professionals in the implementation of Lean Manufacturing (LM), a tool called Value Stream Map (VSM) stands out due to the level of detail of the process that it can represent [4]. However, manufacturing companies face volatile market changes, shorter product life cycles and increased production complexity [5].

Digitisation is already a necessary reality for process improvement, for example. With the advent of Industry 4.0 (I4.0), a more digital process environment is possible, where companies can better take advantage of opportunities for continuous improvement [6]. Cyber-physical systems (CPS) are technologies widely used in the context of I4.0, which contribute to improving communication and dynamically transferring information [7] in production processes. Wireless networks, sensors and cloud

N. C. F. Teles · J. M. Abe (✉)
Paulista University, São Paulo, Brazil
e-mail: jairabe@uol.com.br

N. C. F. Teles
e-mail: nilton.teles@aluno.unip.br

K. Nakamatsu et al. (eds.), *Advanced Intelligent Technologies for Information and Communication*, Smart Innovation, Systems and Technologies 365,
https://doi.org/10.1007/978-981-99-5203-8_1

computing technologies also offer competitive advantages to companies [9]. These technologies, when combined with the LM approach, increase the ability of companies to deliver more excellent value to customers [10], because they help to reduce high inventory levels, improve transportation, reduce labour movement and obtain better delivery times. Continuous improvement is a vital feature of LM.

Traditional VSM is a hand-generated multi-purpose tool focussing on overall process productivity and showing consistent production flows [12]. However, it is questionable whether traditional VSM approaches can still significantly improve the efficiency and productivity of lean manufacturing systems without the proposed I4.0 technologies [13] as process digitisation offers many opportunities to improve business outcomes [14]. In this context, the need arises to develop a more robust VSM, which, although well applied in the manufacturing industry, has obvious drawbacks such as lack of flexibility in scenarios with high volatility, viewing only one product or a series of products, difficulties when using it in more complex processes, custom production systems and lack of visualisation of equipment efficiency, among others [12–16]. Thus, this article aims to identify the benefits of traditional VSM with the increment of I4.0 called VSM 4.0.

2 Methodology

A review of the literature served as support to identify the state of the art of the current VSM. The following research question guided this work: "What changes can be made in the current VSM to dynamically meet production needs (Q)?". Therefore, the paper is composed as follows: Sect. 3.1 deals with the procedure for starting a bibliographic search. Thus, Sect. 3.2 reviews and highlights the advantages and disadvantages of traditional VSM. Section 3.3 reviews the digital integration of lean methods. Figure 1 summarises the systematic structure of this article [6].

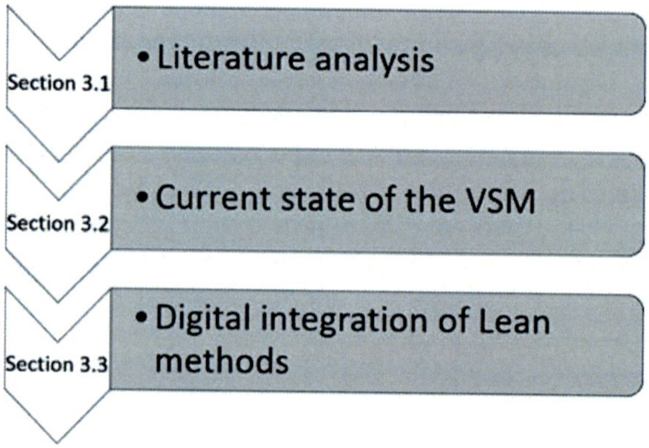

Fig. 1 Steps of the systematic structure. *Source* Authors

3 Literature Review

3.1 *Literature Review*

The steps proposed by [17] were followed in the literature analysis in search of the state of the art of VSM and digitalised approaches of Lean methods. The Web of Science and Scopus databases were chosen for their reliability recognised by the academic community and for their content relevant to the topic presented. The search for scientific articles was limited to 2007–2022, as it is a current topic [6].

3.2 *Current Status of the VSM*

When searching for scholarly articles, use the keyword "value stream", and this term should appear first in the title. In the Web of Science, 67 publications were identified, while in Scopus, 68 publications match the research proposal. Duplicate articles were removed, leaving 95 titles. After reading the titles and abstracts, 28 publications were considered suitable. Preliminary results show that conventional VSM helps reduce waste and contributes to qualitative and quantitative process improvements, as shown in [4].

In [18], the quality and efficiency of production are improved for [19–21] inventory and lead time reduction, as well as better production performance. In the study by [22], VSM provides better team communication. Some shortcomings are also noted. For example, [23, 25] show the problems when using VSM in complex manufacturing systems. In the study by [24], the process lacks detail, and the dynamic factors are neglected, leading to inefficient results. For [26], the traditional VSM is unsuitable for pushed systems and [27] shows the problem with the range of product options.

On the other hand, [28] points out the static representation of the method as a significant disadvantage. The works of [29, 30] suggest that poorly trained professionals can lead to errors in collecting production data. Finally, [31] points to the gap between the theory and practical application of VSM. Table 1 presents the systematic literature review results and identifies the advantages and disadvantages of the VSM method [6].

3.3 *Digital Integration of Lean Methods*

The study by [33] describes that Lean approaches were successful in the past but are no longer sufficient to solve companies' current problems. The authors suggest automation in lean manufacturing and combining lean methods with I4.0 technologies. For [34], Lean is limited in complex manufacturing processes and states that automation with CPS is essential for companies to achieve greater agility.

Table 1 Traditional VSM, advantages and disadvantages. *Source* Authors

Author	Benefits	Disadvantages	Comments
[4]	x		Waste and the potential for improvement
[18]	x		Quality and increased efficiency
[19, 20]	x		Reduction in throughput parameters, cycle times and inventories
[21]	x		Improved delivery performance and increased value-added component
[22]	x		Means of communication: management and employees
[32]	x		It was possible to improve inventories by 70.2%, production times by 56.2% and cycle times by 52.6% through the analysis of 91 publications
[23]		x	Show difficulties in a complex production environment
[24]		x	Dynamic influence parameters would not be objects of consideration
[25]		x	Inappropriate methodology for complex value streams
[26]		x	Adequacy of the methodology only for purely pull systems
[27]		x	Identify difficulties among a wide range of variants
[28]		x	Considering the static picture of the method is a weakness
[29, 30]		x	Obtaining production data and poorly trained people
[31]		x	Establishing a gap between VSM theory and the use of the method in practice

In [35], the relationship between the production system and other subsystems, such as maintenance, quality management and planning, is discussed and suggests the integration of these areas. For [36–38], the combination of Lean and I4.0 is necessary to improve the use of the concepts of these two methods, reduce the complexity of production communication and improve process control. In [39, 40], the use of digital solutions between suppliers, operations and customers is proposed. Table 2 presents the results of the systematic literature review [6].

4 Results and Discussion

The results show that traditional VSM in companies is still little researched. Those who use it can optimise processes qualitatively and quantitatively within limits provided by this tool. Scanning in combination with VSM provides even better results [27]. However, this combination is still not widely explored by companies in practice.

For [36], introducing I4.0 technologies further supports lean principles, although I4.0 is not yet mature [38]. These two approaches have common goals of increasing productivity and flexibility; however, Lean proposes to reduce complexity through

Table 2 Lean approach in the context of I4.0. *Source* Authors

Author	Comments
[33]	The proposition of Lean automation in the face of being outdated with today's challenges despite Lean being widely distributed and successful
[34]	Preference for a lean automation approach, development of an intelligent Jidoka system and decentralisation based on CPS with a commitment to greater flexibility
[35]	Modelling lean implementation with a systems approach, intertwining production with maintenance, quality, planning and control subsystems
[36]	Lean methods combined with Industry 4.0 with real-time data allow companies to increase production speed, reduce waste and inventory and improve one-piece flow
[37]	Digital innovation and applications in innovation management
[38]	Lean reduces complexity, while Industry 4.0 controls complexity with IT solutions. Combining the two concepts indicates a better proposal
[39]	Digital solutions for suppliers, customers, processes and control
[40]	Consideration of Lean and Industry 4.0 tools in the industrial sector, without a clear connection between the two proposals
[41]	Enterprise VSM digitisation is cost-effective downstream with short-term continuous improvement or long-term optimisation projects
[42]	Reduction of total internal lead time from 6.5 h to 15 min with VSM 4.0 due to implementing an online configurator and direct connection of almost all information flows
[43]	Workers reduced search time from 7.2 min/basket to less than 2 min/basket in real-time RFID system data tracking in a case study at an apparel company
[6]	A holistic treatment that includes simulation, real-time data integration and consistency between systems is needed

ongoing standardisation, while I4.0 proposes to reduce complexity with technological increments without worrying about standardisation. This factor is paramount, showing how much one approach complements the other.

Concerning the VSM tool, [39] points to the need for more research on integrating I4.0 technologies. In the work of [29], for example, he highlights the advantages and disadvantages of the method, the transparency of information and material flows, and the fragility in complex environments.

Another crucial factor is the static view that the VSM presents, and the authors consider this its greatest weakness. Therefore, the traditional value stream mapping method must migrate to digitised models as LM does not contradict I4.0. For example, real-time data can be collected with RFID systems, so the value stream, bottlenecks and process improvements can be dynamically monitored.

This new model, called VSM 4.0, can provide even more stable results than the previous model [41]. For example, in the study [42] using VSM 4.0, the lead time was reduced from 6.5 h to 15 min. In another study, the material location was reduced from 7.2 min to 2 min [43]. Finally, integrating real-time data in the day-to-day activities of companies also require changes in management and adjustments in organisational structures to receive the dynamic flow of information for decision-making [6].

5 Final Considerations

From the study presented, it is possible to point out some answers from the investigation: "What changes can be made in the current VSM to meet production needs (Q) dynamically?". Integrating RFID systems and CPS technologies can support VSM development for fast and dynamic detection of process changes. With the advent of I4.0 technologies, it is possible to reduce this tool's fragility significantly. On the other hand, companies now have new challenges: How to make fast and intelligent decisions with VSM 4.0 updating process information in real time?

Acknowledgements This study was supported partially by the Coordenação de Aperfeiçoamento de Pessoal de Nível Superior-Brazil (CAPES)-Finance Code 001 Number Process 88887.663537/2022-00.

References

1. Ghosh, M.: Lean manufacturing performance in Indian manufacturing plants. J. Manuf. Technol. Manag. **24**(1), 113–122 (2013)
2. Womack, J.P., Jones, D.T.: Lean thinking: banish waste and create wealth in your corporation. J. Oper. Res. Soc. **48**(11), 1148–1148 (1997)
3. Bhasin, S.: Performance of Lean in large organisations. J. Manuf. Syst. **31**(3), 349–357 (2012)
4. Grewal, C.: An initiative to implement lean manufacturing using value stream mapping in a small company. Int. J. Manuf. Technol. Manag. **15**(3–4), 404–417 (2008)
5. Adolph, S., Tisch, M., Metternich.J.: Challenges and approaches to competency development for future production. J. Int. Sci. Publications–Educational Altern. **12**(1), 1001–1010 (2014)
6. A. Lugert, A. Batz, H. Winkler.: Empirical assessment of the future adequacy of value stream mapping in manufacturing industries. J. Manuf. Technol. Manag. **29**(5), 886–906 (2018)
7. Zheng, P., Wang, H, Sang, Z., Zhong, R.Y., Liu, Y., Liu, C., Mubarok, K., Yu, S., Xun Xu, X.: Smart manufacturing systems for Industry 4.0: Conceptual framework, scenarios, and future perspectives. Front. Mech. Eng. **2**(13), 137–150 (2018)
8. Weyer, S., Schmitt, M., Ohmer,M., Gorecky, D.: Towards Industry 4.0—Standardization as the crucial challenge for highly modular, multi-vendor production systems. IFAC-Pap. **3**(48), pp.579–584. Elsevier, Kaiserslautern, Germany (2015)
9. Ahuett-Garza, H., Turfess, T.: A brief discussion on the trends of habilitating technologies for Industry 4.0 and Smart manufacturing. Manuf. Lett. (15), 60–63 (2018)
10. Mayra, A., Weigelta, M., Kühla, A., Grimmb, S., Erlla, A., Potzela, M., Frankea, J.: Lean 4.0: A conceptual conjunction of lean management and Industry 4.0. Procedia CIRP. **72**, 622–628 (2018)
11. Rother, M., Shook, J.: Learning to see: value stream mapping to add value and eliminate muda. Version 1.3.,1st edn. Productivity Press, Cambridge (2003)
12. Huang, Z., Kim, J., Sadri, A., Dowey, S., Dargusch, M.S.: Industry 4.0: Development of a multi-agent system for dynamic value stream mapping in SMEs. J. Manuf. Syst. (52), 1–12 (2019)
13. Kolberg, D., Zühlke, D.: Lean automation enabled by industry 4.0 technologies. IFAC-Pap. **48**(3), 1870–1875 (2015)
14. Bennett, D.: Future challenges for manufacturing. J. Manuf. Technol. Manag. **25**(1), 2–6 (2014)

15. Schoenemann, M., Kurle, D., Herrmann, C., Thiede, S.: Multi-product EVSM simulation. In: Research and Innovation. Manufacturing: key enabling technologies for the factories of the future, pp. 334–339. Proceedings of the 48th cirp conference on manufacturing systems, Braunschweig (2016)
16. Khaswala, Z.N., Irani, S.A.: Value network mapping (VNM): visualisation and analysis of multiple flows in value stream maps. In: Proceedings of the lean management solutions conference, pp. 1011. Department of Industrial, Welding and Systems Engineering the Ohio State University, St. Louis (2001)
17. Fink, A.: Conducting research literature reviews: From the internet to paper, 5ª Sage publications, UCLA, Los Angeles (2019)
18. Lacerda, A.P., Xambre, A.R., Alvelos, H.M.: Applying value stream mapping to eliminate waste: a case study of an original equipment manufacturer for the automotive industry. Int. J. Prod. Res. **54**(6), 1708–1720 (2016)
19. Rekha, R.S., Periyasamy, P., Nallusamy, S.: An optimised model for reduction of cycle time using value stream mapping in a small scale industry. Int. J. Eng. Res. Afr. **27**, 179–189 (2016)
20. Seth, et al.: Application of value stream mapping (VSM) for minimisation of wastes in the processing side of supply chain of cottonseed oil industry in Indian context. J. Manuf. Technol. Manag. **19**(4), 529–550 (2008)
21. Vinodh, S., Selvaraj, T., Chintha, S.K., Vimal, K.E.K.: Development of value stream map for an Indian automotive components manufacturing organisation. J. Eng. Des. Technol. **13**(3), 380–399 (2015)
22. Singh, B., Sharma, S.K.: Value stream mapping as a versatile tool for lean implementation: an Indian case study of a manufacturing firm. Meas. Bus. Excell. **13**(3), 58–68 (2009)
23. Seth, D., Nitin, S., Pratik, D.: Application of value stream mapping (VSM) for lean and cycle time reduction in complex production environments: a case study. Prod. Plan. & Control. **28**(5), 398–419 (2017)
24. Schmidtke, D., Heiser, U., Hinrichsen, O.: A simulation-enhanced value stream mapping approach for optimisation of complex production environments. Int. J. Prod. Res. **52**(20), 6146–6160 (2014)
25. Berndt, R., Sunk, A.: Value stream mapping with vasco-from reducing lead time to sustainable production management. Ercim News **41**, 26–27 (2016)
26. Bertolini, M., Braglia, M., Romagnoli, G., Zammori, F.: Extending value stream mapping: the synchro-MRP case. Int. J. Prod. Res. **51**(18), 5499–5519 (2013)
27. Braglia, M., Frosolini, M., Zammori, F.: Uncertainty in value stream mapping analysis. Int. J. logistics—Research Appl. **12**(6), 435–453 (2009)
28. Lian, Y.H., Van Landeghem, H.: Analysing the effects of Lean manufacturing using a value stream mapping-based simulation generator. Int. J. Prod. Res. **45**(13), 3037–3058 (2007)
29. Forno, A.J.D., Pereira, F.A., Forcellini, F.A., Kipper, L.M.: Value Stream Mapping: a study about the problems and challenges found in the literature from the past 15 years about application of Lean tools. Int. J. Adv. Manuf. Technol. **72**, 779–790 (2014)
30. Lasa, I.S., Laburu, C.O., Castro Vila, R.: An evaluation of the value stream mapping tool. Bus. Process. Manag. J. **14**(1), 39–52 (2008)
31. Serrano Lasa, I., Castro, R., Laburu, C.O.: Extent of the use of Lean concepts proposed for a value stream mapping application. Prod. Plan. & Control. **20**(1), 82–98 (2009)
32. Shou, W., Wang, J., Wu, P., Wang, X., Chong, H.Y.: A cross-sector review on the use of value stream mapping. Int. J. Prod. Res. **55**(13), 3906–3928 (2017)
33. Kolberg, D., Knobloch, J., Zühlke, D.: Towards a lean automation interface for workstations. Int. J. Prod. Res. **55**(10), 2845–2856 (2017)
34. Ma, J., Wang, Q., Zhao, Z.: SLAE–CPS: Smart lean automation engine enabled by cyber-physical systems technologies. Sensors **17**(7), 1500–1500 (2017)
35. Mahmood, A., Montagna, F.: Making lean smart by using system-of-systems' approach. IEEE Syst. J. **7**(4), 537–548 (2013)
36. Netland, T.: Industry 4.0: where does it leave lean. Lean Manag. J., (5), 22–23 (2015).

37. Nicoletti, B.: Optimising innovation with the lean and digitise innovation process. Technol. Innov. Manag. Rev. **5**(3), 29–38 (2015)
38. Rüttimann, B.G., Stöckli, M.T.: Lean and Industry 4.0: twins, partners, or contenders? A due clarification regarding the supposed clash of two production systems. J. Serv. Sci. Manag. **9**(6), 485–500 (2016)
39. Sanders, A., Elangeswaran, C., Wulfsberg, J.P.: Industry 4.0 implies lean manufacturing: Research activities in industry 4.0 function as enablers for lean manufacturing. J. Ind. Eng. Manag. **9**(3), 811–833 (2016)
40. Veza, I., Mladineo, M., Gjeldum, N.: Selection of the basic lean tools for development of croatian model of innovative smart enterprise. Tehnički vjesnik **23**(5), 1317–1324 (2016)
41. Meudt, T., Metternich, J., Abele, E..: Value stream mapping 4.0: Holistic examination of value stream and information logistics in production. CIRP Ann.-Manuf. Technol. (66), 413–416 (2017)
42. Hartmann, L., Meudt, T., Seifermann, S., Metternich, J.: Value stream method 4.0: holistic method to analyse and design value streams in the digital age. In: 6th cirp global web conference—envisaging the future manufacturing, design, technologies and systems in innovation era, pp. 249–254. Procedia CIRP, Darmstadt (2018)
43. Phuong, N.A., Guidat, T.: Sustainable value stream mapping and technologies of Industry 4.0 in manufacturing process reconfiguration: A case study in an apparel company. In: Phuong, N.A., Guidat, T. (eds) IEEE International conference on service operations and logistics and informatics, pp. 85–90. Institute of Electrical and Electronics Engineers, Singapore (2018)

Regular Papers

Passive Detection Beamforming Based on Arc Array

MengLei Li, YueLin Xu, ShiRui Wang, and YiQing Li

Abstract Linear arrays are widely used in underwater acoustic detection equipment, but they often have the problem of port and starboard ambiguity, and the azimuth resolution will decrease rapidly with the increase of the deviation from the natural pointing angle. In order to solve this problem, the conventional beamforming method is adopted in this paper, and the beamforming performance of non-linear arrays is simulated and studied. The simulation results show that the ring array can have consistent beam response and azimuth resolution in any direction. Further considering that the actual application scenario of the array is the shore, the actual required detection range is from −90° to 90°. In order to avoid the waste of array elements, the beamforming performance of the arc array is studied. The simulation results show that for a semi-circular arc array with a beam radius product of 21π, the azimuth resolution can be maintained between 2° and 4° in the range of −70° to 70°, which is beneficial to the azimuth estimation of small targets and meets the requirements of usage requirements.

1 Introduction

Linear array is widely used in underwater acoustic detection equipment such as towed array sonar and shore-based sonar, but its use often has the problem of port and starboard ambiguity. There are two technical ways to solve this problem, one is to use vector transducers and the other is to use non-linear array methods, such as ring arrays, arc arrays, and so on. In this paper, the beamforming performance of the arc array is studied and simulated. The simulation results show that the arc array can solve the problem of port and starboard ambiguity very well [1, 2].

M. Li · S. Wang · Y. Li
CSSC Ocean Exploration Technology Institude Co., Ltd, Wuxi, Jiangsu, China

M. Li · Y. Xu (✉)
Science and Technology On Near-Surface Detection Laboratory, Wuxi, Jiangsu, China
e-mail: 1140860425@qq.com

K. Nakamatsu et al. (eds.), *Advanced Intelligent Technologies for Information and Communication*, Smart Innovation, Systems and Technologies 365,
https://doi.org/10.1007/978-981-99-5203-8_2

2 The Principle of Passive Directional Beamforming

The basic idea of the conventional beamforming algorithm is to artificially compensate the signals propagating to each basic array element according to the relative sound path difference, so as to synchronize the compensated signals, and then perform summation processing, so that the output of the array can have the largest response to the signal incident in the specified direction, and it can be considered that there is a target signal source in this direction. In practical engineering applications, conventional beamforming algorithms have the following advantages:

(1) Less computation, easy to implement in real time.
(2) Applies to both wideband and narrowband signals, coherent and incoherent signals.
(3) Spatial filtering is compatible with bearing estimation.
(4) Has the highest robustness, i.e., very little sensitivity to errors.

However, the azimuth resolution of the conventional beamforming algorithm is low, limited by the Rayleigh limit, and it cannot distinguish two signal sources that are very close and located in the same beam width, so it is not suitable for the multi-signal source situation [3].

For CW signals, narrowband beam scanning can be used to estimate the azimuth spectrum directly. A cross array of M array elements is set up. The coordinates of the array elements are (x_m, y_m, z_m), and the far-field narrowband signal whose center frequency is f_0 is from (θ_0, ϕ_0). If the azimuth is incident on the array, the signal received by the array is

$$\boldsymbol{x}(t) = \boldsymbol{p}(\theta_0, \phi_0)s(t) + \boldsymbol{n}(t) \tag{1}$$

In the formula:

$\boldsymbol{p}(\theta_0, \phi_0)$—$\boldsymbol{p}(\theta_0, \phi_0) = e^{-j2\pi f_0 \boldsymbol{\tau}(\theta_0, \phi_0)}$, $M \times 1$-dimensional beam response vector. Calculate according to formula (2), where $\boldsymbol{\tau}(\theta_0, \phi_0)$ is the time delay vector relative to the reference array element, when τ is a negative number, it means that the received signal of the array element arrives earlier than the reference point signal.

$s(t)$—signal source.

$\boldsymbol{n}(t)$—$M \times 1$-dimensional noise vector.

Then, the beam scanning method is used to estimate the azimuth spectrum. Assuming that the observation direction of the beam scanning is (θ, ϕ), the corresponding conventional beamformer weighting vector is

$$\mathbf{w}(\theta, \phi) = \frac{\boldsymbol{p}(\theta, \phi)}{M} \tag{2}$$

The corresponding beam output power is

$$\boldsymbol{P}(\theta, \phi) = \mathbf{w}^{\boldsymbol{H}}(\theta, \phi)\boldsymbol{R}_{\boldsymbol{x}}\mathbf{w}(\theta, \phi), (\theta, \phi) \in \Omega \tag{3}$$

$$\boldsymbol{R}_x = E\left[\boldsymbol{x}\boldsymbol{x}^H\right] = \sigma_s^2 \boldsymbol{p}(\theta,\phi)\boldsymbol{p}^H(\theta,\phi) + \sigma_n^2 \boldsymbol{\rho}_n \tag{4}$$

In the formula:

Ω—The range of viewing azimuths of interest or viewing area.

σ_s^2—The power of the signal $s(t)$.

σ_n^2—Noise power.

$\boldsymbol{\rho}_n$—White Gaussian noise normalized cross-spectral matrix. Because the noise received by each point in space is not correlated with each other, $\boldsymbol{\rho}_n$ is equal to the identity matrix.

$\boldsymbol{R}_x$—Data covariance matrix. Equation (5) is the ideal data covariance matrix, but in practical applications, since the ideal data covariance matrix cannot be obtained, the data sample covariance matrix $\widehat{\boldsymbol{R}}_x$ is often used instead, that is, a snapshot sample is used for estimation:

$$\widehat{\boldsymbol{R}}_x = \frac{1}{L}\sum_{n=1}^{L}\left[\boldsymbol{x}(n)\boldsymbol{x}^H(n)\right] \tag{5}$$

where L is the length of the data sample.

For the LFM signal, the reception time series of the mth element of the array can be expressed as

$$x_m(t) = s[t - \tau_m(\theta_0,\phi_0)] + n_m(t) m = 1,\cdots,M \tag{6}$$

Take the received data of appropriate length and perform discrete Fourier transform of K points, then one M × 1-dimensional matrix snapshot complex vector can be obtained on each sub-band:

$$\boldsymbol{X}(f_k) = \boldsymbol{p}(f_k,\theta_0,\phi_0)S(f_k) + \boldsymbol{N}(f_k), k = 1,\cdots,K \tag{7}$$

In the formula:

f_k—The kth frequency in the working frequency band.

$p(f_k,\theta_0,\phi_0) - M \times 1$-dimensional frequency domain matrix response vector.

$S(f_k)$—Signal spectrum.

$N(f_k)$—M × 1-dimensional frequency domain noise vector.

Then, by analogy with the narrowband beamforming algorithm, the data covariance matrix corresponding to each frequency sub-band f_k is calculated in the frequency domain, and the beam observation direction (θ,ϕ) is changed in the observation view area Ω of interest, and the beam scanning azimuth spectrum $P(f_k,\theta,\phi)$ is calculated, and finally average the azimuth spectrum of each frequency to obtain the broadband azimuth spectrum:

$$\boldsymbol{P}_{brd}(\boldsymbol{\theta},\boldsymbol{\phi}) = \frac{1}{\boldsymbol{K}}\sum_{\boldsymbol{k}=1}^{\boldsymbol{K}}\boldsymbol{P}\left(\boldsymbol{f}_k,\boldsymbol{\theta},\boldsymbol{\phi}\right) \tag{8}$$

3 Array Design

Since we only care about the target azimuth information in the horizontal direction, the linear array can be used to meet the target detection in the plane range. The following will analyze the beamforming of the uniform linear array, ring array, and arc array, respectively.

3.1 Uniform Linear Array

Consider first a uniform linear array. Assume that the number of array elements of the uniform linear array is M_1, the spacing of the array elements is d_1, and the array is arranged along the y-axis. The schematic diagram of the layout is shown in Fig. 1.

Set the array element spacing to the half wavelength corresponding to 20 kHz, and the number of array elements to 54. At this time, the beam pattern of the linear array is shown in Fig. 2, and the −3 dB beam width in the 0° direction is about 1.9°. The −3 dB beamwidth in the 70° direction is about 5.6°, and the azimuth resolution decreases rapidly as the deviation from the natural pointing angle increases.

3.2 Ring Array

Then consider the ring array. Let the radius of the ring array be r_2, the number of array elements be M_2, the ring array is placed on the xoy-plane, and the center of the circle is the origin of the coordinates, assuming that the orientation of the mth array element on the array is

$$\vartheta_m = \frac{2\pi(m-1)}{M_2} - \pi, m = 1, \ldots, M_2 \tag{9}$$

The layout of the ring array is shown in Fig. 3.

Fig. 1 Uniform linear array

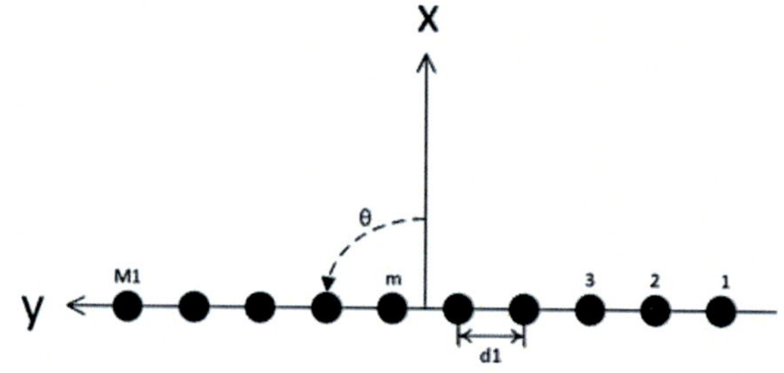

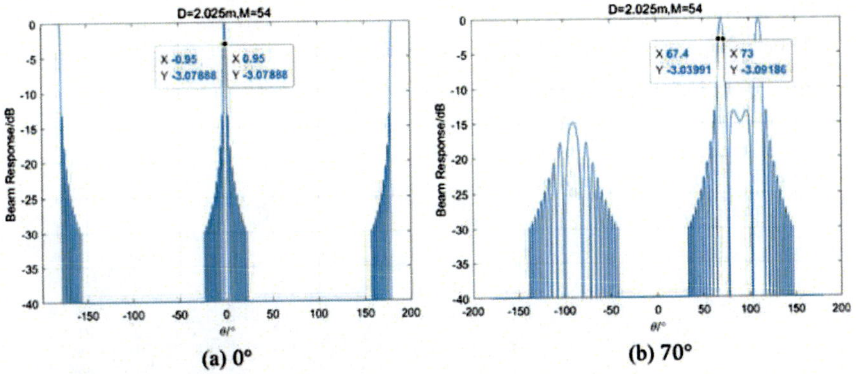

Fig. 2 Conventional beam pattern in horizontal direction of linear array

Fig. 3 Uniform ring array

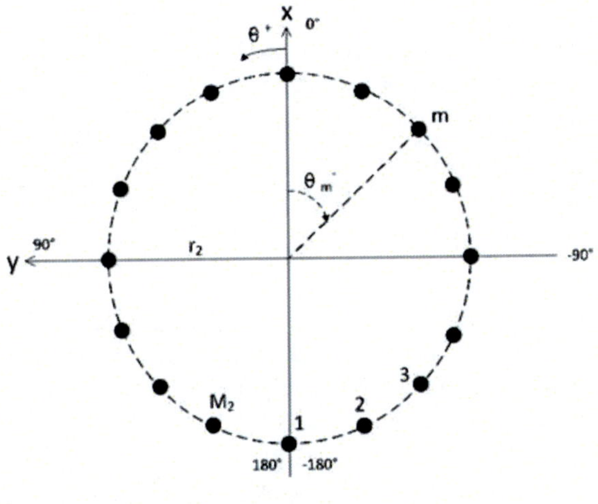

Referring to the linear array, the array manifold vector of the circular array can be deduced as

$$p(\Omega) = \left[e^{ikr_2\cos(\vartheta_1-\theta)}, e^{ikr_2\cos(\vartheta_2-\theta)}, \ldots, e^{ikr_2\cos(\vartheta_M-\theta)}\right]^T \quad (10)$$

After simulation analysis, taking the wavenumber radius product $kr_2 = 21\pi$ and the number of array elements $M_2 = 132$ can meet the requirement of -3 dB beam width within 2°, and there is no grating lobe. The beam pattern of the ring array is shown in Fig. 4.

Compared with the linear array, the ring array has no problem of ambiguity of port and starboard, and can form beams with the same response in multiple directions, that is to say, the circular array has the same target detection and orientation estimation performance [4, 5].

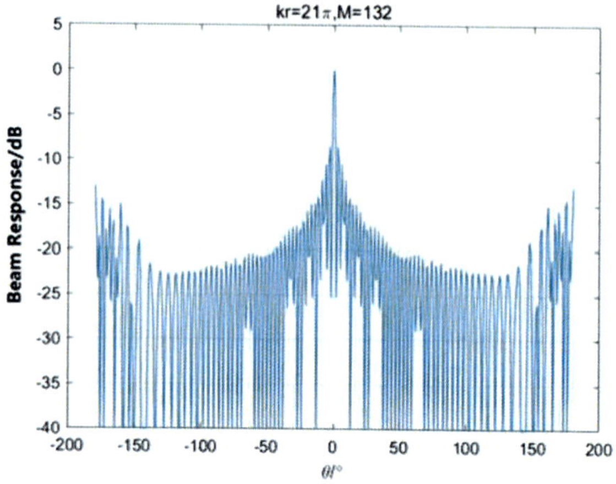

Fig. 4 Conventional beam pattern in horizontal direction of uniform circular array, 0°

However, because the hydrophone array is placed in the waters near the shore, it only needs to detect the surrounding 180° range, and the number of elements of the ring array is somewhat redundant.

3.3 Arc Array

Consider arc arrays on the basis of circular arrays. Suppose the radius of the arc array is r_3, the number of array elements is M_3, the arc array is placed on the xoy-plane, the center of the arc is the origin of the coordinates, and the angle of the arc is θ_a, then the angle of the mth array element on the arc is

$$\vartheta_m = \frac{\theta_a \cdot (\mathrm{m} - 0.5)}{M_3} - \frac{\theta_a}{2}, m = 1, \ldots, M_3 \tag{11}$$

The arc array layout is shown in Fig. 5.

Taking the beam radius product $kr_3 = 21\pi$, the number of array elements $M_3 = 45$, and the arc angle $\theta_a = 180°$. Simulation result shows that no side lobes appear

Fig. 5 Uniform arc array

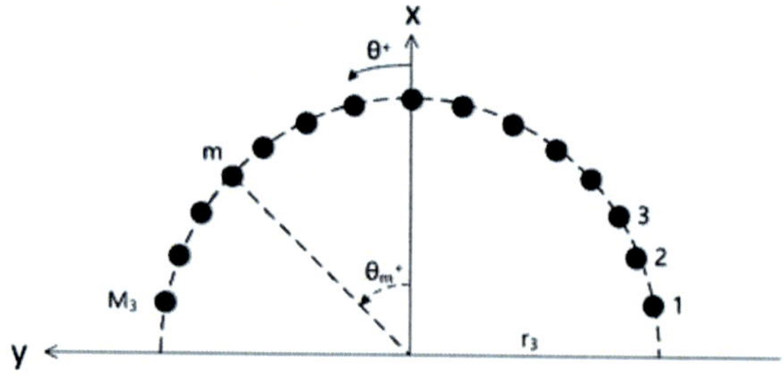

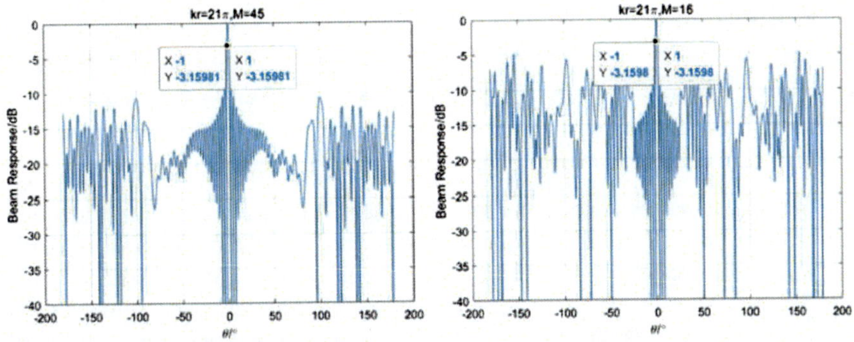

(a) 45 elements (b) 16 elements

Fig. 6 Conventional beam pattern in horizontal direction of arc array, 0°

in the range of $-90° \sim 90°$, and they are below -10 dB, and the -3 dB beamwidth is within 2°. The arc array beam simulation result is shown in Fig. 6a.

When the number of array elements is 16, the beam pattern is as shown in Fig. 6b, side lobes are below -5 dB, and the azimuth resolution is about 2°.

The beam response to the signal in the $-70°$ direction is shown in Fig. 7, and the azimuth resolution of 45 elements and 16 elements is about 4°. Therefore, the detection range of the arc array can be determined as $-70°$~$70°$.

Compared with the ring array, although the arc array cannot have the same target detection performance and azimuth estimation performance at any angle; however, there is neither the starboard and starboard ambiguity problem like linear array, nor does it have more array elements than ring array, and the detection angle range is larger than that of the linear array [6–10].

Comprehensively considered, the arc array is chosen as the hydrophone array.

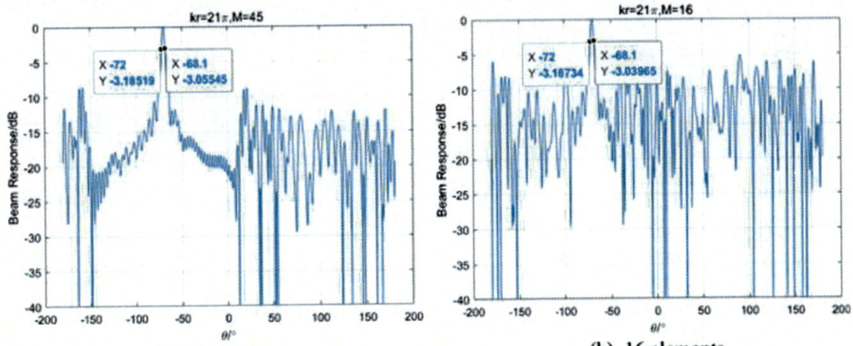

(a) 45 elements (b) 16 elements

Fig. 7 Conventional beam pattern in horizontal direction of arc array, $-70°$

4 Simulation

As shown in Table 1, the relevant indicators of the arc hydrophone array are set below.

Then, set the sound field environment and basic parameters:

(1) Set up two narrowband signal sources. Signal 1 is incident from the direction of $-10°$, and signal 2 is incident from the direction of $10°$. The SNR is 10 dB and 5 dB, respectively, the noise is white noise, and the power is 0 dB;
(2) Set the center of the arc as the origin of the coordinates. Each array element is regarded as a point element, the array element is isotropic, and the mutual influence between the array elements is not considered.

The azimuth spectrum of the arc array is shown in Fig. 8a.

When the SNR is -5 dB and -10 dB, respectively, the azimuth spectrum is shown in Fig. 8b.

It can be seen from the figure that the arc array can still maintain a good azimuth estimation performance in the case of low SNR.

Table 1 Arc array parameters

Parameter	Value
Beam radius product	21 π
Number of array elements	45
Arc angle	180°

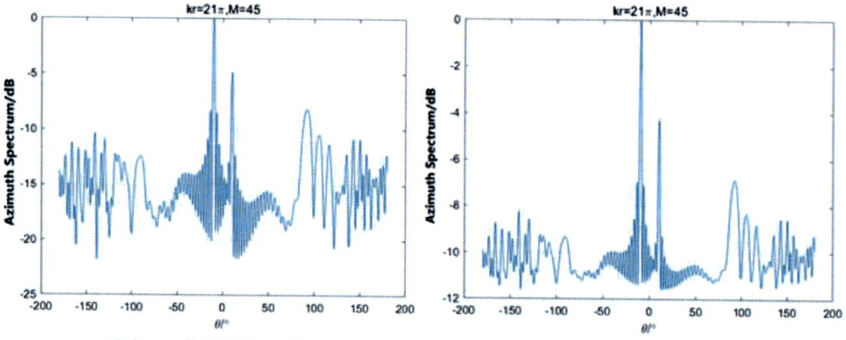

(a) $\theta_1 = -10°, SNR_1 = 10dB,$
$\theta_2 = 10°, SNR_2 = 5dB$

(b) $\theta_1 = -10°, SNR_1 = -5dB,$
$\theta_2 = 10°, SNR_2 = -10dB$

Fig. 8 Arc array azimuth spectrum

5 Conclusion

The arc array solves the problem of port and starboard ambiguity of the linear array very well. At the same time, according to the actual needs, 45 array elements are selected to form a semi-circular array. Under the same azimuth resolution, compared with the ring array, it uses fewer array elements. However, it also must be recognized that under the same array size, the smaller the number of array elements, the smaller the field of view. Therefore, if the number of array elements is small, when estimating the target orientation, it is necessary to pay attention to the influence of side lobes and eliminate false targets.

Acknowledgements This research work was funded by Science and Technology on Near-Surface Detection Laboratory pre-research fund, funded project number: 6142414190101.

References

1. Cheng, Y.S., Li, H.T., Wang, S.: Analysis of port/starboard discrimination capability using vector sensor uniform line array and scalar sensor arc array. J. Appl. Acoust. **036**(003), 276–282 (2017)
2. Sun, G.Q., Zhang, C.H., Huang, H.N., et al.: Left-right resolution of acoustic vector sensor line arrays. J. Harbin Eng. Univ., Harbin, China **31**(7), 848–855 (2010)
3. Yan, Sh.F., Li, Q.H.: Optimizing array signal processing (Volume 1 and 2), 1st edn. Science Press, Beijing (2018)
4. Xie, Y., Huang, M., Zhang, Y., Duan, T., Wang, C.: Two-stage fast DOA estimation based on directional antennas in conformal uniform circular array. Sensors **21**(1), 276 (2021). https://doi.org/10.3390/s21010276
5. Djigan, v.: Circular adaptive antenna array. In: 2021 IEEE East-West Design & Test Symposium (EWDTS), pp. 1–5. (2021). https://doi.org/10.1109/EWDTS52692.2021.9580987
6. Tian, Y., Mei, R., Huang, Y., Tang, X. and Cui, T.: 2D-DOA estimation in arc-array with a DNN based covariance matrix completion strategy. In: IEEE Access, vol. 10, pp. 57608–57620. (2022). https://doi.org/10.1109/ACCESS.2022.3172478
7. Li, X., Zhang, W., Yuan, Q., Cheng, Y.: Angles estimation for wideband signals using arc nested array. In: IET International Radar Conference (IET IRC 2020), pp. 1352–1355. (2020). https://doi.org/10.1049/icp.2021.0483
8. Gao, Z.H., Wang, X., Huang, P., Xu, W., Zhang, Z.: Arc array radar IAA-STAP algorithm based on sparse constraint. In: 2020 International conference on information science, parallel and distributed systems (ISPDS), pp. 277–282. (2020). https://doi.org/10.1109/ISPDS51347.2020
9. Gao, Z.H., Wang, X., Huang, P., Xu, W., Zhang, Z.H.: Iterative adaptive approach STAP algorithm based on arc antenna array. J. Phys.: Conf. Ser., v1607 012070(2020)
10. Kim, S., Kim, J.W., Cho, H., Ahn, B., Kim, K.S., Yu, J.W.: Squint-less arc array for near-field focusing in wideband systems. J. Electromagn. Waves Appl. 1890642, (2021). https://doi.org/10.1080/09205071.2021.

Shoal Target Recognition Based on Sample Data Enhancement and Target Detection

Jian Jun Wang, Yue Lin Xu, Shi Rui Wang, and Xing Yue Du

Abstract For shoal targets (boats, cars, people, etc.), the data is scarce and fuzzy, and it is difficult to detect and identify. A data enhancement technology based on generative adversarial network and target detection method based on multi-scale feature fusion and knowledge distillation are proposed. First, in the absence of paired data, it is not necessary to establish a one-to-one mapping between training data, and learn the method CycleGAN from the source data domain X to the target data domain Y, so as to realize the migration between the source domain and the target domain. So as to obtain shoal target images in different environments, construct shoal target datasets of different categories, and solve the problem of scarcity and blur of shoal target data. Then, by studying the use of atrous convolution to replace the general convolution kernel, the small target detection performance is improved and the spatial feature pyramid pooling structure based on the attention mechanism is used to realize the fusion of local features and global features, enrich the feature information of the target itself, and suppress the influence of environmental factors such as illumination on target detection improves the detection performance of the model in dark night scenes, so as to achieve target detection and recognition.

J. J. Wang · S. R. Wang · X. Y. Du
CSSC Ocean Exploration Technology Institute Co., Ltd, Wuxi, Jiangsu, China

Y. L. Xu (✉)
Science and Technology On Near-Surface Detection Laboratory, Wuxi, Jiangsu, China
e-mail: 1140860425@qq.com

K. Nakamatsu et al. (eds.), *Advanced Intelligent Technologies for Information and Communication*, Smart Innovation, Systems and Technologies 365,
https://doi.org/10.1007/978-981-99-5203-8_3

1 Introduction

In this paper, under the condition of limited resources, it is the main content of this paper to achieve effective sample data enhancement and target detection and recognition of scarce and ambiguous shoal targets. Neural network learning and machine image vision [3] have been widely used [4]. Through the construction of datasets, data enhancement, target detection, and other technical means, the accurate positioning and identification of shoal targets (people, vehicles, ships, etc.) are achieved. First, collect and label shoal target data to construct a dataset containing different shoal targets; augment and enrich the data of each category through data enhancement technology; and then train the target detection model to locate and identify the specific location and category of the target.

2 GAN-Based Data Augmentation

In view of the fact that the shoal target dataset is not rich enough, a method of learning from the source data domain X to the target data domain Y is proposed to use CycleGAN to realize the migration between the source domain and the target domain, so as to get images in a variety of styles [2].

The generator consists of an encoder, a converter, and a decoder. Encoder: The first step utilizes a convolutional neural network to extract features from the input image. Compress the image into 256 64*64 feature vectors. Converter: Convert the feature vector of the image in the DA domain to the feature vector in the DB domain by combining the dissimilar features of the image, using a six-layer Reset module, each A Reset module is a neural network layer composed of two convolutional layers, which can achieve the goal of preserving the original image features during conversion. The decoder uses the deconvolution layer to complete the work of restoring low-level features from the feature vector. Finally, get the generated image.

The generator structure of CycleGAN follows the residual network used in fast neural style transfer, which includes downsampling convolutional layers, some residual modules, and upsampling convolutional layers. This type of network has shown good performance in image generation tasks by virtue of the characteristics of residual network being easy to train and strong feature extraction ability, but it has the defects of lack of diversity and insufficient quality of generated images [10]. In response to this problem, we have partially improved the generative model. Under the condition that the depth of the generator network remains unchanged [9], we use the first three of the downsampling convolutional layer and the residual module of the generator as the encoder, and the last three. A residual module and an upsampling convolutional layer are used as decoders, and the three residual modules in the middle are used as feature transfer modules. This module has different residual modules for different mapping tasks, instead of only one residual module. Different mappings: The encoder and decoder parts can be shared. In addition, the sub-pixel convolution

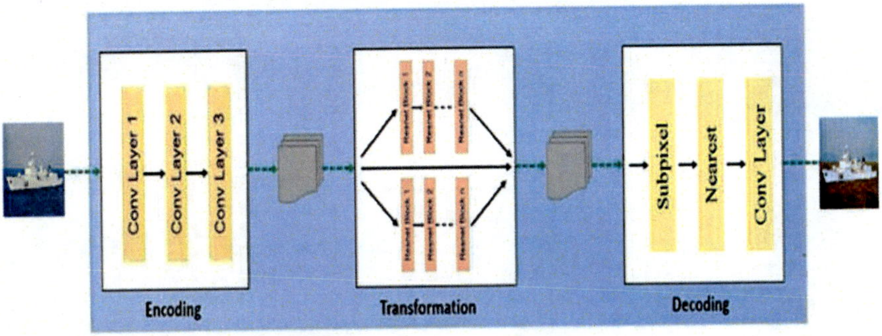

Fig. 1 Generator structure diagram

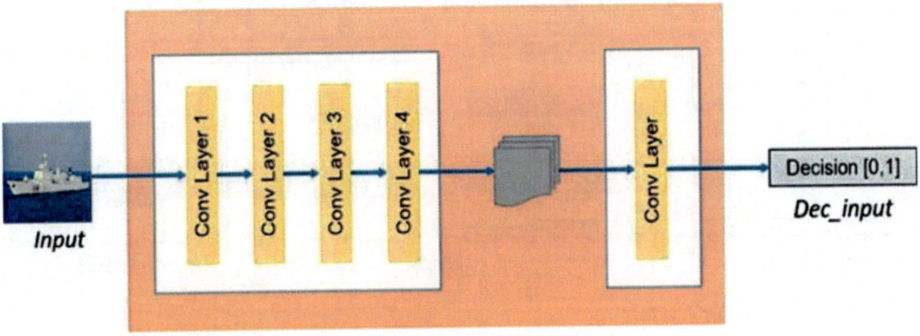

Fig. 2 Schematic diagram of the discriminator structure

and nearest neighbor upsampling methods will be used in the upsampling part to improve the quality of image upsampling, and then the original BN will be replaced by IN (instance norm), and the domain consistency loss function will be added to further improve the generation of ships. The quality improved generator is shown in Fig. 1. The discriminator takes an image as input and tries to predict whether it is the original image or the output image of the generator. The discriminator extracts features from the image, and then determines whether the extracted features belong to a specific category by adding a convolutional layer that produces a one-dimensional output [5]. The structure of the discriminator is shown in Figs. 1 and 2.

3 Object Detection Method

Solve the difficult problem of small target detection. Compared with other target detection networks [1], YoloV5 divides the output of the backbone network into three checkerboard grids with different sparsity by upsampling [8], and uses the densest grid to generate a large number of prediction frames for small target detection. The training method improves the detection ability of small targets to a certain extent.

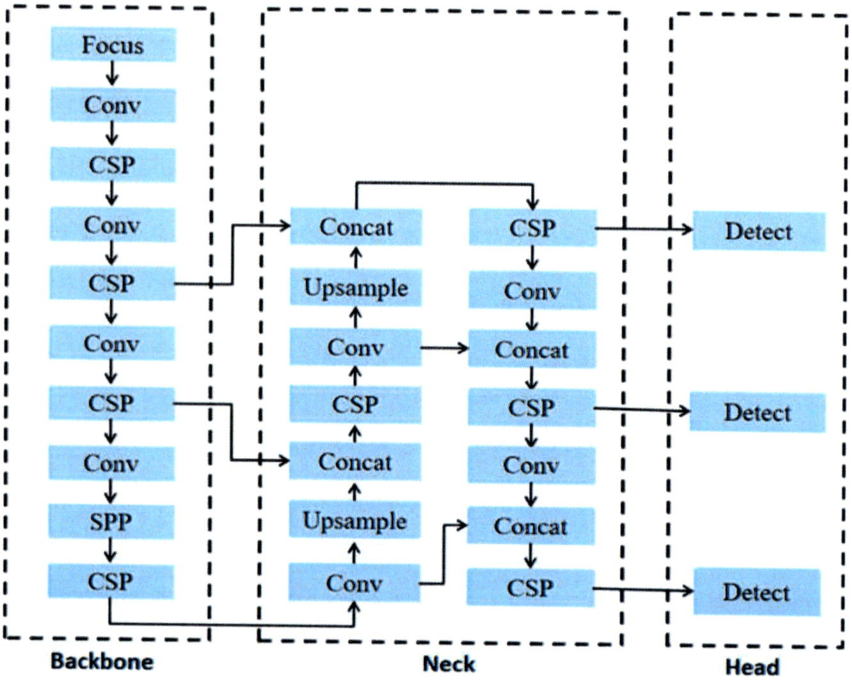

Fig. 3 Model structure diagram

The YoloV5 model consists of three parts: the feature extraction backbone network: Backbone, the feature fusion Neck, and the detection head. The Model structure is shown in Fig. 3.

The Neck part is the PANet (path aggregation network). On the basis of the FPN (feature pyramid network), the PANet adds a bottom-up branch, which transfers the location information of the shallow layer in the FPN to the high-level through this branch, and enhances the location information in the high level. The Neck part enhances the position information and semantic information of each output level and further strengthens the model feature extraction ability through two branches and unilateral connection, namely, top-down and bottom-up. The Structure Diagram of Main Modules is shown in Fig. 4.

The core of the Head part is prediction. For each grid of the three different scale feature maps output from Neck, predict boxes, each box contains four coordinates, one confidence level, and three conditional category probabilities. The Detect module mainly uses one 11 convolution to modulate channels into * (5 + C) channels, corresponding to the regression parameters, confidence level, and category of the target. Finally, the redundant detection frames are eliminated through Non-maximum Suppression (NMS) to obtain the final prediction frame.

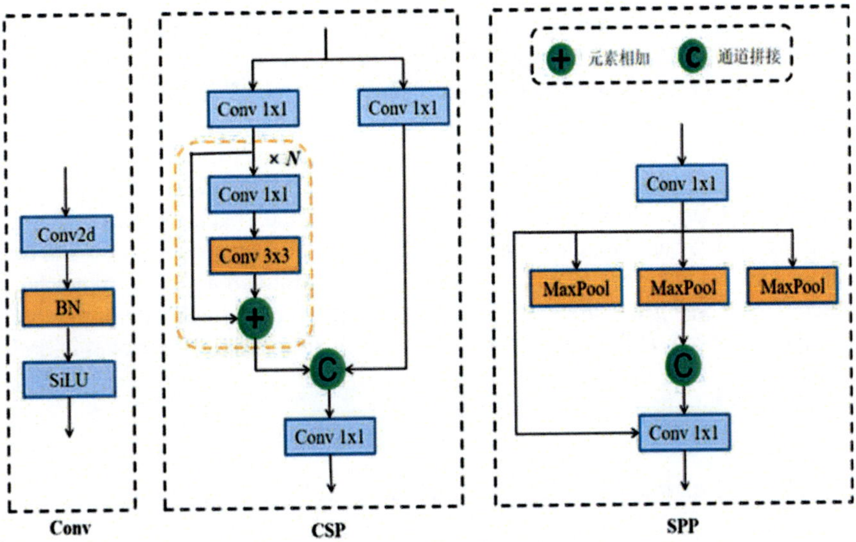

Fig. 4 Structure diagram of main modules

This scheme adopts the data enhancement method of mosaic splicing, and splices multiple training pictures into one according to a certain proportion, and increases the number of pictures in each training step in disguise. The samples generated by different splicing methods enhance the robustness of the model. The original image will be cropped during the stitching process. After the large-scale target is cropped, only part of the features may be retained, while the small-scale target is less affected by cropping and will generally be completely preserved. Therefore, mosaic stitching can encourage the model to learn small-scale targets. Characteristics. In the model training process, the traditional convolutional neural network needs to convolve the image first and then do the pooling to reduce the image size and increase the receptive field. However, since the multi-scale learning part of YoloV5 adopts an upsampling structure, the process of upsampling the size feature map to the large-size feature map will inevitably result in the loss of key information. This scheme uses a hole convolution to replace the general convolution kernel, expands the receptive field range of the convolution layer, and captures multi-scale global information. Reduce the loss of key information in the upsampling step, thereby improving small object detection performance [6].

Improve target detection performance for shallow targets in dark environments. The performance of target detection is greatly affected by the lighting environment, especially in the scenario of this project, such as ships sailing at night, how to ensure the detection effect in the period of poor lighting conditions has become an urgent problem to be solved. By means of domain transfer, this scheme is based on the backbone network trained on low-light data, transfers the YoloV5 model including the target in the night scene, and adopts the spatial feature pyramid pooling structure based on the attention mechanism to achieve local features and the fusion of global features further enriches the feature information of the target itself, thereby suppressing the influence of environmental factors such as illumination on target detection, and improving the detection performance of the model in dark night scenes [7].

4 Experimental Results and Performance Analysis

4.1 Data Augmentation Experiment

The proposed loop-consistent adversarial automatic colorization scheme in this paper can produce reasonable colorization results that convince most people. For the color style conversion of ships, the ship output effect based on the improved CycleGAN is shown in Fig. 5. It can be seen that the black-and-white ships are converted to the colors in the actual scene.

4.2 Shoal Target Detection and Recognition Test

We finally established a total dataset including 16,523 images, of which 14,870 were used as training sets and 1,653 were used as test sets. Each image contained several targets of individuals, cars, and boats. It is 14,000, the number of cars is about 6,000, and the number of ships is about 15,000. Based on the existing data, the team used YoloV5 and the current mainstream YoloV3 target detection model to train and test the shoal target detection model. Under the Yolo5 method, the indicator loss graph results are shown in Fig. 6 and Table 1.

The indicator Map@0.5 is shown in the table.

Compared with YoloV3, YoloV5 has a great improvement in mAP@0.5. We selected 10 images containing small ship targets, including 50 large and small targets, of which 48 were detected by YoLoV5 and 43 were detected by YoLoV3. It shows that this solution has greatly improved the problem of missing detection of small targets. As shown in the figure below, YoloV5 successfully detected small ships that were missed by the YoloV3 model, and the comparison of YoloV5 and YoloV3 results is shown in Fig. 7.

For shoal targets in different scenarios, the detection results of the method in this paper are shown in the following figure. The reasoning time of each picture is about 50 ~ 80 ms, the reasoning time can meet the needs of use, and the picture accuracy is also high. In the picture, the shallow targets in multiple scenes, including large targets and small targets, can be well detected and recognized come out. The Shoal target detection results are shown in Fig. 8.

Fig. 5 Ship color conversion renderings, the left side is before conversion, the right side is after conversion

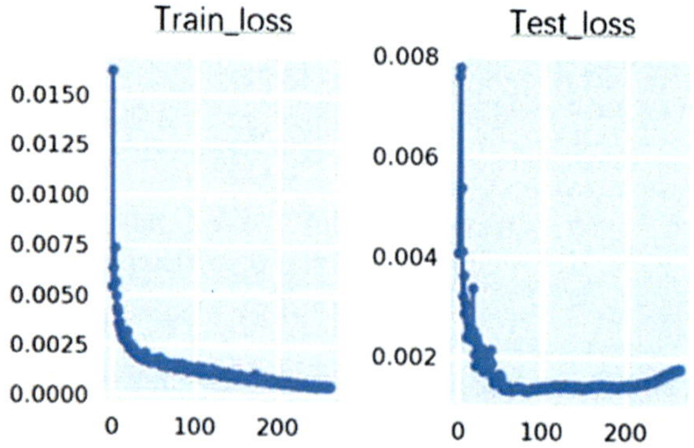

Fig. 6 The indicator loss graph results

Table 1 The indicator map

Model	mAP@0.5
YoloV3	0.69
YoloV5	0.77

Fig. 7 Comparison of YoloV5 (left) and YoloV3 (right) results

Fig. 8 Shoal target detection results

Acknowledgements This project was supported by Stable Foundation Project of the Science and Technology on Near-Surface Detection Laboratory with funding number 6142414190101.

References

1. Ren, S., He, K., Girshick, R., et al.: Faster R-CNN: Towards real-time object detection with region proposal networks. IEEE Trans. Pattern Anal. Mach. Intell. **39**(6), 1137–1149 (2017)
2. Xie, S., Girshick, R., Dollar, P., Tu, Z., He, K.: In: Proceedings of the IEEE conference on computer vision and pattern recognition (CVPR), pp. 1492–1500. (2017)
3. Liu, W., Anguelov, D., Erhan, D., et al.: SSD: Single shot multibox detector. Eur. Conf. Comput. Vis.. Springer, Cham, (2016)
4. Szegedy, C., Liu, W., Jia, Y., Sermanet, P., Reed, S., Anguelov, D., Erhan, D., Vanhoucke, V., Rabinovich, A.: In: Proceedings of the IEEE conference on computer vision and pattern recognition (CVPR), pp. 1–9. (2015)
5. Sayed, M., Brostow, G.: Improved handling of motion blur in online object detection. In: Proceedings of the IEEE/CVF conference on computer vision and pattern recognition, (2021)
6. Girshick, R., Donahue, J., Darrell, T., Malik, J.: Proceedings of the IEEE Conference on Computer Vision and Pattern Recognition (CVPR), pp. 580–587. (2014)
7. He, K., Zhang, X., Ren, S., et al.: Spatial pyramid pooling in deep convolutional networks for visual recognition. Pattern Anal. & Mach. Intell. IEEE Trans. On **37**(9), 1904–1916 (2015)
8. Redmon, J., Divvala, S., Girshick, R., et al.: You only look once: unified, real-time object detection. Comput. Vis. & Pattern Recognit., (2016)
9. He, K., Zhang, X., Ren, S., Sun, J.: Deep residual learning for image recognition. CVPR (2016).
10. Huang, G., Liu, Z., van der Maaten, L., Weinberger, K.Q.: Proceedings of the IEEE Conference on Computer Vision and Pattern Recognition (CVPR), pp. 4700–4708. (2017)

Research on the Influence of Aging Information System Optimization Based on Big Data on China's Rural Revitalization

Lina Wang, Qiuyue Zhang, Jian Cui, Yitong Wang, and Yongqiang Pan

Abstract The strategy of rural revitalization is an important part of resolving the main contradictions of China's current development, and it is of great significance to China's rural development. However, the long-term urban–rural dual structure has led to serious loss of rural population, and the gradual transformation of rural areas to hollow and aging is an important factor in the steady implementation of the rural revitalization strategy. Rural revitalization has five connotations, namely, industrial prosperity, ecological livability, rural civilization, effective governance, and affluent life. This paper applies advanced intelligent technologies such as cloud computing and big data and green computing, selects 20 indicators to evaluate the five dimensions of rural revitalization, and carries out data analysis. At the same time, a system dynamics model is constructed to simulate rural revitalization and the interactive relationship between the level of information system on aging and predict the five dimensions of rural revitalization from 2020 to 2030. The results found that the deepening of the aging level has a positive effect on rural revitalization; it has different degrees of promotion effect on the five dimensions, and the basic sequence from fast to slow is: living affluent > rural culture > industrial prosperity > ecological livability > governance effective; the overall rating of rural revitalization has also shown rapid growth. In the context of big data, discuss the impact of aging information system on rural revitalization from a quantitative perspective, hoping to provide a certain reference for the implementation of rural revitalization.

1 Introduction

Looking back on the past, the Chinese people of all ethnic groups to win the battle against poverty. 98.99 million rural poor people have been lifted out of poverty, 832 poverty-stricken counties have taken off their hats, millions of poverty alleviation cadres have devoted themselves to the cause of poverty alleviation, and more than

L. Wang (✉) · Q. Zhang · J. Cui · Y. Wang · Y. Pan
School of Economy and Management, Hainan Normal University, Haikou 571158, Hainan, China
e-mail: lina1976113@126.com

K. Nakamatsu et al. (eds.), *Advanced Intelligent Technologies for Information and Communication*, Smart Innovation, Systems and Technologies 365,
https://doi.org/10.1007/978-981-99-5203-8_4

1800 have lost their lives in the cause of poverty alleviation. Rural revitalization is an important strategic transformation in China's rural construction after poverty alleviation. With the introduction of the rural revitalization strategy, a series of important documents such as the Rural Revitalization Strategic Plan have been issued one after another to ensure the orderly connection between rural revitalization and poverty alleviation, further promote China's rural development and strengthen the endogenous driving force of the domestic circulation.

The research on rural revitalization has a long history, and scholars at home and abroad have rich research results. Gladwin and others [1] emphasized the importance of rural entrepreneurship for rural revitalization. Through the consideration of agricultural development, Greene [2] believes that the government plays an irreplaceable role in rural revitalization. A large number of foreign scholars have also carried out research on a case-by-case basis to summarize the practical experience of rural revitalization around the world. Chinese scholars mostly explore three aspects, they are theoretical logic, practical problems, and implementation paths. Xiang Jiquan, PanJiaen, and others [3–5] pointed out the basic and theoretical research problems of agricultural and rural, and sorted out the history of rural construction in China; Zhang Yong, Pan Jiaen, and others [6, 7] discussed rural decline, rural renaissance, urban–rural contradictions, rural imbalances, and other issues; Xiwen, Liu Yansui, and others [8, 9] studied China's urban and rural development in the new era; Wei et al. [10, 11] analyzed the successful cases of rural revitalization, summarized their methods and achievements, and provided path choices for rural revitalization. In terms of rural aging, foreign scholars have studied the mental health of the elderly in rural areas, elderly service problems, and pension models. Monteso and others [12] believe that it is necessary to strengthen the mental health problems of the elderly in rural areas and actively intervene; Watanabe-Galloway et al. [13, 14] have studied the problem of nursing services and proposed to provide meticulous and targeted care services for the elderly in rural areas. Hatano, Bertuzzi, and other scholars have explored the elderly care model according to the main characteristics of the region. Chinese scholars mostly discuss rural aging from a qualitative perspective. Bai et al. [15–17] discussed the plight of poverty among the elderly in rural areas in China, and proposed to solve the problem of poverty among the elderly in rural areas from the aspects of support mechanism, social security system, and elderly service system. Li and others [18] believe that rural revitalization can effectively solve the plight of rural aging in all aspects.

To sum up, the research on aging information system and rural revitalization mostly adopts qualitative research, and is mostly analyzed from the perspective of theoretical discussion and current situation. There is little quantitative research on the influencing factors of rural revitalization. Based on this, this paper applies advanced intelligent technologies such as cloud computing and big data and green computing. By building a system dynamics model for data analysis, the impact of rural aging information system on rural revitalization is discussed. The evaluation dimension of rural revitalization adopts five aspects: industrial prosperity, ecological livability, rural civilization, effective governance, and affluent life. By comparing the evaluation indicators that have been studied, 20 secondary evaluation indicators are selected, and a system dynamics model is constructed. In the context of big data, this paper obtains the impact of rural information system aging on the comprehensive indicators of rural revitalization, and explores the impact path of rural aging information system on rural revitalization.

2 Conceptual Stage and Research Assumptions

With the deepening of the aging of rural areas in China, the negative effects of rural development have gradually become significant. Especially under the long-term dual economic development pattern of China, the outflow of rural labor force is serious, which directly leads to a rapid increase in the proportion of the rural elderly population and seriously threatens the implementation of the rural revitalization strategy. The rural revitalization strategy is the new deployment to agriculture, rural areas, and farmers. At the same time, it puts forward the five general requirement of "industrial prosperity, ecological livability, rural civilization, effective governance and affluent life".

2.1 Conceptual Stage

Industrial prosperity, as the primary problem of rural revitalization, is also the foundation and key to solving the rural agricultural problem [19]. Rural industrial development depends on a large number of factors. As a regional complex of nature, society, and economy, rural areas have all aspects of industrial development. For most rural areas, the revitalization of rural industry is mainly agricultural revitalization, and the industrial prosperity is mainly agricultural prosperity [20].

Ecological livability is a key link in the implementation of rural revitalization, which is of great significance and profound connotation. Ecological livability not only includes the beautiful ecological environment, but also has high requirements in terms of complete infrastructure, perfect public services, and harmonious social environment, which are realized by relying on the corresponding economic foundation.

Rural civilization mainly refers to the continuous improvement of the ideological, cultural and moral level of the peasant masses, the advocating scientific civilization, healthy social atmosphere, vigorous development of education, culture, health, sports, and other undertakings, and adapting to the cultural life needs of farmers. It is the guarantee of rural revitalization and contains rich cultural connotations, including political civilization, moral civilization, behavioral civilization, and ecological civilization [21].

Rural governance refers to strengthening and improving the leadership of the rural work and giving full play to the leading role of rural organizations, so as to unite and promote the consensus and strength of rural governance, improve the efficiency of public goods and public services, and finally maximize and sustainable public interests in the countryside and even society as a whole [22]. It is an important guarantee for social stability and rural development.

Affluent life is the fundamental goal of rural revitalization. It aims to improve the income level and quality of life of residents on the basis of comprehensive poverty eradication, narrow the gap between urban and rural development, and truly realize the integrated development of urban and rural areas and common prosperity [23]. Its connotation includes not only material attributes, but also spiritual and cultural attributes, with two layers of material wealth and spiritual life [24].

2.2 Research Assumptions

Due to the long-term outflow of rural labor, the supply of rural labor has been seriously affected. A large proportion of the poverty-stricken villages are due to the shortage of labor and the high proportion of the elderly population, especially the rural left-behind elderly [15], which directly leads to the shortage of labor factors for the development of rural industries. The reduction of the agricultural-age population has affected agricultural production and the serious degradation of cultivated land. In the context of urban–rural population mobility, the relatively high-educated young labor force goes out, and the proportion of left-behind elderly, left-behind children, and left-behind women with lower education levels is relatively large, resulting in lower education levels of the rural residents [25]. At the same time, compared with urban areas, rural areas have their own complexities in terms of economic development, modernization, traditional cultural influence, neighboring interpersonal relations, etc., which requires that effective rural governance cannot be achieved without talents with scientific knowledge, and the universal participation and support of the rural population are also essential [26]. However, rural population outflow and population aging restrict the supply of local talents and external talents. At this stage, the forms of rural childlessness and aging are severe. In the future, due to the aging of the rural labor force, the agricultural industry will even have a low-level, part-time, and extensive development trend, which is not conducive to the realization of industrial revitalization, especially the improvement of rural per capita income. For a post-industrialized country with a small per capita arable land area and a serious rural overpopulation, only by gradually absorbing the surplus rural population and expanding the average agricultural arable land through industrialization and urbanization can we improve farmers' income and narrow the urban–rural income gap. This is also the main viewpoint of the dual structure theory [27].

Rural aging information system also has a positive effect on rural revitalization, mainly manifested in consumer savings changing capital investment and human capital changing production technology. Savings is the source of investment, and consumption is the ultimate goal of economic activities. Macroscopically, according to the life-cycle theory, rational consumers will arrange consumption and savings at all stages of life expectancy according to the standard of maximizing utility, indicating that the intensification of rural aging will change the age structure, leading to the marginal tendency of rural residents to save. And changes in marginal consumption tendencies [28]; microscopically, the impact of aging on micro-savings comes from the life-cycle consumption effect and the prevention effect, but the prevention motivation is greater than the consumption effect, which has a significant positive impact on whether to save and the scale of savings [29]. Human capital is an important factor in the production process, and it has a higher and long-term rate of return compared with material capital. On the one hand, it has contributed to an increase in the number of years of service for future generations, and the present value of the return on investment in education will also increase. The length of education of children in the family will be extended, and families tend to increase their investment in education, thus promoting the improvement of human capital and positively increasing the level of human capital of enterprises related to productive services [30]. On the other hand, the aging information system of the population will increase people's preference for health investment. Increased health investment is conducive to improving the health status of the elderly, and the improvement of health will enable the elderly to have better physical conditions to receive more education and training, thus increasing the level of human capital accumulation of the elderly [31].

It can be seen that the impact of aging on rural revitalization is multifaceted, with both positive and negative effects. Based on this, this paper puts forward the following assumptions:

Assumption 1: The aging information system of the rural population has positive and negative effects on industrial prosperity and affluence life, and the overall effect depends on the size of the two.

Assumption 2: The aging information system of the rural population has a negative effect on ecological livability by affecting the use of cultivated land, and has a positive effect through the circulation of land management rights.

Assumption 3: The aging information system of the rural population has an impact on the effectiveness of rural governance and affluence through human capital, life-cycle consumption effect, and savings level.

Generally speaking, the "five-in-one" of rural revitalization is by no means fragmented, but closely linked. It needs to be considered with systematic thinking, but each requirement has its own focus. Industrial prosperity and affluence life are the foundation of rural revitalization, and ecological livability, rural civilization, and effective governance are the soft environment for rural revitalization. Affluence life is the fundamental goal of rural revitalization.

3 Indicator Selection and Model Building

3.1 Indicator Selection

The related report pointed out that rural revitalization should take rural "industrial prosperity, ecological livability, rural civilization, effective governance and affluent life" as the construction goal, requiring comprehensive overall planning of rural construction. Therefore, this paper selects this five aspects as the first-level indicators. As the same time, by comparing the existing research, 20 s-level indicators are selected (Table 1). Among them, the data are mainly derived from the China Statistical Yearbook, the China Rural Statistical Yearbook, the National Bureau of Statistics, and the census data and existing research results [32–35]. Due to the limitations of data availability, the effective governance is calculated by Delin [34] based on two secondary indicators: rural medical service level and rural living security level. Effective evaluation factors for governance, the average weight of indicators is used, and the level of rural aging is replaced by the national level [36].

3.2 Relationship Determined

The OLS regression and table function methods are used to determine the functional relationship between aging information system and various indicators, and the systematic simulation of the development of aging information system and rural revitalization is carried out dynamically with the systematic simulation of the system dynamics model with each index. By changing the parameters of aging and setting up reference groups to compare the dynamic impact of aging information system change on rural revitalization.

3.3 Relationship Determined

System boundary. Determining the system boundary is an important step in establishing a system dynamics model. The impact of aging information system on rural development is multifaceted. However, the complexity of China's regional environment and the difference in the actual situation lead to great differences in the actual situation in various places. In the existing studies, the relationship between the two has been discussed from a qualitative perspective, and there is little quantitative analysis. However, due to its complexity, this paper further considers the interaction between the evaluation indicators related to rural revitalization and aging information system, and the rest of the factors are not considered for the time being.

Table 1 Evaluation index system and function relationship of rural revitalization

First-level indicators	Second-level indicators (Yi)	Unit	Functional relationship	R-squared	Adjusted R-squared
Industrial prosperity	Per capita agricultural, forestry, animal husbandry, and fishery output value	Yuan	$Y1 = -1.64 *** +31.95 *** X$	0.9926	0.9917
	Proportion of effective irrigated area	%	$Y2 = 0.314 *** +0.779 *** X$	0.9033	0.8912
	Grain output per hectare of cultivated land	Ton/ha	$Y3 = 3.215 *** +6.318 *** X$	0.6881	0.6491
	Agricultural mechanical power per unit area	Kilowatt/ha	Table function		
Ecological livability	Per capita national financial poverty alleviation and investment in rural construction	Yuan/person	$Y4 = -2899.99 *** +33108.45 *** X$	0.959	0.9530
	Length of village roads per capita	Meter/person	$Y5 = -4.582 *** +82.919 *** X$	0.9776	0.9731
	Peasant toilet penetration rate	%	Table function		
	Fixed asset investment per capita for rural households	Yuan/person	Table function		
	Number of health workers per 10,000 people in rural areas	One/10,000 people	$Y6 = -3.425 *** +236.33 *** X$	0.9926	0.9911
Rural civilization	Number of teachers per capita	One/person	Table function		
	TV broadcast coverage	%	$Y7 = 93.779 *** +43.882 *** X$	0.9422	0.9349

(continued)

Table 1 (continued)

First-level indicators	Second-level indicators (Yi)	Unit	Functional relationship	R-squared	Adjusted R-squared
	Number of rural cultural stations per 10,000 people	One/10,000 people	$Y7 = 0.294***$ $+2.568***X$	0.9569	0.9507
	Per capita expenditure on cultural, recreation, and education for rural residents	Yuan/person	$Y8 =$ $-1584.504***$ $+24331.98***X$	0.9945	0.9931
Effective governance	Number of village committees per 10,000 people	One/10,000 people	$Y10 = 7.269***$ $+20.712***X$	0.7461	0.7038
	Number of village committee directors per 10,000 people	One/10,000 people	$Y11 =$ $3.273+59.845**X$	0.6693	0.5866
Affluent life	Per capita disposable income of rural residents	Yuan/person	$Y11 = -1.1708*$ $**+22.099***X$	0.9970	0.9963
	Cost of unit residential area	10,000yuan/ square meter	$Y12 =$ $0.7736***X$	0.8190	0.8190
	Mobile phone volume per 100 households	One/100 households	$Y13 = 34.307 +$ $1844.523***X$	0.9232	0.9040
	Engel's coefficient	%	$Y14 = 49.333***$ $-157.496***X$	0.9503	0.9379
	Car ownership per 100 households	Vehicle/100 households	$Y15 = 43.942***$ $+552.728***X$	0.9644	0.9555

Note X stands for the level of aging, and Y_i stands for the second-level indicators of the i; *, **, *** represent the significant levels of 0.1, 0.05, 0.01, respectively

Causal circuit diagram. Causal circuit diagram can visualize the interaction between real variables, mainly including arrows and plus or minus signs [37]. The arrow represents the causal relationship between each circuit, and the positive and negative signs indicate causal polarity, that is, the change of one variable and the increase or decrease of another variable [38]. The cycle is divided into positive cycle and negative cycle. The positive cycle represents the enhancement loop, and the negative cycle represents the equilibrium loop [37]. Figure 1 shows the causal relationship between the level of aging information system and the indicators of rural revitalization. Expect for Engel's coefficient, the level of aging information system has a positive effect on each indicator.

Flowchart and simulation. Based on the causal circuit diagram, a system dynamics flow diagram is constructed to determine the initial value of variables and their mutual mapping relationship, so as to dynamically simulate the interaction between aging information system and rural revitalization. The dynamic simulation flowchart is shown in Fig. 2, which includes a simple dynamic simulation of the total rural population. The number of births is calculated by the product of the total population and the annual rate of birth, and the birth rate is calculated by the national average; the emigrated population is calculated by the product of the total population and the average annual urbanization growth rate; and the dead population is calculated by the product of the total population and the mortality rate. Reference Table 1 for functional relations between variables.

Carry out historical inspection of the model. Set the step length to 1 year and the simulation time to 2011–2020. It is obtained that the simulation value is compared with the real value, and the error rate is calculated [39]:

$$r = \frac{R_1 - R_0}{R_0} \times 100\%$$

R represents the error rate, R_1 represents the analog value, and R_0 represents the real value.

After historical testing, the error rate is less than 10%, indicating that the model is more reasonable for system simulation, and the simulation results are credible, which can simulate and predict aging information system and rural revitalization.

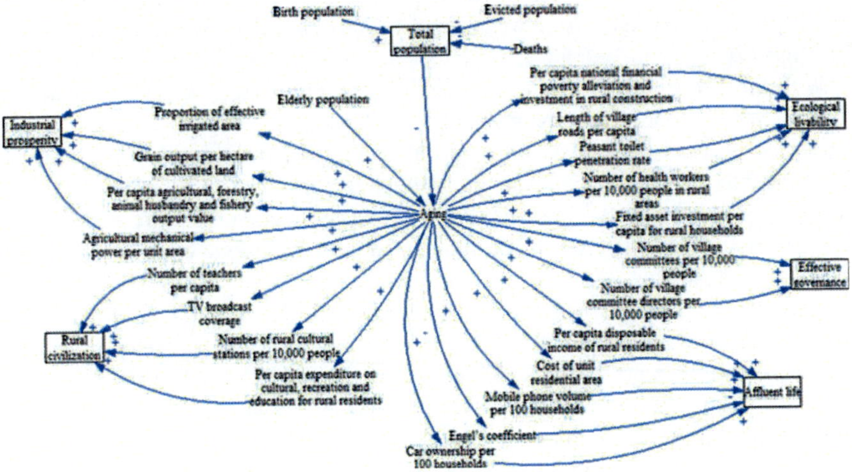

Fig. 1 Causal circuit diagram of evaluation indicators of aging information system and rural revitalization

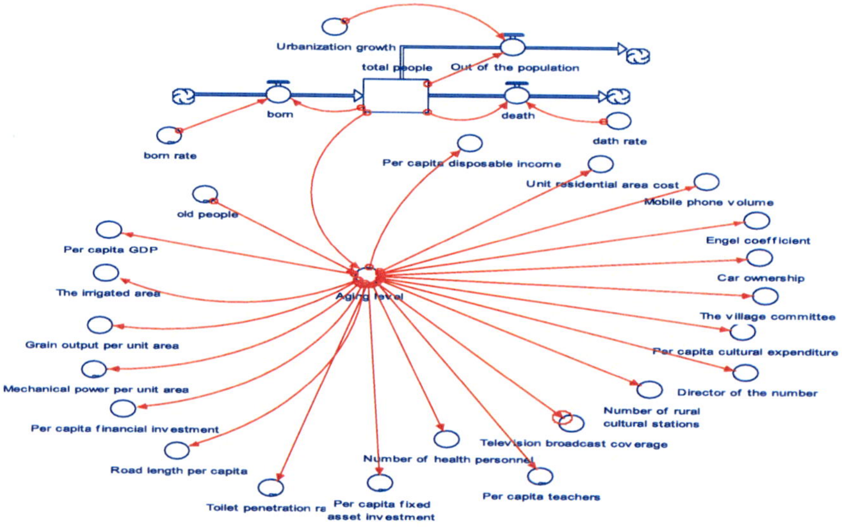

Fig. 2 Flowchart of the aging information system and rural revitalization system

4 Result Analysis

Under the premise that the model error test is acceptable, the development trend of rural revitalization in industrial prosperity, ecological livability, rural civilization, effective governance, and affluent life from 2020 to 2030 is simulated and evaluated one by one. Finally, the rural revitalization is comprehensively evaluated. The entropy method is used to determine the weight of each index. The specific calculation method refers to the calculation process of He [40]. Table 2 shows the scoring results. The score value of each project is between 0 and 1. The closer the score is to 1, the better the project is.

With the deepening of aging, the number and proportion of the rural elderly population will continue to increase, but the deepening of aging will not have a significant negative impact on rural revitalization, and rural development can be accelerated to a certain extent. According to the forecast results from 2020 to 2030, the industrial prosperity, ecological livability, rural civilization, effective governance, and affluent life have increased to a certain extent, and the growth trend has remained stable. Specifically:

First, the score of industrial prosperity rose from 0.095 in 2020 to 0.182 in 2030, and the foundation of rural industry has been continuously laid. It can be seen from the secondary indicators that agriculture, as a priority industry for rural development, has greatly promoted agricultural development with effective irrigated area and agricultural mechanization. At the same time, financial support for rural construction has been increasing, and the rural industrial structure has been adjusted and improved

Table 2 Rural revitalization score sheet from 2020 to 2030

Year	Level of aging	Industrial prosperity	Ecological livability	Rural civilization	Effective governance	Affluent life	Comprehensive score
2020	0.14	0.095	0.108	0.098	0.056	0.139	0.495
2021	0.14	0.104	0.115	0.106	0.061	0.158	0.544
2022	0.14	0.112	0.123	0.113	0.066	0.168	0.583
2023	0.15	0.121	0.131	0.121	0.071	0.179	0.623
2024	0.15	0.128	0.139	0.129	0.076	0.199	0.671
2025	0.16	0.137	0.147	0.138	0.082	0.210	0.713
2026	0.16	0.146	0.156	0.149	0.087	0.221	0.759
2027	0.17	0.155	0.164	0.157	0.093	0.242	0.812
2028	0.17	0.164	0.174	0.166	0.099	0.254	0.857
2029	0.17	0.173	0.183	0.175	0.106	0.276	0.912
2030	0.18	0.182	0.192	0.187	0.112	0.289	0.962
Average annual growth	0.004	0.0087	0.0084	0.0089	0.0056	0.015	0.0467

[34]. Many measures have brought positive effects on the development of rural industries. Although there are certain negative effects on rural aging information system, generally speaking, China's existing policies and measures can offset its negative impact.

Second, ecological livability is an environmental element of rural development and an important part of rural revitalization. The ecological livability score rose from 0.108 in 2020 to 0.192 in 2030, and the growth trend is slow. Rural ecological protection and rural construction are important drivers of ecological livability. The deepening of rural aging requires both ecology and livability. In recent years, with the severization of environmental control in China, the problem of rural environmental pollution has been gradually improved, infrastructure and rural construction investment have been continuously increased, and the level of livability in rural areas is constantly improving.

Third, the score of rural civilization increased from 0.098 in 2020 to 0.187 in 2030, and the growth rate is basically the same as that of ecological livability. At this stage, the older the population in China, the lower their average education level [41], and the lack of human capital makes them an economically "vulnerable group". The theory of cultural capital in Bourdieu (1986) shows that families lead to social reproduction through cultural reproduction [42], with the result that the rural elderly population will increase spending on cultural, recreation, and education for their offspring. At the same time, with the narrowing of the gap between urban and rural public services, the number of teachers per capita in rural areas has steadily increased, the education level has been continuously improved, and the increase of cultural stations and radio and television has greatly enriched the spiritual civilization of the residents.

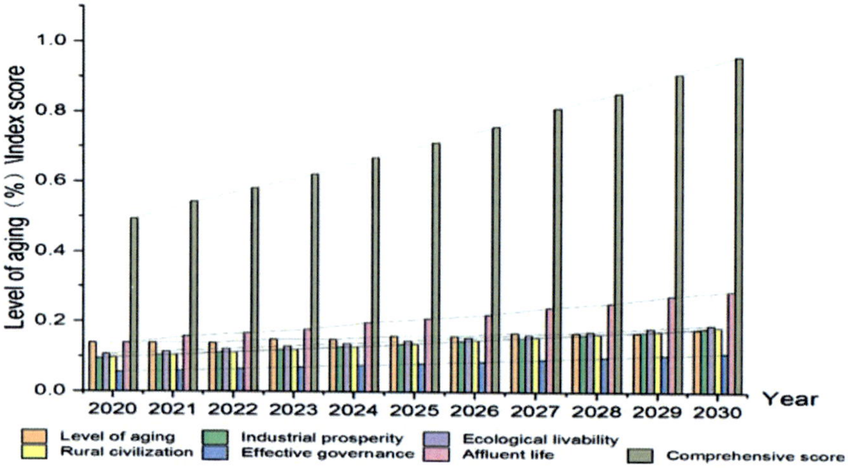

Fig. 3 Growth trend chart of rural revitalization index from 2020 to 2030

Fourth, the score of effective governance increased from 0.056 in 2020 to 0.112 in 2030, ranking at the end of the five indicators. The complexity and intractability of rural governance have made it a weak link in rural revitalization [43]. The deepening of aging has expanded the demand for community pension. While the development of rural communities has laid the foundation for community pension, it has also promoted the improvement of the level of rural governance, which is reflected in the increase in the number of village committees and the number of village committee directors.

Fifth, the score of affluent life increased from 0.139 in 2020 to 0.962 in 2030, with the fastest growth rate. According to the life-cycle theory, the deepening of aging will lead to a higher level of savings, and the individual consumption structure is also more inclined to food, transportation, communication, and health care [44]. The proportion of consumption expenditure to food expenditure has decreased, and Engel's coefficient decreases as the level of aging deepens. At the same time, China's social security system has been continuously improved, and the rural living environment and living level have improved significantly, which is reflected in the increasing level of housing prices, automobile, and telephone popularization (Fig. 3).

5 Conclusion

Based on the practical problems of rural aging information system and the five goals of "industrial prosperity, ecological livability, rural civilization, effective governance and affluent life" in the rural revitalization strategy, this paper builds a dynamic system dynamics model linked to the level of rural aging information system by selecting evaluation indicators for rural revitalization. On this basis, this paper predicts the evaluation indicators of rural revitalization from 2020 to 2030 and uses the entropy right method to quantitatively evaluate the five goals of rural revitalization, explore the role mechanism of rural aging information system in the rural revitalization strategy, and improve the policy reference for solving the problems of hollowing and aging in rural revitalization. The main conclusions are as follows:

First, the role of aging in the "five-in-one" of rural revitalization has both positive and negative effects, but it generally shows a positive effect. Since 2020, the level of rural aging in China has continued to deepen. The scores of "industrial prosperity, ecological livability, rural civilization, effective governance and affluent life" have not been negatively affected by aging, and they still guarantee a steady rise in the overall trend.

Second, the impact of aging on all levels is large, and the growth rate varies significantly. The basic order from fast to slow is: affluent life > rural civilization > industrial prosperity > ecological livability > effective governance. The average annual increase in the score of affluent life has reached 0.015, while the average annual growth in the score of effective governance is only 0.0056, and the average annual growth of the remaining three indicators is about 0.0085. It can be seen that the "five-in-one" of rural revitalization lacks balance in the future development, and the gap between individual indicators is large. It is necessary to coordinate the five elements, make up for the shortcomings, and achieve the comprehensive and balanced development of rural revitalization.

Third, under the background of rural aging, the overall score growth rate of rural revitalization is relatively fast. The continuous deepening of rural aging has a role in promoting rural development. The overall score of rural revitalization construction in 2020 was 0.459, reaching 0.962 by 2030, and the level of rural development has been rapidly improved.

6 Prospects

The deepening of aging is an inevitable phenomenon with economic development, and its impact is extensive and complex. Due to China's special urban–rural dual structure, the form of rural aging is more severe and has a more far-reaching impact on rural development. Therefore, it is necessary to properly deal with rural aging on the basis of consolidating and expanding the achievements of poverty alleviation and comprehensively promoting the rural revitalization strategy, innovate the pension system, expand the way for the rural elderly population to enjoy the dividends of socialist modernization, and actively introduce and encourage talents to return home and talents to go to the countryside. At the same time, it improves rural infrastructure, implements mergers with conditions, expands the rural population base, improves the efficiency of basic social services in rural areas, and realizes high-quality rural development.

By constructing a system dynamics model of aging information system and rural revitalization indicators, it is found that aging information system plays a positive role in promoting the rural industrial prosperity, ecological livability, rural civilization, effective governance, and affluent life, which is of great significance to the implementation of the rural revitalization strategy. However, this article also has some shortcomings: first of all, aging information system also has an indirect impact on rural revitalization, and the indirect impact has a role that cannot be ignored, considering the indirect impact in the model makes the study more perfect; second, there is also interaction between the five overall requirements of rural revitalization, not fragmentation. In a word, the interaction between population aging information system and quantitative indicators of rural revitalization in a systematic model at the time level is a new attempt, which will provide a new perspective to examine the impact of rural aging information system on rural revitalization, which is of great theoretical and practical significance. The next step will be to increase the variables involved in the model, improve and supplement unconsidered factors, and improve the accuracy of the model.

Acknowledgements This work was financially supported by (1) Hainan Provincial Natural Science Foundation of China in 2022 (722MS064): Research on resource and environment coordination mechanism and realization path driven by county innovation in Hainan Province under the background of Rural Revitalization; (2) Haikou Federation of Social Sciences 2022 Social Science Planning Project (2022—ZCKT—22): The integration mechanism and implementation path of international carbon emissions trading and supervision sandbox on Hainan Free Trade Island; (3) Key topics of Hainan Free Trade Port Finance Society in 2022 (Qiongjin Xuezi [2022] Item 43 of Project No. 13): Hainan Free Trade Port establishes a carbon trading financial innovation research focusing on "regulatory sandbox"; (4) 2018 National Social Science Fund Project (18BJL017), Project Name: Study on Marx's Theory of Ground Rent and Deepening the Reform of Rural Land System; (5) Talents project of Hainan Normal University of 2021; (6) The 2022 United Front Project of Hainan Normal University (Haishi Dang Tong Han [2002] No. 003, Item 10): The Historical Experience of China's Silk Road and the Construction of a Community with a Shared Future for Mankind.

References

1. Gladwin, C.H., Long, B.F., Babb, E.M., et al.: Rural entrepreneurship: one key to rural revitalization. Am. J. Agr. Econ. **71**(5), 1305–1314 (1989)
2. J. G.M.: Agriculture diversification covernment roles in rural revitalizatio. Rural. Econ. Altern., (1988)
3. Jiquan, X.: Rural construction in China: A century-old exploration and path conversion. J. Gansu Adm. Inst. **02**, 87–94 (2009)
4. Pan, J., Wen, T.: Tree hundred years: the context and development of China's rural construction. Open Times, (04), 126–45+7 (2016)
5. Huang, J.: The past and future of Chinese agriculture. J. Manag. World, (03), 95–104+11 (2004)
6. Jiaen, P.: Returning home to write: the problem of hometown or the dilemma of urban and rural? Beijing Cult. Rev. **03**, 16 (2017)
7. Yong, Z.: Through the Doctor's Return to Home during the Spring Festival to see rural problems, urban-rural contradictions and urban-rural integration. Theor. Explor. **04**, 86–93 (2016)
8. Xiwen, C.: Rural reform and development in China. Forum Lead.Ship Sci. **06**, 3–15 (2017)
9. Yansui, L., Bin, Y., Yanfei, W.: The main problems and transformation countermeasures of China's urban and rural development in the new era. Econ. Geogr. **36**(07), 1–8 (2016)

10. Wei, G., Cui, Y.: Revitalization of decline villages: Research on rural communities in Zhengding County, Hebei Province and their regenerative construction strategies. Mod. Decor. (Theory), (09), 256 (2016)
11. Chen, Z.: Reshaping the factor flow environment and an the active rural rejuvenation-the empirical evidence of Dashan Village, Gaochun County, International Slow City. Urban Planning Forum **03**, 28–35 (2013)
12. Montesó, P., Ferre, C., Lleixa, M., et al.: Depression in the elderly: study in a rural city in southern Catalonia. J. Psychiatr. Ment. Health Nurs. **19**(5), 426–429 (2012)
13. Watanabe-Galloway, S., Zhang, W., Watkins, K., et al.: Quality of end-of-life care among rural medicare beneficiaries with colorectal cancer. J. Rural Health **30**(4), 397–405 (2014)
14. Wu, C., Gao, L., Chen, S., et al.: Care services for elderly people with dementia in rural China: a case study. Bull. World Health Organ. **94**(3), 167–173 (2016)
15. Bai, Z., Wang, S., Zhou, Y.X.: Poverty management in rural areas for the elderly from the perspective of relative poverty. J. Nanjing Agric. Univ. (Soc. Sci. Ed.), **20**(04), 68–77 (2020)
16. Wang, S., Zeng, X.: Preliminary exploration of poverty in past-2020. J. Hohai Univ. (Philos. Soc. Sci.), **20**(02), 7–13+89 (2018)
17. Du, Z., Fang, K.: Practical dilemma and optimization path of the accurate supply of home-based elderly care services in rural communities. Chongqing Social Sci. (09), 130-140 (2020)
18. Li, Z., Li, Y.: Farmers' "resource pension" and its realization form under the background of rural revitalization strategy. Qilu J **06**, 99–109 (2020)
19. Xing, C., Haishan, C., Jiazhang, Z.: Empirical study on the impact of financial development on rural revitalization and development. Contemp. Econ. Manag. **41**(08), 91–97 (2019)
20. Xiaolei, X., Huifang, Z.: Factors influencing mountain agricultural modernization from the perspective of rural industrial revitalization: an empirical study based on the survey data of 315 administrative villages in Guizhou Province. Rural. Econ. Sci.-Technol. **32**(01), 3–8 (2021)
21. Zhenqin, S., Lu, D.: In-depth thinking on the construction of rural civilization. J. Yan'an Univ. (Soc. Sci. Ed.), **42**(02), 78–82+7 (2020)
22. Ruishan, C., Mingming, R.: The implication and consideration of effective governance in rural areas. Sci. Social. **03**, 116–121 (2019)
23. Yongqing, Z.: Exploration and practical prospects of the connotation of rural revitalization strategy under the background of the new era. Hubei Agric. Mech. **02**, 3–5 (2019)
24. Dong, L., Wu, C.: Realistic basis, internal logic and risk foresight of rural revitalization and young farmers returning home to start businesses. Mod. Econ. Res. (05), 125-132 (2019)
25. Mao, C., Xiong, H.: Rural population dualization under the strategy of rural revitalization and risk foresight. J. Southwest Minzu Univ. (Hum.Ities Soc. Sci.), **40**(08), 57–63 (2019)
26. Liuwen, L., Xiaolu, G.: Research progress and prospects on the impact of population aging on rural development. Prog. Geogr. **37**(05), 617–626 (2018)
27. Wenlong, W.: Five major issues should be clarified in the implementation of the rural revitalization strategy. J. Jishou Univ. (Soc. Sci.), **41**(02), 82–90 (2020)
28. Wu, X.: Study on the impact of population aging on China's economic growth. Shaanxi Normal University, (2018)
29. Li, C., Rundong, L.: Aging, prevention motivation and household savings rate: an empirical study on China's second demographic dividend. Popul. & Econ. **02**, 104–113 (2018)
30. Zhou, Y., Deng, F., Zhuo, C.: Has the decline of the demographic dividend hindered the rise of China's manufacturing value chain?. Inq. Econ. Issues, (04), 13–26 (2020)
31. Qi, H., Yan, H.: Has the aging population inhibited China's economic growth?, (2019)
32. Jin, J., Li, X., Yun, S.: Construction and empirical analysis of the indicator system of rural revitalization strategy. Financ. & Econ. **11**, 70–82 (2018)
33. Xu, L.B., Chunlong, M.X.: Research on the measurement and spatiotemporal heterogeneity of the development speed of inter-provincial rural revitalization in China-comprehensive evaluation method based on combined weighted principal component analysis. J. Appl. Stat. Manag. **40**(02), 205–221 (2021)
34. Delin, L., Qian, Z.: Research on the spatial and temporal evolution and influencing factors of the construction level of beautiful villages in China. East China Econ. Manag. **34**(01), 1–8 (2020)

35. Chengchao, L., Yue, C.: Rural revitalization and development: indicators evaluation system, regional gap and spatial polarization. Issues Agric. Econ., (05), 20–32 (2021)
36. Hongxia, W.: Rural population aging and rural space evolution: an exploration on the process of population aging from the perspective of rural microcosmic space. Popul. Res. **43**(05), 66–80 (2019)
37. Li, P., Lijun, T., Wenbin, H.: Improvement and empirical testing of population migration gravity model based on system dynamics. Syst. Eng.-Theory & Pract. **41**(07), 1722–1731 (2021)
38. Shi, H., Sun, G., Zhang, B.: Basic model construction and policy analysis of platform ecosystem evolution. Forum Sci. Technol. China, (07), 113–123 (2018)
39. Xinyu, G.A., Huihe, D.O., Lunian, G.A. et al.: Dynamic simulation of Changchun urban economy-society-land use system based on SD model. Resour. & Ind., 1–21
40. He, R., Lei, X., Wei, S.: Comprehensive evaluation of urbanization quality of Beijing-Tianjin-Hebei urban agglomeration. AR. Res. Dev. **35**(06), 42–47 (2016)
41. Hu, A., Hu, L., Xiong, Y.: Vigorously develop human resources and actively respond to the elderly society, F. (2012)
42. Yali, J.: Summary of research on family background and access to education. High. Educ. Rev. **4**(01), 199–213 (2016)
43. Yongzhi, G., Qiuxi, Z.: The implementation of the rural revitalization strategy needs to improve the level of rural governance as a whole. Res. Agric. Mod. **39**(05), 717–724 (2018)
44. Xin, Y., Peng, X.: New understanding of ole-age consumption and reflection on public policy. Fudan J. (Soc. Sci. Ed.)., **58**(02), 126–134 (2016)

Analysis of Hainan Tourism Carbon Balance Information System Based on Big Data

Lina Wang, Qiuyue Zhang, Guoqin Cai, Yitong Wang, and Yongqiang Pan

Abstract In the context of big data, this paper applies advanced intelligent technologies such as cloud computing and big data and green computing, based on the time series data related to tourism in Hainan Province 2010–2020, a carbon emission information system measurement model of tourism and tourism ecosystem carbon absorption information system is constructed, and the carbon balance information system of tourism in Hainan Province is preliminarily calculated to clarify the dilemma of carbon emission reduction in tourism, and explore the low-carbon development mode of tourism. The results show that: (1) from 2010 to 2020, the carbon emissions of tourism in Hainan Province increased from 3.02972 million tons to 4.61407 million tons, of which tourism transportation, warehousing, and post and telecommunications are the main sources of carbon emissions information system. (2) From 2010 to 2020, the total carbon absorption of the tourism ecosystem in Hainan Province increased from 2.78928 million tons to 3.52312 million tons, of which the tourism waters and tourism forest ecosystems are the main carbon sinks, and the capacity of tourism farmland and grassland ecosystems is relatively weak. (3) The estimated annual carbon deficit of the tourism industry is 693600 tons. It presents a serious carbon imbalance; Hainan tourism industry has become a significant carbon source, and energy conservation, emission reduction, and ecological exchange rate increase will be an effective way for future development.

1 Introduction

The traditional concept is that tourism is a "low energy consumption and low pollution" industry. However, recent studies have shown that tourism accounts for 5% of global carbon emissions, and tourism carbon emissions will grow at an average annual rate of 2.5% by 2035 [1]. Tourism has become one of the important sources of carbon emissions information system [2]. In view of this, the Opinions of the State Council

L. Wang (✉) · Q. Zhang · G. Cai · Y. Wang · Y. Pan
School of Economy and Management, Hainan Normal University, Haikou, Hainan 571158, China
e-mail: lina1976113@126.com

K. Nakamatsu et al. (eds.), *Advanced Intelligent Technologies for Information and Communication*, Smart Innovation, Systems and Technologies 365,
https://doi.org/10.1007/978-981-99-5203-8_5

on Accelerating the Development of Tourism strongly advocates low-carbon tourism and puts forward new requirements for energy conservation and emission reduction in tourism. China's tourism industry is an important force in the global tourism economy. It is of practical significance to clarify the carbon emissions generated by tourism energy consumption and the carbon absorption of terrestrial ecosystems, which is an important prerequisite for exploring the carbon balance information system status of regional tourism and clarifying future emission reduction goals.

The carbon balance information system analysis of tourism involves two aspects: tourism carbon emission information system measurement and ecosystem carbon absorption information system measurement. In terms of tourism carbon emission information system measurement, the measurement at different spatial scales is mainly based on the "top-down" method [3], the "bottom-up" method [4], the combination of "top-down" and "bottom-up" methods [5], and the "carbon footprint" method [6], including global [7], national [8], regional [9], provincial and municipal [10], tourism destination [12], etc. In terms of ecosystem carbon absorption information system measurement, scholars mainly measure the carbon absorption of ecosystems such as grasslands [11], forests [12], farmland [13], and waters [14] at different spatial regional scales. Academics have made a lot of research results in these two aspects, but because carbon emissions from tourism and ecosystem carbon absorption belong to different academic fields, only a few scholars [15] combined the two and analyzed them. The research scale also covers only a few provinces and tourist destinations, which shows that academic achievements in related fields are insufficient. In previous studies, it has been found that the emission reduction pressure of regional tourism cannot be clearly judged based on the measurement results of carbon emission information system alone. Carbon balance information system analysis of tourism, as an exploratory work for the development of green and low-carbon tourism, will help solve this problem.

In view of this, this paper applies advanced intelligent technologies such as cloud computing and big data and green computing, builds a tourism carbon emission and ecosystem carbon absorption model based on the tourism-related time series data of Hainan Province from 2010 to 2020. Though the tourism carbon-deficit measurement model, it combines tourism carbon emissions with ecosystem carbon absorption, and analyzes the carbon balance information system state of tourism in Hainan Province, with a view to enriching the theoretical system of low-carbon tourism, clarifying the responsibility of Hainan tourism to the local environment, and providing a scientific basis for relevant departments to formulate low-carbon paths.

2 Research Methods and Data Sources

2.1 *Tourism Carbon Emission Information System Measurement Model*

Based on the "tourism consumption divestiture coefficient", this paper separates the energy consumption of tourism from industries closely related to tourism, and then calculates the carbon emissions of tourism according to the carbon emission coefficient method [16]. At present, this method has been widely studied in the study of carbon emissions in tourism.

$$R_i = S_i / M_i \tag{1}$$

$$C = \sum_{i=1}^{n} C_i = \sum_{i=1}^{n} \sum_{j=1}^{n} (E_{ij} \times R_i \times F_j \times CE_j) \tag{2}$$

In the formula, R_i represents the tourism consumption divestiture coefficient of i industry; S_i represents the added value of tourism in i industry; M_i represents the added value of i industry. i can be obtained by multiplying the value-added rate of i industry by i industry tourism revenue, where the i industry value-added rate refers to the ratio of the value-added value of i industry to the total output value of i industry and i industry tourism revenue can be obtained by multiplied by the composition of i industry tourism expenditure by tourism revenue; C indicates the total carbon emissions of tourism; C_i represents the carbon emissions of the i industry involved in the tourism; i = 1,2, represents the wholesale, retail, accommodation and catering industries and transportation, warehousing and postal industries, respectively; E_{ij} represents the amount of j energy consumed by the i industry; F_j represents the standard coal conversion coefficient of class j energy; and CE_j represents the carbon dioxide emissions per unit of standard coal. Based on the existing literature, the CE_j value is 2.45 [17].

2.2 *Tourism Ecosystem Carbon Absorption Information System Measurement Model*

Based on the existing relevant research results and the actual situation in Hainan Province, this paper first divides the ecosystem into four ecosystems: forests, grasslands, waters, and farmland. Secondly, the carbon absorption of these four ecosystems is measured, and finally the measure results are summarized to obtain carbon absorption of the entire ecosystem. On this basis, the proportion of tourism output in GDP is multiplied by the carbon absorption of the entire ecosystem to obtain the carbon absorption of the tourism ecosystem. The specific measurement formula is as follows:

$$Y_t = \frac{I}{GDP}(Y_{at} + Y_{bt} + Y_{ct} + Y_{dt}) \tag{3}$$

In the formula, Y_t represents the carbon absorption of the tourism ecosystem, I represents the total income of the regional tourism industry, GDP represents the gross regional product, Y_{at} represents the carbon absorption of the forest ecosystem, Y_{bt} represents the carbon absorption of the grassland ecosystem, Y_{ct} represents the carbon absorption of the aquatic ecosystem, and Y_{dt} represents the carbon absorption of the farmland ecosystem.

Carbon absorption information system measurement of forest ecosystem Carbon absorption in forest ecosystem includes forest vegetation carbon absorption and forest soil carbon absorption, the specific formula is as follows:

$$Y_{at} = Y_A + Y_a \tag{4}$$

$$Y_A = A \times a \tag{5}$$

$$Y_a = (Y_A \times 30\% + \frac{2}{3Y_A}) \times 0.5 \tag{6}$$

In the formula, Y_{at} represents the carbon emissions of forest ecosystem, Y_A represents the carbon absorption of forest vegetation, Y_a represents the carbon absorption of forest soil, A represents the area covered by forest vegetation, and a represents the average annual carbon sequestration intensity of forest vegetation, with a value of $3.81 t/hm^2$ [18].

Carbon Absorption Information System Measurement of Grassland Ecosystem

$$Y_{bt} = B \times b \tag{7}$$

In the formula, Y_{bt} represents the carbon absorption of grassland ecosystem, B represents the grassland area, and b represents the carbon sequestration coefficient per area of grassland. The average annual carbon sequestration coefficient is $0.21 t/hm^2$ [19].

Carbon Absorption Information System Measurement of Aquatic Ecosystem

$$Y_{ct} = C \times c \tag{8}$$

In the formula, Y_{ct} represents the carbon absorption of aquatic ecosystem, C represents the water area, and c represents the carbon sequestration coefficient per unit area of the water, with a value of 5t/hm^2 [20].

Carbon Absorption Information System Measurement of Farmland Ecosystem

$$Y_{dt} = D - E \tag{9}$$

In the formula, Y_{dt} is the net carbon absorption of farmland ecosystem, D is the carbon absorption of farmland ecosystem, and E is the carbon emissions of farmland ecosystem.

Among them, the formula for measuring the carbon absorption of farmland ecosystem is as follows:

$$D = \sum_{i=1}^{n} S_i = \sum_{i=1}^{n} Q_i \times d_i \times (1 - f_i)/E_i \tag{10}$$

In the formula, D represents the carbon absorption of regional farmland ecosystem, i represents the types of different crops, S_i represents the carbon reserves of category i crops, Q_i is the yield of category i crops (t), d_i represents the carbon content rate of category i crops (%), E_i represents the economic coefficient of category i crops (%), and f_i represents the moisture coefficient of category i crops (%). The various coefficients of this part refer to the research results of Tian Yun, etc. [21].

According to the research results of Zhao Rongqin and others [22] on farmland carbon emissions, the carbon emissions of major farmland ecosystems are divided into three categories: fertilizer, agricultural machinery, and agricultural irrigation. The specific formula is

$$E = E_1 + E_2 + E_3 \tag{11}$$

In the formula, E represents the carbon emissions of farmland ecosystem, E_1 represents the carbon emissions from fertilizer use, E_2 represents the carbon emissions from agricultural irrigation, and E_3 represents the carbon emissions from agricultural machinery use.

$$E_1 = G_i \times g_i \tag{12}$$

In the formula, E_1 represents the carbon emissions generated during the use of fertilizers, G_i is the amount used in category i fertilizers; and g_i represents the carbon emission conversion factors of category i fertilizers.

$$E_2 = H \times h \tag{13}$$

In the formula, E_2 represents the carbon emissions generated in the agricultural irrigation process, H is the agricultural irrigated area, and h is the carbon emission conversion coefficient of the agricultural irrigated area.

$$E_3 = I \times i + J \times j \tag{14}$$

In the formula, E_3 represents the carbon emissions generated during the use of agricultural machinery, I represents the area under crop cultivation, i represents the carbon emission conversion coefficient of crop cultivation area, J represents the total power of agricultural machinery, and j represents the carbon emission conversion factor of the total power of agricultural machinery.

2.3 *Tourism Carbon-Deficit Model*

Carbon balance information system in tourism is an important indicator to evaluate the impact of tourism on the local environment on the basis of comprehensive consideration of the level of regional tourism economic development and carbon absorption (carbon sequestration) capacity of local ecosystems. At present, the academic research results on carbon balance information system in tourism are insufficient [23]. This study's assessment of the carbon balance information system status in tourism is mainly reflected by comparing the difference between tourism carbon emissions and tourism carbon sequestration (i.e., tourism carbon deficit). This difference (if positive) can indicate the degree of carbon imbalance in tourism, is an important basis for measuring the impact of regional tourism development on global environmental change, and is also the current responsibility of tourism to reduce emissions. The specific calculation formula is as follows:

$$K = C - Y_t \tag{15}$$

In the formula, K is the carbon deficit of tourism, C is the carbon emission of tourism, Y_t is carbon absorption of tourism. If the carbon emission of tourism is more carbon of tourism than that of tourism, that is, the corresponding carbon offset of the total carbon absorption of the regional ecosystem for tourism is not enough to offset the carbon emissions generated by closely related industries in the region, then the tourism carbon deficit in the region is positive, which is reflected in the carbon imbalance of tourism, which shows that tourism activities in the region have significantly contributed to climate warming and have a serious adverse impact on the local ecological environment. If the carbon emissions from tourism are less than that of tourism, the regional tourism carbon deficit is negative, indicating that the tourism has achieved carbon-balanced development.

2.4 Data Source and Processing

The data sources involved in this article are mainly as follows: first, the energy consumption data of tourism-related industries, from the energy balance table in the 2011–2021 Hainan Statistical Yearbook; second, the economic data of tourism and closely related industries, including the composition of various consumption of tourism, the added value, total output, and total tourism revenue of the three industries are mainly from the 2011–2021 Hainan Statistical Yearbook, the 2011–2021 China Tourism Statistics Yearbook, and the Tourism Sample Survey Data; third, it is the basic data for ecosystem carbon absorption information system measurement, including forest and grassland coverage area, crop yield, fertilizer application, total power of agricultural machinery, agricultural sowing area, irrigation area, etc., it comes from the 2011–2021 Hainan Statistical Yearbook, and the other calculation coefficients are from the relevant research results of scholars at home and abroad. Some of the missing data are linearly interpolated with the values of adjacent years. At the same time, in order to eliminate the impact of the analysis results of price factors, the relevant data are unified in 2020 as the base period and the GDP Deflator is used for deduction.

3 Results and Analysis

3.1 Analysis of Carbon Emissions Information System from Tourism in Hainan Province

Measurement of carbon emissions from tourism in Hainan Province

Based on the added value, total output value, tourism expenditure ratio of tourism wholesale, retail, accommodation and catering industry and tourism transportation, warehousing and post and telecommunications industries in Hainan Province, and the total tourism revenue data in Hainan Province from 2010 to 2020, the tourism consumption divestiture coefficient of tourism wholesale, retail, accommodation and catering industry and tourism transportation, warehousing, and post and telecommunications industries in Hainan Province are calculated using Formula (1). The results are shown in Table 1.

Based on the energy consumption and tourism consumption divestiture coefficient data of the wholesale, retail, accommodation and catering industry and tourism transportation, warehousing, and post and telecommunications industries in Hainan Province from 2010 to 2020, Formula (2) is used to calculate the overall carbon emissions of various tourism industries and tourism in Hainan Province. The results are shown in Table 2.

Table 1 Calculation of the divestiture coefficient of tourism consumption in Hainan Province 2010–2020

Year	Industry 1 value-added ratio (L_1)	Industry 2 value-added ratio (L_2)	Industry 1 proportion of travel expenses	Industry 2 proportion of travel expenses	Total tourism revenue	Industry 1 tourism revenue (J_1)	Industry 2 tourism revenue (J_2)	Industry 1 tourism consumption divestiture coefficient (R_1)	Industry 2 tourism consumption divestiture coefficient (R_2)
2010	0.483	0.397	0.571	0.429	197.122	112.557	84.565	0.243	0.455
2011	0.465	0.378	0.549	0.451	228.130	125.244	102.887	0.246	0.505
2012	0.438	0.366	0.526	0.474	257.916	135.664	122.252	0.235	0.528
2013	0.476	0.329	0.542	0.458	286.060	155.044	131.015	0.227	0.477
2014	0.485	0.327	0.574	0.426	331.716	190.405	141,311	0.254	0.471
2015	0.483	0.350	0.467	0.553	373.286	174.324	198.961	0.212	0.604
2016	0.456	0.333	0.463	0.537	430.094	199.134	230.961	0.220	0.637
2017	0.434	0.337	0.481	0.519	505.629	243.207	262.421	0.246	0.663
2018	0.399	0.312	0.591	0.409	573.314	338.829	234.486	0.313	0.540
2019	0.410	0.315	0.584	0.416	624.622	364.779	259.843	0.316	0.561
2020	0.432	0.330	0.563	0.437	511.909	288.205	223.704	0.242	0.516

Note Industry 1 and Industry 2 refer to the wholesale, retail, accommodation and catering industries and transportation, warehousing, and post and telecommunications industries, respectively

Table 2 Calculation of carbon emissions from tourism in Hainan Province 2010–2020 (ten thousand tons)

Year	Industry 1 energy consumption	Industry 2 energy consumption	Tourism industry 1 energy consumption	Tourism industry 2 energy consumption	Tourism industry 1 carbon emissions	Tourism industry 2 carbon emissions	Carbon emissions from the tourism industry	The proportion of carbon emissions in the total carbon emissions of the tourism industry
2010	51.140	244.550	12.414	111.248	30.414	272.558	302.972	0.094
2011	56.290	263.770	13.867	133.148	33.947	326.213	360.187	0.095
2012	65.050	272.280	15.287	143.828	37.453	352.378	389.831	0.097
2013	69.720	288.470	15.794	137.736	38.694	337.453	376.147	0.089
2014	76.130	280.980	19.368	132.332	47.452	324.214	371.667	0.083
2015	79.440	261.700	16.864	158.154	41.316	387.478	428.794	0.091
2016	84.260	256.990	18.559	163.756	45.469	401.201	446.670	0.092
2017	92.640	273.620	22.817	181.378	55.901	444.376	500.278	0.098
2018	111.810	300.640	34.999	162.448	85.748	397.998	483.746	0.091
2019	126.770	307.530	40.056	172.653	98.137	423.000	521.137	0.094
2020	123.540	307.090	29.958	158.372	73.397	388.010	461.407	0.083

Current Situation of Carbon Emissions from Tourism in Hainan Province

At present, the contribution rate of tourism carbon emissions to total carbon emissions in Hainan Province is relatively high, and carbon emissions from tourism transportation, warehousing, and post and telecommunications are the main sources of carbon emissions from tourism. Table 2 shows that the total carbon emissions of tourism in Hainan Province in 2020 were 4.61407 million tons, and the carbon emissions of tourism wholesale, retail, and accommodation industries and tourism transportation, warehousing, and post and telecommunications were 733,970 tons and 3.8801 million tons, respectively. It can be calculated that the carbon emissions of tourism in Hainan Province account 8.294% of the total carbon emissions, and the wholesale, retail and accommodation industries and tourism transportation, warehousing and post and telecommunications account for 15.907% and 84.093% of the carbon emissions of the tourism industry, respectively. Global tourism carbon emissions account for only about 5% of the total carbon emissions [24]. It can be seen that at present, the proportion of carbon emissions from tourism in Hainan Province is 3.294% higher than 5%. Therefore, the contribution of carbon emissions to total carbon emissions in Hainan Province is relatively high. Carbon emissions from tourism transportation, warehousing, and post and telecommunications in Hainan Province account for 84.093% of the carbon emissions from tourism, which is the main source of carbon emissions from tourism, which is basically in line with the findings released by the UNWTO in 2009.

Trends in carbon emissions from tourism in Hainan Province

Since 2010, the total carbon emissions from the tourism industry in Hainan Province have generally shown a trend of fluctuating growth, and its dynamic growth is relatively slow. As can be seen from Fig. 1, from 2010 to 2020, the total carbon emissions from the tourism industry increased from 3.02972 million tons to 4.61407 million tons, an increase of 1.58435 million tons, with an average annual growth rate of 3.749%. Among them, it increased significantly in 2017 compared with 2016 and decreased significantly in 2020 compared with 2019.

Since 2010, the changes in carbon emissions from tourism transportation, warehousing, and post and telecommunications in Hainan Province are basically the same as the total carbon emissions from tourism, while the carbon emissions from the wholesale, retail, accommodation, and catering industries of tourism have increased slowly in fluctuations. As can be seen from Fig. 1, from 2010 to 2020, the carbon emissions of tourism transportation, warehousing, and post and telecommunications increased from 2.72558 million tons to 3.8801 million tons, an increase of 1.424 times in just 10 years; carbon emissions from the wholesale, retail, accommodation, and catering industries of tourism increased from 304,140 tons to 733,970 tons, with an average annual growth rate of 11.059%.

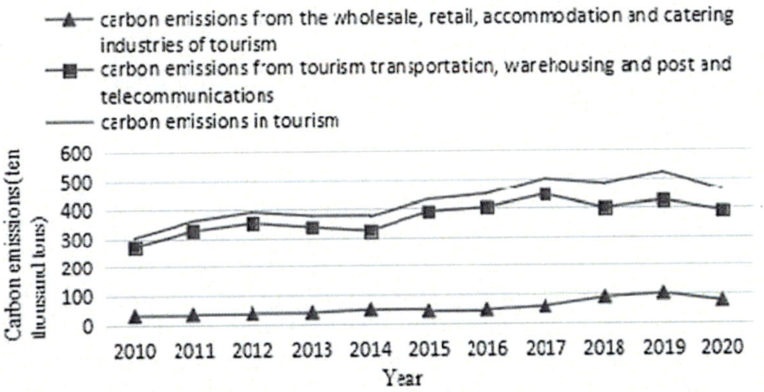

Fig. 1 Carbon emissions from tourism and its industries in Hainan Province 2010–2020

3.2 *Carbon Absorption Information System Analysis of Tourism in Hainan Province*

Measurement of carbon absorption in tourism in Hainan Province

Based on the forest cover area, grassland area, main river area, farmland fertilizer use, pesticide use, farmland effective irrigation area, crop cultivation area, total power of agricultural machinery, various crop yield data, and corresponding carbon sequestration coefficients in Hainan Province from 2010 to 2020, the carbon absorption of the four ecosystem of forests, grassland, waters, and farmland in Hainan Province is measured using Formula (4)–(14). On this basis, the carbon absorption of tourism industry is measured by Formula (3). The results are shown in Table 3.

Current situation of carbon absorption in tourism in Hainan Province

At present, the carbon absorption of the tourism aquatic ecosystem in Hainan Province contributes the most to the carbon absorption of tourism, followed by the tourism forest ecosystem, then the tourism farmland ecosystem, and finally the tourism grassland ecosystem, that is, the tourism aquatic ecosystem has the strongest carbon sequestration capacity, followed by the tourism forest ecosystem, then the tourism farmland ecosystem, and finally the tourism grassland ecosystem. Table 3 shows that in 2020, the total carbon absorption of tourism in Hainan Province was 3.52312 million tons, the carbon absorption of the tourism forest ecosystem was 1.47 658 million tons, the carbon absorption of the tourism grassland ecosystem was 1520 tons, carbon absorption of the tourism aquatic ecosystem was 1.85225 million tons, and the net carbon absorption of the tourism farmland ecosystem was 192,770 tons. It can be calculated that the tourism forest ecosystem, the tourism grassland ecosystem, the tourism aquatic ecosystem, and the tourism farmland ecosystem account for 41.911%, 0.043%, 52.574%, and 5.472% of the carbon absorption of tourism, respectively.

Table 3 Carbon absorption in tourism in Hainan Province (ten thousand tons)

Year	Carbon absorption in forest ecosystems	Carbon absorption in grassland ecosystems	Carbon absorption in aquatic ecosystems	Net carbon absorption in farmland ecosystems			Total tourism revenue	Regional GDP	Carbon absorption in tourism
				Carbon emissions	Carbon absorption	Net carbon absorption			
2010	754.281	0.966	1174	133.969	392.288	258.319	197.122	1545.980	278.928
2011	759.063	0.966	1174	114.220	377.973	263.754	228.130	1734.590	289.408
2012	766.662	0.966	1174	132.799	404.402	271.603	257.916	1897.662	300.812
2013	772.284	0.966	1174	139.408	437.373	297.965	286.060	2079.799	308.812
2014	926.512	0.966	1174	144.236	402.879	258.664	331.716	2258.682	346.614
2015	934.399	0.966	1174	148.127	339.876	191.749	373.286	2434.838	352.785
2016	934.399	0.966	1174	140.412	299.006	158.595	430.094	2617.426	372.670
2017	934.399	0.966	1174	142.416	274.203	131.787	505.629	2800.632	404.620
2018	935.899	0.966	1174	129.970	247.868	117.898	573.314	2963.047	431.237
2019	935.899	0.966	1174	122.257	239.079	116.822	624.622	3147.818	442.038
2020	935.899	0.966	1174	115.533	273.713	122.181	511.909	3244.598	352.312

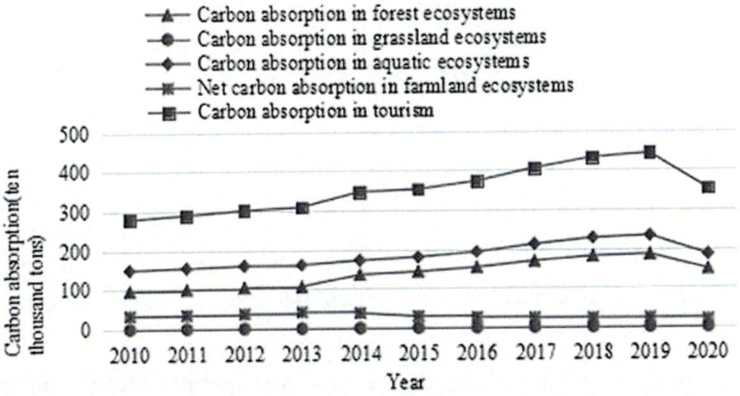

Fig. 2 Carbon absorption of tourism and ecosystems in Hainan Province 2010–2020

The trend of carbon absorption in tourism in Hainan Province. Since 2010, the total carbon absorption in tourism in Hainan Province has generally shown a slow growth trend (except 2020). As can be seen from Fig. 2, from 2010 to 2020, the total carbon absorption in tourism increased from 2.78928 million tons to 3.52312 million tons, an increase of 733,840 tons, with an average annual growth rate of 2.737%. During this period, the carbon absorption in tourism increased significantly in 2014 and decreased significantly in 2020.

Since 2010, the dynamics of carbon absorption in tourism forest ecosystems and tourism aquatic ecosystems in Hainan Province have been basically consistent with the total carbon absorption in tourism. The carbon absorption of the tourism grassland ecosystem fluctuates steadily, and the tourism farmland ecosystem shows a steady decline. As can be seen from Fig. 2, from 2010 to 2020, the carbon absorption of the tourism forest ecosystems increased from 961,760 tons to 1.47658 million tons, with an average annual growth rate of 4.996%; the carbon absorption of the tourism grassland ecosystems increased from 1230 to 1520 tons, with an average annual growth rate of 2.514%; the carbon absorption of the tourism aquatic ecosystems increased from 1.49692 million tons to 1.8225 million tons, with an average annual growth rate of 2.514%; and the net carbon absorption of the tourism farmland ecosystems decreased from 329,370 tons to 192,770 tons, with an average annual decline rate of 4.657%.

3.3 *Carbon Balance Information System Analysis of Tourism in Hainan Province*

Since 2010, the tourism industry in Hainan Province has been in a carbon deficit and has fluctuated steadily. As can be seen from Fig. 3, although the carbon emissions of tourism and the carbon absorption of tourism are generally on the rise, the carbon

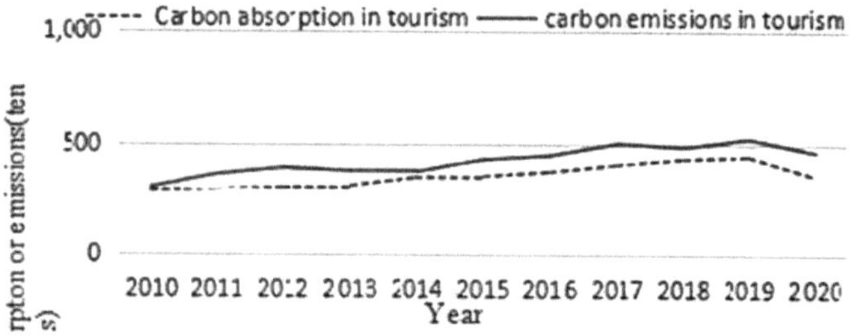

Fig. 3 Dynamic changes in carbon emissions and carbon absorption in tourism in Hainan Province

emissions of tourism have always been greater than those of tourism, showing a carbon deficit. From 2010 to 2020, the average annual carbon emissions and annual carbon absorption of tourism were 4.22076 million tons and 3.52715 million tons, respectively, showing an average annual carbon deficit of 693,600 tons. It shows that the carbon sinks of the tourism ecosystem cannot completely offset the carbon emissions released by tourism development. Therefore, Hainan Province tourism is a significant carbon source. As an increasingly important economic pillar industry in modern society, tourism is one of the future development trends of low-carbon tourism. Therefore, the tourism industry in Hainan Province must pay attention to energy conservation, emission reduction and ecological exchange rate increase, strengthen the construction of tourism low-carbon development strategy, and ensure the sustainable and healthy development of tourism.

4 Results and Analysis

4.1 Main Research Conclusions

The assessment of the carbon balance information system of tourism is an important basis for quantitatively assessing the impact of tourism on global climate change and formulating energy conservation and emission reduction policies to improve carbon efficiency. Through a comprehensive assessment of carbon emissions, carbon absorption, and carbon balance information system in tourism in Hainan Province, the following main conclusions can be drawn:

(1) From 2010 to 2020, the overall carbon emissions of tourism in Hainan Province showed a volatile growth trend, from 3.02972 million tons in 2010 to 4.61407 million tons in 2020. Among them, tourism transportation, warehousing, and post and telecommunications are the main sources of carbon emissions from tourism, accounting for about 84% of the total carbon emissions in tourism every year.

(2) From 2010 to 2020, the carbon absorption of the tourism ecosystem in Hainan Province showed a slow growth trend, from 2.78928 million tons in 2020 to 3.52312 million tons in 2020. Among them, the tourism aquatic ecosystem has the highest carbon absorption capacity, followed by the tourism forest ecosystem, then the tourism farmland ecosystem, and finally the tourism grassland ecosystem. The contribution rate to the total carbon absorption of the tourism ecosystem is about 52.574%, 41.911%, 5.472%, 0.043%, respectively.
(3) From 2010 to 2020, the carbon emissions of tourism in Hainan Province have been in a carbon deficit. The average annual deficit of the tourism industry in Hainan Province from 2010 to 2020 was 693,600 tons, showing a serious carbon imbalance. The tourism industry should take some responsibility for local environmental risks. At the same time, it also shows the necessity and urgency of changing energy-intensive economic growth and increasing the carbon sinks of regional ecosystems.

4.2 *Related Recommendations*

Energy conservation, emission reduction, and exchange rate increase are the "dual engines" to promote the development of low-carbon tourism. Only when both work together can the low-carbon transformation of tourism in Hainan Province be realized. Specifically, it includes.

(1) In terms of energy conservation and emission reduction, tourism transportation, warehousing, and post and telecommunications dominate the carbon emissions of tourism. Controlling carbon emissions from transportation is the focus of developing low-carbon tourism in Hainan Province. Therefore, we should actively promote the low carbonization of transportation modes, build a public service system for tourism transportation, improve the transportation turnover efficiency of units, and vigorously advocate low-carbon travel modes such as railways, electric vehicles, and bicycles. Secondly, change the focus of marketing, increase publicity in and around the city, and focus on developing the short- and medium-distance tourism source market.
(2) In terms of increasing exchange rates, relevant departments should actively implement forestry policies, do a good job in afforestation, disease and pest control, water and soil conservation, etc., effectively improve forest vegetation coverage, and give full play to their carbon sink potential. Secondly, we should strengthen farmland water conservancy construction, adjust crop structure according to local conditions, improve the technical conditions of agricultural production, and strive to increase crop carbon sinks and reduce carbon emissions in agricultural production. Third, strengthen the restoration and management of grassland, scientifically calculate the productivity of the grassland system, and reasonably control the grassland load. Finally, attention

should be paid to protecting forest vegetation, increasing water conservation, moderately expanding the area of waters and surrounding wetlands, and focusing on improving their carbon absorption capacity.

Acknowledgements This work was financially supported by (1) Hainan Provincial Natural Science Foundation of China in 2022 (722MS064): Research on resource and environment coordination mechanism and realization path driven by county innovation in Hainan Province under the background of Rural Revitalization; (2) Haikou Federation of Social Sciences 2022 Social Science Planning Project (2022—ZCKT—22): The integration mechanism and implementation path of international carbon emissions trading and supervision sandbox on Hainan Free Trade Island; (3) Key topics of Hainan Free Trade Port Finance Society in 2022 (Qiongjin Xuezi [2022] Item 43 of Project No. 13): Hainan Free Trade Port establishes a carbon trading financial innovation research focusing on "regulatory sandbox"; (4) 2018 National Social Science Fund Project (18BJL017), Project Name: Study on Marx's Theory of Ground Rent and Deepening the Reform of Rural Land System; (5) Talents project of Hainan Normal University of 2021; (6) The 2022 United Front Project of Hainan Normal University (Haishi Dang Tong Han [2002] No. 003, Item 10): The Historical Experience of China's Silk Road and the Construction of a Community with a Shared Future for Mankind.

References

1. The UN Tourism Organization (UNWTO): Towards a Low Carbon Travel and Tourism Sector [R]. The World Economic Forum (2009)
2. Dogan, E., Seker, F., Bulbul, S.: Investigating the impacts of energy consumption, real GDP, tourism and trade on CO_2 emissions by accounting for cross-sectional dependence: a panel study of OECD countries. Curr. Issue Tour. **20**(16), 1701–1719 (2017)
3. Huang, H.P., Qiao, X.Z., Zhang, J., Li, Y.L., Zeng, Y.: Research on spatiotemporal differentiation and influencing factors of regional tourism in the context of green development: Taking the Yangtze River Economic Belt as an example. Econ. Geogr. **39**(11), 214–224 (2019)
4. Zhong, Y.D., Shi, S.Y., Li, S.H., Luo, F., Luo, W.L.: Construction and empirical research on carbon emission measurement methods in China's tourism industry: based on the perspective of input and output. J. Central South Univ. For. Technol. **35**(01), 132–139+144 (2015)
5. Han, Y.J., Wu, P.: Calculation and comparison of carbon emissions from tourism in Beijing-Tianjin-Hebei. Hum. Geogr. **31**(04), 127–134 (2016)
6. Wiedmann, T., Minx, J.: A definition of 'carbon footprint.' Ecol. Econ. Res. Trends **1**, 1–11 (2008)
7. Peeters, P., Dubois, G.: Tourism travel under climate change mitigation constraints. J. Transp. Geogr. **18**(3), 447–557 (2009)
8. Shi, P.H., Wu, P.: Preliminary estimates of energy consumption and CO_2 emissions from China's tourism industry. Acta Geogr. Sin. **66**(2), 235–243 (2011)
9. Huang, H.P., Qiao, X.Z., Zhang, J.: Analysis of the spatial and temporal evolution of carbon emissions from tourism in the Yangtze River Economic Belt. Guizhou Soc. Sci. **2**, 143–152 (2019)
10. Zhao, X.C., Teng, J.: Decomposition of factors affecting carbon emissions from tourism in Hunan Province based on LMDI. Environ. Sci. Technol. **41**(09), 192–199 (2018)
11. Ding, Y.L., Liu, D.W., Zhang, F.Q., Shen, J.F.: Study on the estimation of net carbon emissions from tourism in Chaji ancient villages in southern Anhui Province. Resour. Dev. Mark. **34**(07), 935–940 (2018)

12. Xing, P.F., Li, G., Chen, X.P., Li, D.Y., Wang, C.H., Dong, K.H., Zhao, X.: The influence of grazing on carbon exchange in the grassland ecosystem in the agricultural and pastoral interlaced zone in northern Jin. Acta Prataculturae Sin. **28**(10), 1–11 (2019)
13. Yin, J.P., Zhang, Y.X., Fu, Y., Wang, X.J.: Study on carbon emissions and the carbon absorption potential of forest vegetation in China. For. Resour. Manag. **3**, 53–61 (2021). https://doi.org/10.13466/j.cnki.lyzygl.2021.03.009
14. Li, M.Q., Liu, S.L., Wu, X., Sun, Y.X., Hou, X.Y., Zhao, S.: Spatiotemporal changes of the carbon footprint of farmland ecosystems in Yunnan Province and their influencing factors. Acta Ecol. Sin. **38**(24), 8822–8834 (2018)
15. Li, S.X., Chen, Z.H.: Carbon balance analysis and low-carbon development of tourism in Yangquan City, Shanxi Province. Journal of Shaanxi Normal University (Natural Science Edition) **49**(06), 64–74 (2021). https://doi.org/10.15983/j.cnki.jsnu.2021.04.019
16. Han, Y.J., Wu, P., Lin, T.: Calculation and comparative analysis of the efficiency of tourism industry in representative provinces based on carbon emissions. Geogr. Res. **34**(1 0), 1957–1970 (2015)
17. Chen, F., Da, J.: Theoretical methods of low-carbon urban research and empirical analysis in Shanghai. Urban Dev. Stud. **16**(10), 71–79 (2009)
18. Xie, H.Y., Chen, X.S., Lin, K.R., Hu, A.Y.: Fossil energy and electricity ecological footprint based on carbon cycle. Acta Ecol. Sin. **4**, 1729–1735 (2008)
19. Fang, J.Y., Guo, Z.D., Piao, S.L., Chen, A.P.: Estimate of carbon sinks of terrestrial vegetation in China from 1981 to 2000. Scientia Sinica (Terrae) **6**, 804–812 (2007)
20. Yan, G.A., Liu, Y.D.: Carbon cycle of aquatic ecosystems and sinks of atmospheric CO_2. Acta Ecol. Sin. **5**, 827–833 (2001)
21. Tian, Y., Zhang, J.B., Wu, X.R., Cheng, L.L.: Dynamic changes and regional differences in carbon sink surpluses in China's planting industry - based on panel data from 2000 to 2012 in 31 provinces (municipalities and districts). J. Nat. Resour. **30**(11), 1885–1895 (2015)
22. Zhao, R.Q., Huang, A.M., Qin, M.Z., Yang, H.: Research progress in carbon sink increase/emission reduction technology in farmland ecosystems in China. J. Henan Univ. (Nat. Sci.) **1**, 60–65 (2004). https://doi.org/10.15991/j.cnki.411100.2004.01.014
23. Zhou, N.X., Huang, Z.F., Liang, Y.Y.: Measurement and equilibrium of carbon sources and sinks in Lushan scenic spots. Acta Ecol. Sin. **33**(13), 4134–4145 (2013)

Power Data Sharing Model Based on Blockchain and IPFS Storage

Kaiqiang Xian, Xiaolong Wang, Qianjun Wu, Caijun Zhang, and Jiayi Lang

Abstract With the rapid development of information technology, the electric power industry has also accelerated informatization process. To improve the efficiency of information management and optimize customer service, it is necessary to realize data sharing among electric power enterprises. Due to the privacy, polymorphism, and huge amount of power data, it is very difficult for power enterprises to connect with each other, and it is difficult to realize safe sharing and data storage in the process of power data sharing. To solve the security problem of data sharing, this paper introduces blockchain technology and designs a power data sharing model. Using traditional blockchain technology to store a large amount of data will seriously affect its performance, resulting in its slow running speed. Therefore, this paper designs a data storage network based on IPFS technology, which improves the efficiency of data transmission and storage. Experiments show that this model has good effect.

1 Introduction

In recent years, electric power enterprises continue to optimize customer service system by informational zed management. Remarkable achievements have been made in data fusion and business capability improvement. However, the existing data sharing channels in many industries have not been really opened, and there

K. Xian
Marketing Service Centre, State Grid Jiangsu Electric Power Co., LTD. Nanjing, Jiangsu 211000, China

X. Wang (✉) · Q. Wu
State Grid Electric Power Research Institute Co. LTD, Nanjing 211106, Jiangsu, China
e-mail: wblanksky@hotmail.com

C. Zhang · J. Lang
Customer Service Centre, State Grid Co., LTD, Tianjin 300309, China

K. Nakamatsu et al. (eds.), *Advanced Intelligent Technologies for Information and Communication*, Smart Innovation, Systems and Technologies 365,
https://doi.org/10.1007/978-981-99-5203-8_6

are problems such as low real-time data exchange, difficult data life-cycle management, and low data correctness and security. Information connectivity is difficult, and "information islands" still exist. Both internal and external enterprises have not formed a mature and reliable trusted data sharing mechanism, and some businesses that require internal and external data trust and sharing are difficult to complete. Due to the confidentiality of enterprise data, enterprises cannot use mainstream cloud servers for data sharing, and cloud servers with third-party organizations as central nodes have problems such as system crash and data leakage.

To prevent security risks brought by centralized services, ensure data security, and realize trusted data sharing between nodes, we can use blockchain technology to realize decentralized secure storage of data and third-party trust. Blockchain technology has developed rapidly in recent years due to its decentralized and tamper-proof characteristics. Several scholars have proposed the use of blockchain for data sharing. Wang et al. proposed a blockchain-based data sharing scheme to solve problems such as distrust, privacy issues, data abuse, and asymmetric valuation of shared data between entities in the supply chain [1]. Wei et al. proposed a cross-Internet of Things data access control framework based on blockchain [2]. Tanwar et al. solved the recording and sharing of medical data using blockchain [3]. Ziteng et al. implemented the international trade L/C payment system based on the Fabric framework of alliance chain [4].

In this paper, combined with distributed file system protocol IPFS (InterPlanetary File System) [9], we propose a power data sharing model based on blockchain and IPFS storage to solve the problem of privacy information protection in power data sharing [5, 6] and the problem of slow blockchain speed caused by large amount of data storage [7, 8], so as to realize a mutual trust, transparent and reliable power data sharing mechanism.

2 Shared Architecture and Storage Methods

This section describes the data sharing architecture and the IPFS network storage method in the data sharing architecture in detail. In Sect. 2.1, the structure division and specific functions of the data sharing method are analyzed, and in Sect. 2.2, the IPFS storage network in the data sharing model is described in detail.

3 Data Sharing Method

To realize the efficient sharing of power data, the role of data is divided into system administrator, institutional node, and institutional user. The alliance chain is used to build the data sharing platform. Institutions access the alliance chain in the way of nodes, and the system administrator manages the institutional nodes and builds IPFS network for data storage. The data sharing architecture is shown in Fig. 1.

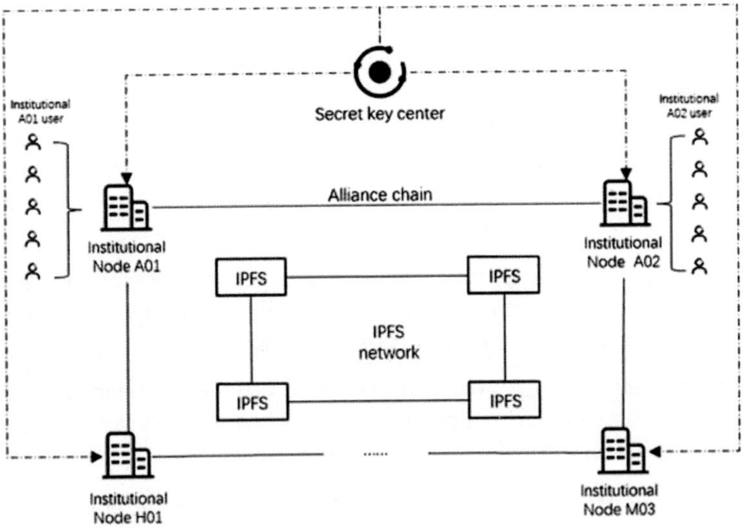

Fig. 1 Data sharing architecture

System role division. The data sharing system is mainly divided into two parts:

One is the intra-institutional system, which consists of institutional node, institutional users, and institutional databases. Institutional users are the main users of the data sharing system, responsible for uploading data from the organization and applying for data sharing. The institutional node is responsible for reviewing the operations of institutional users, and after the review is passed, it is used as a blockchain node to access and execute these operations, and manage institutional users and institutional databases.

The other is the blockchain sharing system, which consists of system administrators, institutional nodes, and IPFS network. The institutional nodes use metadata to perform sharing operations on the blockchain. The IPFS network is responsible for storing the complete data, and the system administrator manages the institutional nodes. Key roles in the system include the following:

System administrator: The system administrator is the supervisory node of the system, responsible for registering institutional nodes, setting institutional information, executing smart contracts and other operations, and dynamically constructing a data sharing list to ensure that authorized nodes can share specified information.

Institutional node: As the basic node of blockchain, it is connected to the blockchain, responsible for managing institutional users of the institution, reviewing the application sharing and uploading of institutional users. After passing the operation, the data processing request will be initiated from the institutional node, and the data sharing record on the chain will be recorded by this node. The data storage will be transmitted to the IPFS system for processing and storage with the organization node as the core, and the returned address will be encrypted and linked to complete data uploading and sharing.

Institutional user: An institutional user who is dependent on the institutional node and is responsible for initiating data sharing applications and uploading data. The operation is audited by the institutional node, and the shared data is stored and managed by the institutional node.

Secret key center: This model uses the proxy re-encryption method [10, 11] to secure the system. The secret key center is responsible for distributing keys to institutional nodes and system administrators and encrypting data.

Data sharing function. Power data sharing mainly includes data entry, request data sharing, review data sharing, transfer sharing data, sharing data storage, and other functions, some data sharing function is shown in Fig. 2.

(1) **Data entry:** After reviewing the data upload application, the organization will generate the metadata of the input data, including the key information needed for data sharing, and input the metadata into the blockchain. According to the data storage mechanism proposed in this paper, data is passed into IPFS and the Hash address returned is stored waiting for data sharing application.

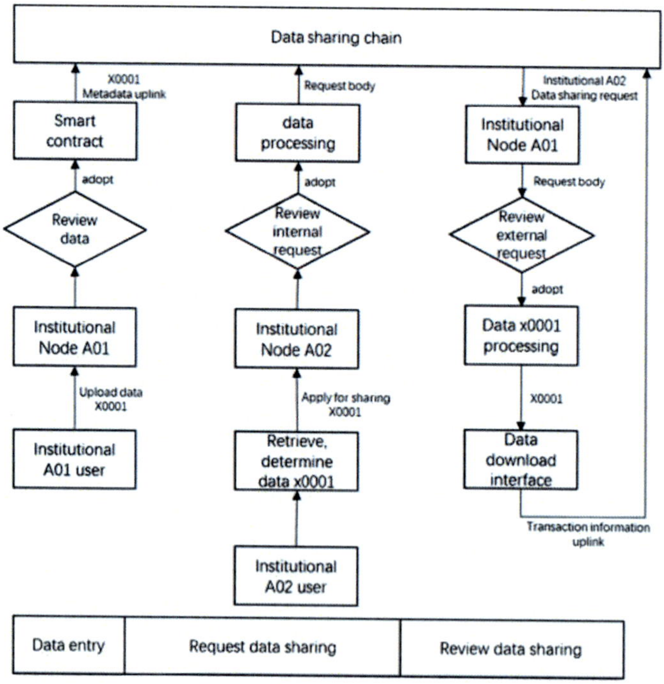

Fig. 2 Data sharing function

(2) **Data retrieval:** Retrieves shared data in the system through the data information carried by metadata.
(3) **Request for data sharing:** The data sharing request is initiated by the institutional user, and the request body is generated after the review by the institutional node. The request body is sent to the institutional node holding the data in the form of blockchain transaction, including shared application node, target node, time stamp, data information, remarks information, etc.
(4) **Review data sharing requests:** The institutional node that holds data reviews the data sharing requests of other nodes and determines whether to share data with this node. If yes, use the public key of the target node to encrypt the Hash address of IPFS data according to the smart contract to complete the data sharing.
(5) **Data download:** After the data sharing is completed, the application node can obtain the IPFs address of the target data from the data sharing system, and use the account private key to obtain information from the IPFs storage network.

4 IPFS Storage Network

In the grid data sharing model, there will be a huge amount of sharing files, which uses different formats and types [12]. If all data need to be stored on the blockchain, the operation efficiency of the blockchain will be seriously affected, so the system is difficult to meet the requirements of storing shared data. IPFS (InterPlanetary File System) is a technology that manages files in a distributed way [13]. It is a point-to-point protocol that addresses data content. Large files are divided into blocks of fixed size and stored in distributed nodes. The files generate unique hash values as the addressing addresses.

Combined with IPFS, this paper designs a shared data storage method to distribute shared data to the off-chain storage system, which solves the problem of massive data storage on the blockchain.

IPFS storage module design. The main function of IPFS storage module is to transfer and store encrypted shared data. It is based on the Hash address value and the content addressing feature of the data binding, so that nodes can get the corresponding data based on the Hash value.

Institutional nodes own ownership of their data and interact with the IPFS storage system until shared transaction data is generated in the blockchain network. After agreeing to share the new data with other institutional nodes, the institutional node encrypts the data with the generated symmetric key, and the encrypted shared data is uploaded by the institutional node to the IPFS storage network. After the encrypted data is uploaded to IPFS, the system generates a hash address for the shared data, which is used as the address of the encrypted data uploaded by the accessing institutional node. Once an institutional node obtains an IPFS address, it uses the IPFS

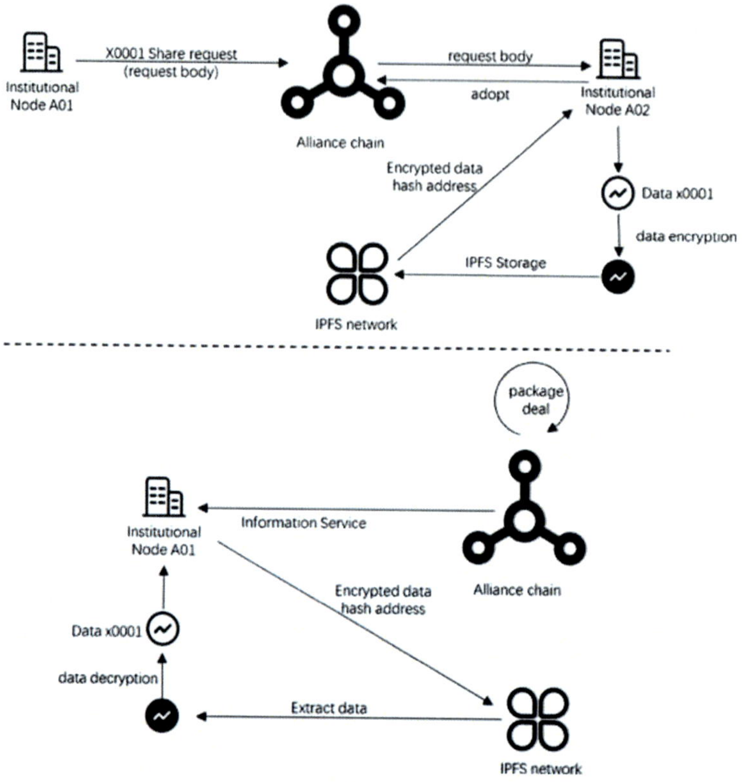

Fig. 3 Interaction between institutional nodes and IPFS network

address to generate a shared data transaction, which is broadcasted to other nodes in the blockchain network. The main interaction process between institutional nodes and IPFS network is shown in Fig. 3.

The main process of data storage in IPFS storage network is as follows: Merkel directed acyclic graph interface organizes data into a Merkel directed acyclic graph structure and returns the IPFS hash value of the following nodes of the directed acyclic graph.

5 Experiments and Analysis

This section describes the experiment and initial results of the system. To test the performance of the data sharing system, Ganache2.1.2 and GO IPFS 0.6.0 are used to simulate the system alliance chain nodes and IPFS storage network, and 100 data sharing nodes are generated with Ganache.

Transmission efficiency test. In this test, three hosts were used to build IPFS nodes to form IPFS storage system, and an FTP server was set up as a comparison test. In the IPFS storage system, upload and download speeds of files of different sizes are tested and compared with those of FTP. The results are as follows (Figs. 4 and 5):

In Table 1, the calculation formula of the optimization rate is

$$O_c = (t_{FTP} - t_{IPFS})/t_{FTP} \tag{1}$$

The O_c is transmission efficiency, t_{FTP} is the time required for FTP transmission, and t_{IPFS} is the IPFS transmission time required. In the test results, IPFS storage network has an average optimization rate of 93.16% in data entry and an improvement of 37.52% in data download, which greatly optimizes the storage efficiency of data sharing.

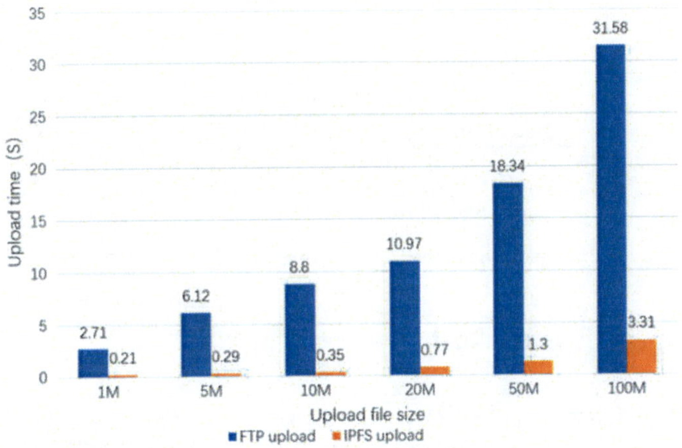

Fig. 4 Comparison of data upload time

Fig. 5 Comparison of data download time

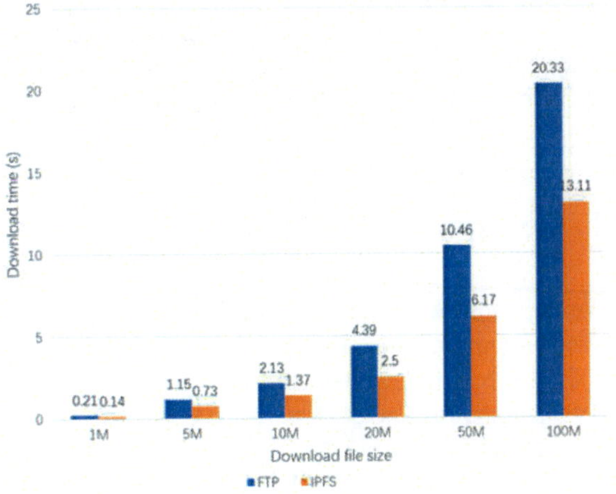

Table 1 Time optimization rate

	Min. optimization rate (%)	Avg. optimization rate (%)	Max. optimization rate (%)
Upload	89.52	93.16	96.02
Download	33.33	37.52	43.05

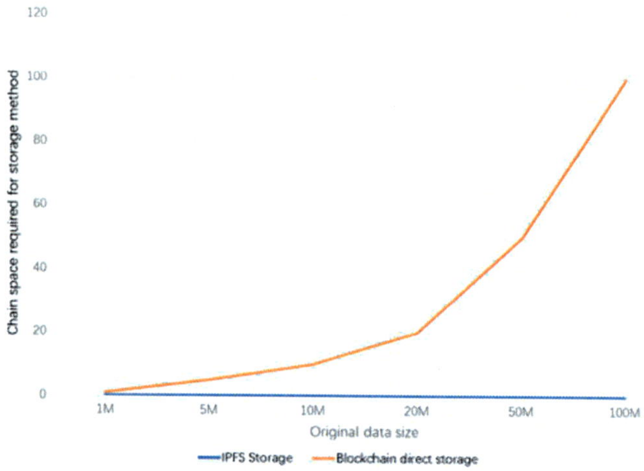

Fig. 6 Storage space comparison between IPFS method and blockchain direct storage method

6 Storage Efficiency Test

To calculate the storage space data required by the two storage methods on the chain, the on-chain storage space required by the IPFS storage method is the metadata storage space and the Hash value returned by the IPFS system after storage. However, the storage space required by the direct storage on the chain method is the space required by the complete data (Fig. 6).

As can be seen from Fig. 6, the storage space required by IPFS storage method is much smaller than that of blockchain direct storage. And the larger the original data is, the more efficient the IPFS storage method is.

7 Retrieval Efficiency Test

We compare the centralized storage with the storage method proposed in this paper in terms of retrieval time. Compare using file sizes (KB) of 50, 100, 150, 200, 250, and 300. The experimental results are shown in Fig. 7, the results show that the proposed IPFS storage for file retrieval requires less time.

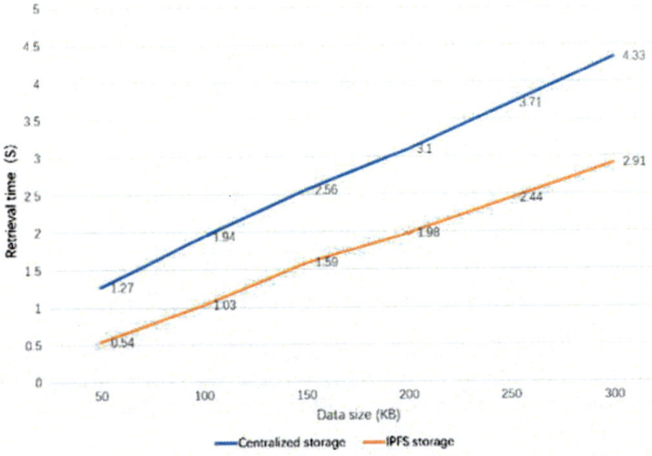

Fig. 7 Retrieval time comparison between centralized storage and IPFS storage method

8 Conclusion

This paper first proposes a power data sharing model based on blockchain and IPFS storage, and divides the roles in the data sharing blockchain into management nodes, institutional nodes, institutional users, and secret key centers. The institutional nodes and the institutional users establish the institutional internal system; complete the institutional internal data management and operation review; establish blockchain sharing system among institutional nodes, management node. and secret key center; build alliance chain; and use proxy re-encryption mechanism to ensure system security. IPFS storage network is designed to store power sharing data, solve the problem of massive polymorphic data storage in blockchain sharing system, and improve data storage efficiency and transmission efficiency.

The data sharing system and IPFS network are implemented to verify the feasibility of the proposed scheme. The experimental results show that IPFS not only solves the problem of storage, but also has excellent transmission efficiency.

The model proposed in this paper realizes the sharing of power grid data and ensures the security and efficiency of data storage as well as the reliable storage of shared data. However, there is still room for improvement in blockchain data retrieval. We will focus on improving the efficiency of data retrieval in the model in the following research.

Acknowledgements This research was funded by the Science and Technology Project of The Headquarters of State Grid Corporation of China (Contract no. 5700-202153179A-0-0-00).

References

1. Wang, Z., Zheng, Z., Jiang, W., Tang, S.: Blockchain-enabled data sharing in supply chains: model. Prod. Oper. Manag. Prod. Oper. Manag. Soc. **30**(7), 1965–1985 (2021)
2. Wei, X., Yan, Y., Guo, S., Qiu, X., Qi, F.: Secure data sharing: blockchain-enabled data access control framework for IoT. IEEE Internet Things J. **9**(11), 8143–8153 (2022)
3. Tanwar, S., Parekh, K., Evans, R.: Blockchain-based electronic healthcare record system for healthcare 4.0 applications. J. Inf. Secur. Appl. **50**, 102407 (2020). Elsevier, Amsterdam
4. Ziteng, L.: Design and Implementation of International Trade L/C Payment System Based on Fabric [D]. Huazhong University of Science and Technology, Hubei (2018)
5. Moussaoui, D., Kadri, B., Feham, M., Ammar Bensaber, B.: A distributed blockchain based PKI (BCPKI) architecture to enhance privacy in VANET. In: The 2nd International Workshop on Human-Centric Smart Environments for Health and Well-being (IHSH), pp. 75–79 (2021)
6. Miyamae, T. et al.: ZGridBC: zero-knowledge proof based scalable and private blockchain platform for smart grid. In: 2021 IEEE International Conference on Blockchain and Cryptocurrency (ICBC), pp. 1–3 (2021)
7. dos Santos Abreu, A.W., Coutinho, E.F., Bezerra, C.I.M.: Performance evaluation of data transactions in blockchain. IEEE Lat. Am. Trans. **20**(3), 409–416 (2022)
8. Ahmad, A., Saad, M., Al Ghamdi, M., Nyang, D., Mohaisen, D.: BlockTrail: a service for secure and transparent blockchain-driven audit trails. IEEE Syst. J. **16**(1), 1367–1378 (2022)
9. Rahalkar, C., Gujar, D.: Content addressed P2P file system for the web with blockchain-based meta-data integrity. In: 2019 International Conference on Advances in Computing, Communication and Control (ICAC3), pp. 1–4. IEEE, New York (2019)
10. Ge, C., Liu, Z., Xia, J., Fang, L.: Revocable identity-based broadcast proxy re-encryption for data sharing in clouds. IEEE Trans. Dependable Secur. Comput. **18**(3), 1214–1226 (2021)
11. Su, Z., Wang, H., Wang, H., Shi, X.: A financial data security sharing solution based on blockchain technology and proxy re-encryption technology. In: 2020 IEEE 3rd International Conference of Safe Production and Informatization (IICSPI), pp. 462–465 (2020)
12. Dwivedi, S.K., Amin, R., Vollala, S.: Blockchain-based secured IPFS-enable event storage technique with authentication protocol in VANET. IEEE/CAA J. Autom. Sin. **8**(12), 1913–1922 (2021)
13. Ullah, Z., Raza, B., Shah, H., Khan, S., Waheed, A.: Towards blockchain-based secure storage and trusted data sharing scheme for IoT environment. In IEEE Access, vol. 10, pp. 36978–36994. IEEE, New York (2022)

A Steady-State Model on Finger-Vein Recognition Accuracy

Shilei Liu, Qin Li, Geng Yang, Xudong Lin, and Zhenqi Zheng

Abstract Finger-vein recognition is one of the most focused new biometric methods in recent years, with unique advantages such as non-contact, natural liveness detection, difficult to counterfeit, etc. However, like all the biometric recognition technologies, it has two fundamental premises: uniqueness and persistence. Uniqueness guarantees finger-vein from different classes is strictly different and is able to be proved and validated by real usage and its data in certain levels. As for consistency, there are few academic research results that indicate the timely stability of a same finger-vein class and that every sample from the same finger will be surely categorized into the class. In this paper, we will propose a finger-vein steady-state model to explain and to simulate the fluctuation of finger-vein in-class match scores and use finger-vein data across 4 years to verify the model.

1 Introduction

Since the 1970s, biometric authentication has been a rising trend. Biometric features, including (1) biological ones, such as fingerprint, palmprint, iris, face, vein, etc. and (2) behavioral ones, such as voice, signature, gait, etc., are of four shared characteristics [1, 2] on their capabilities for recognition and verification usage: ***universality***, ***measurability***, ***uniqueness,*** and ***consistency*** (i.e., **stability** in this paper). This paper aims to prove the consistency on type of finger-vein features and further provides a model to describe the dynamical perturbation of in-class finger-vein matching. For this purpose, we have done preparation work including (1) choose proper acquisition device and algorithms for image pre-processing, feature enrollment, and matching; (2) build finger-vein datasets with good quality, large scale, and trustworthy time duration; and (3) observe inner-class matching rate statistics and gain insights.

S. Liu · Q. Li (✉) · G. Yang · X. Lin · Z. Zheng
Shenzhen Institute of Information Technology, Shenzhen 518172, China
e-mail: liqin@sziit.edu.cn

S. Liu
The Hong Kong Polytechnique University, Hong Kong 999077, China

K. Nakamatsu et al. (eds.), *Advanced Intelligent Technologies for Information and Communication*, Smart Innovation, Systems and Technologies 365,
https://doi.org/10.1007/978-981-99-5203-8_7

1.1 Why Stability

On theoretical aspects, stability is one of the foundations of biometric verification. It guarantees a specific kind of biometric feature which is capable of proving "I am myself".

In real biometric verification applications, however, the in-class matching scores of finger-vein images/templates are in constant fluctuation, and sometimes the score might even flux to under the threshold, leading to false non-matching occurrences. FNMR[3] (false rejection rate, i.e., FRR) is a critical index on authentication accuracy. In our previous experiments, FNMR has more impact on EER [3] (equal error rate) than FMR (false matching rate) in our implemented finger-vein verification systems, as shown in Fig. 1.

There are a lot of reasons for the oscillation of matching scores within the same finger-vein class. For internal possible factors, (1) the enroll and match algorithms might not be good enough; (2) human health status is unstable and hence projecting to vein images [4]. For external possible factors, changes of acquisition postures [11], illumination conditions, humidity, lens cleanliness, etc. are all commonly seen impact factors in real verification systems. Hence, one of the main purposes of this paper is to study on the inner-class matching oscillation and try to draw conclusions on stability.

As for uniqueness and other biometric characteristics, we might cover them in other following-up experiments with large scale of between-class matching statistics with more diversity on sampling people group, longer sample year duration, and different levels of acquired image quality, during which FMR will play a more critical role compared to FNMR.

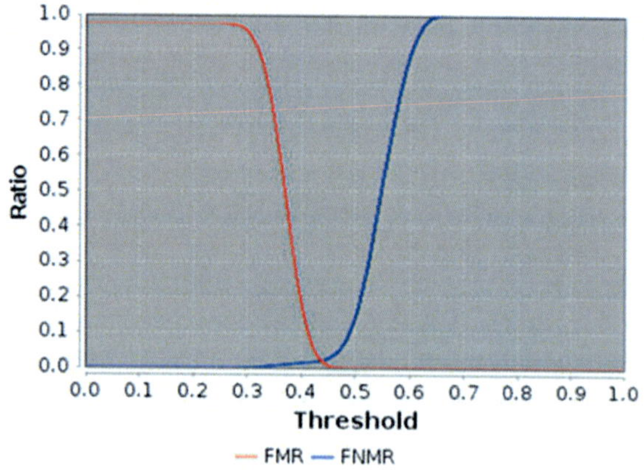

Fig. 1 How FNMR impacts in the matching algorithm

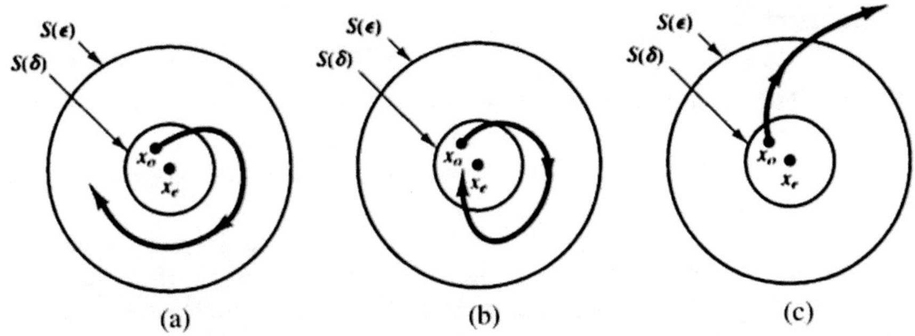

Fig. 2 **a** Lyapunov stable; **b** Asymptotically stable; **c** Instable

1.2 What is Steady State

In modern control theory, Lyapunov defined two types of stable states in dynamical systems[4]: *Lyapunov Stable* and *Asymptotically Stable*, as shown in Fig. 2.

Inspired by Lyapunov's stability analysis, this paper defines three types of stability in Biometrics systems: *Distribution Stability*, *Asymptotical Stability*, and *Convergence Stability*. (We will cover more details in the steady-state model elaboration section.) So, we will conduct corresponding experiments and use results to analysis whether these three types of stability stand.

1.3 Main Contributions

For this paper, our main contributions include the following:

- Choose proper devices and algorithms, collect, and build a finger-vein dataset, with a 4-year span and more than 280,000 registered fingers, over 3,000,000 finger-vein samples and 2,500,000 match records.
- Propose a steady-state model to explain and simulate in-class matching scores.
- Validation on the proposed steady-state model with data analysis using pre-built dataset.

We hope to extend stability on finger-vein into other biometric features by observing stats and building a model upon it. Therefore, in the future, it is possible to come up with universal stability standards, providing guidance on the restrictions on biometric feature's database, matching algorithm, and matching results threshold. If a biometric feature can have expected performance on its selected database, then researcher/developers can claim that the certain biometric feature has already met the corresponding stable standard.

2 Related Work

Stability is first a concept in control theory. It focuses on **dynamical system**, which is a system in which a function describes the time dependence of a point in an ambient space. Previous research on stability expands from mathematics to mechanics, but not on Biometrics.

Although stability (or known as consistency) is a fundamental problem in biometric area, few research has been done to answer it [5]. Some of the key limitations include fast iteration of acquisition device hardware and feature extraction algorithms and restricted time span in acquired data samples [6]. During recent years, some researches show focus on how big major health change impact human biological features, such as diabetes on gait analysis and Parkinson's syndrome on human voice pattern. But all these researches have two shortages: (1) not enough comparison between the same biometric class over long time period. Suitable medical samples are of tens the scale and thus might be unconvincing to some degrees. (2) No experiments have been conducted on healthy groups (e.g., young colleague students).

To summarize here, biometric topics are hot and smart health support systems have been built on breath [7], tongue color [8], vein [9], etc. But more work needs to be done to prove stability does exist for all kinds of biometric features, both behavioral and biological [10]. This paper will try to prove this within large amount of healthy human finger-vein classes.

3 The Steady-State Model

First, there are two assumptions of the finger-vein recognition/verification system for this paper:

1. It is a **nonlinear dynamical system** [13]. At any given time, a dynamical system has a state representing a point in an appropriate state space, often given by a tuple of real numbers or by a vector in a geometrical manifold. In our case here, this state is a one-dimensional float number normalized to interval [0, 1], indicating the distance between two high-dimensional vectors, i.e., normalized finger-vein templates with gray level [0, 255] and size of 128 * 96.
2. It is an **autonomous system** [14]. In mathematics, an autonomous system or autonomous differential equation is a system of ordinary differential equations which does not explicitly depend on the independent variable. We here too take time as the independent variable, thus making the system time-invariant. Use the form $\frac{d}{dt}x(t) = f(x(t))$, where x denotes distance between vectors and t denotes time.

Given these two assumptions, the steady state is then proposed in such a nonlinear autonomous dynamical system. The concept of steady state itself stands for a virtual centralized template, who has the average shortest distance (based on any chosen matching algorithms) to all the templates within the same biometric feature (finger-vein) class. If we can prove for every biometric class this steady state exists, and if we can prove every acquired sample presents a converging trend to it, with observable fluctuation, we then can conclude on stability stands for these biometric features.

Hence the model consists of two parts:

1. For every finger-vein class, we can find a physical sample as the centralized template (steady state) with shortest in-class distances with all sibling samples.
2. During a long enough time, acquired samples will be converging to the steady state, with fast converging speed.

Therefore, we define the steady-state model of finger-vein features as follows:

In a system $\dot{X} = f(x(t)), x(0) = x_0, x(t) \in D \subseteq \mathbb{R}^n$ (state vector, 128 * 96 here). $f : D \to \mathbb{R}^n$ continuous on the open set D (origin included). We call x_e the steady state, if $f(x_e) = 0$.

To calculate this virtual center x_e is a heavy computing task and might require neural network, so we simplify it here with a physical center: one template that has shortest distance with all other templates within same class, defined as steady state in this paper.

We assume this steady state has three properties to prove stability:

1. ***Distribution Stability***. Given a reasonable threshold from EER, the majority of the matching will not see any false rejection situation. Only a very small number of genuine matching targeting on small number of finger-vein classes will see lower than threshold scores. We will use statistics in later experiments to prove if it stands.
2. ***Asymptotical Stability***. all matching that starts out near x_e converge to x_e. in other words, in this recognition system, there will be no recognition series that end up with failure as the final results.

   In our experiments, we make sure every verification (in-class matching) process ends with success within six trials.
3. ***Convergence Stability.*** for a continuing recognition series in 2, the converging speed is high and failure times before success is remarkably small, i.e., there exists $\delta > 0$ such that $\lim_{t\to\infty} \|x(0) - x_e\| = 0$,

   if $\|x(0) - x_e\| < \delta$.

   As for the converging speed, we can define $\alpha > 0, \beta > 0, \delta > 0$, such that:

   if $\|x(0) - x_e\| < \delta$, then $\|x(0) - x_e\| \le \alpha\|x(0) - x_e\|e^{-\beta t}$,
   for all $t \ge 0$.

At this point, the steady-state template x_e is defined and three properties of it proposed: *Distribution Stability, Asymptotical Stability*, and *Convergence Stability*. Now we need to prove these three properties stand so stability is proved. For this purpose, we proposed three parameters for the model:

(1) i for **i**llumination level, which is the gray-level description of the template [0, 255].
(2) d for feature **d**ensity, which is the non-zero pixel proportion in the template (128 * 96 pixels in total).
(3) t for **t**ainted pixel proportion of the original finger-vein image (512*384 pixels in total).

These three parameters i, d, t are calculated using an image quality labeling algorithm which is based on the work from our another paper, so we won't give details of the algorithm in this paper. As a clarification here, they are all normalized float values within [0, 1] interval.

To simplify the model, we use a linear function to predict the fluctuation for each template:

$$f = w_i * i + w_d * d + w_t * t$$

In our experiments, we use a training dataset to train values of w_i, w_d, w_t and evaluate the function in the test dataset. The results indicate that with proper parameter value, the function can predict the matching fluctuation well.

4 Experiment Results

4.1 Dataset and Algorithm Description

As we have mentioned in the introduction section, our collected finger-vein dataset DS-01 has more than 280 k finger-vein classes, 3 million samples, and 2.5 million matching records, with a time span of 5 years. All samples were acquired from young healthy people aged from 17 to 25. To make sure each class is valid, a minimum of three valid templates must be acquired for the class.

108 classes have been picked for timely matching records over the 5 years, with good image quality. And that is DS-02 (Table 1).

The experiments use wide line detector [13] as feature extraction algorithm in enroll and match processes. For the rest of the pre-processing steps, we use Beining Huang's work in 2012 [12], including.

Table 1 Dataset information

Dataset	Classes	Samples
DS-01	263,900	23,257,122
DS-02	108	14,277
Training dataset	179,460	13,955,311
Validation dataset	59,820	4,109,051
Test dataset	59,820	5,192,760

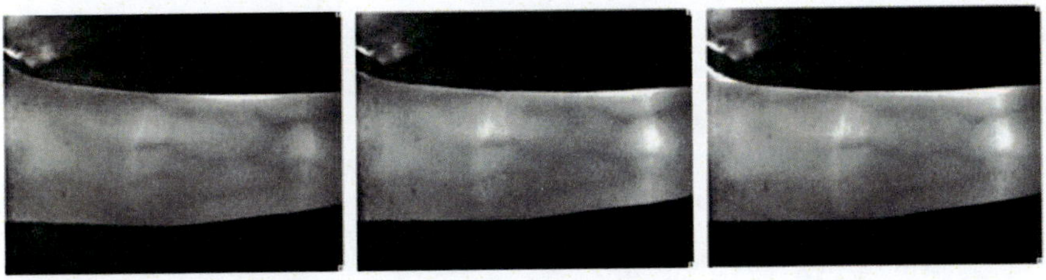

Fig. 3 A set of high within-class matched finger-vein samples

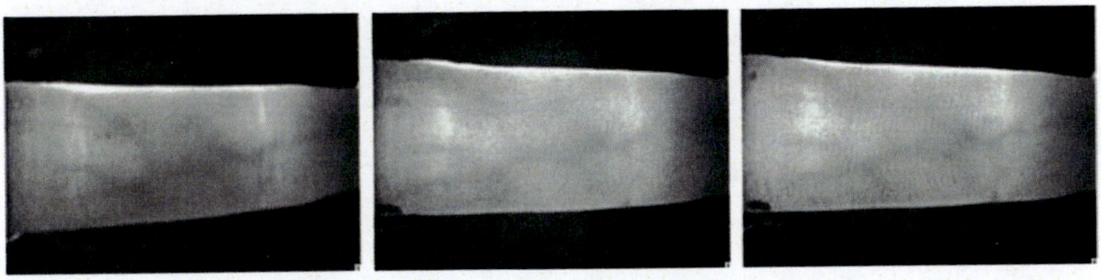

Fig. 4 A set of medium within-class matched finger-vein samples

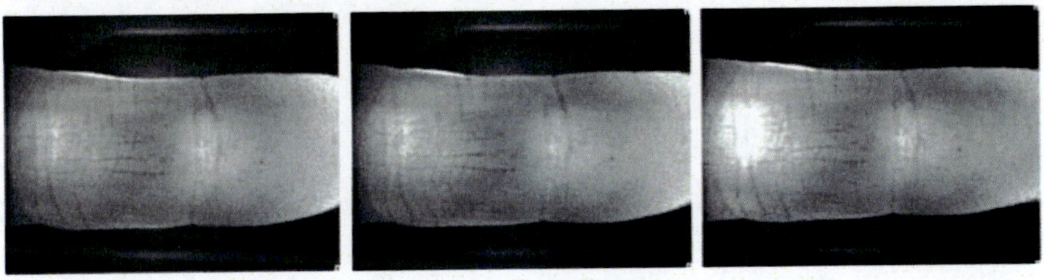

Fig. 5 A set of low within-class matched finger-vein samples

(1) noise reduction,
(2) median filtering (with 8*8 mask),
(3) compression (NNM), and
(4) normalization.

The original quality of finger-vein images is 512 * 384 * 8 bits. The calculated templates are binary 128 * 96 (Figs. 3, 4, 5 and 6).

For matching threshold, we use **0.35**. It is the actual threshold we've been using in real application for years after tuning with EER.

4.2 Experiments and Results

According to statistical results on distribution of genuine match and FR, *Distribution Stability* is satisfied, as is shown in Tables 2 and 3.

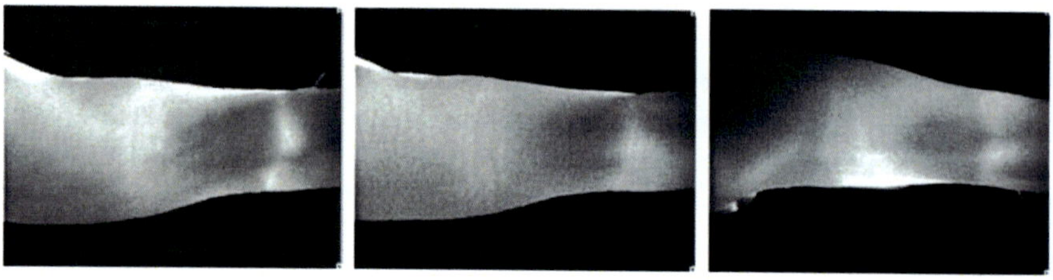

Fig. 6 A set of failed within-class matched finger-vein samples

Table 2 Distribution stability validation: success ratio of in-class matching in DS-01

Match result	Number of genuine match	Ratio (%)
Success	2,695,198	98.20
Failure	49,403	1.80

Table 3 Distribution stability validation: aggregated false rejection times

Times of false rejection in total	Number of classes	Ratio (%)
0	239,384	90.71
1	13,468	5.01
2	7,495	2.84
More than 2	3,512	1.44

According to statistical results on distribution of genuine match and FR, *Asymptotical Stability* is satisfied here, as shown in Table 4.

According to statistical results on FR times for finger classes which has experienced more than two times of FR situation, each finger will be converging to its central template in a fast pace, supporting the *Convergence Stability* assumption, as shown in Table 5.

Table 4 Asymptotical Stability validation: success ratio of in-class matching in DS-02

Match result	Number of genuine match	Ratio
Success	13,321	96.22%
Failure	523	3.78%

Table 5 Convergence stability validation: continuing FR failure series

FR times for a single series	Converging Rate	Case Number
Two times	12.59%	4,946
Three times	7.44%	2,923
>= four times	2.01%	790

Table 6 Steady-state model prediction accuracy on expected and actual fluctuation

Offset diff avg	Var	Prediction number
0.00290	0.00001	5,192,760

We use training set and SVM to train the model and use validation set to tuning. During the process we found w_t has little impact on the result so we removed it and get our best results on prediction accuracy in Table 6.

5 Conclusion

In this paper, the stability of finger-vein features is evaluated. We have collected finger-vein samples from healthy colleague students between 16 years old to 25, from year 2010 to 2015. Preparation includes building testing dataset and choosing corresponding algorithms including matching, enrolling, and quality labeling algorithms. After that we propose a steady-state finger-vein model, that all finger-vein features within same class will have a central sample/template, standing for the "steady state" of the belonging class. The model includes three main characteristics: Distribution Stability, Asymptotical Stability, and Convergence Stability, supported by the experiments. And we use ML method to reveal the impact of illumination level and feature density in the template and get a trained model that indicates we can predict the distance from every template from its steady state. The prediction accuracy is very inspiring.

Two possible directions for the future are improving the trained model and expanding the test subject's aging and health state. But for now, this steady-state model answers the fundamental question on biometric features' consistency and provides a generic method to evaluate how stable different types of biometric features are (e.g., setting up threshold for the converging speed α and β in a biometric finger application depends on business problem statement).

Acknowledgements This work is supported in part by the Intelligent Perception and Computing Innovation Platform of the Shenzhen Institute of Information Technology under Grant PT2019E001, "The Climbing Plan" of special fund for science and technology innovation strategy of Guangdong Province under Grant pdjh202a0982, and the Innovation Team Project of Colleges in Guangdong Province under Grant 2020KCXTD040.

References

1. Cattell, R.B.: Factor analysis: an introduction to essentials. The purpose and underlying models. Biometrics (1965)
2. Jain, A., Bolle, R., Pankanti, S.: Biometrics: personal identification in networked society (2006)

3. Karu, K., Jain, A.: Fingerprint classification. Pattern Recogn. (1996)
4. Lyapunov, A.M.: The General Problem of the Stability of Motion, translated by A. T. Fuller. Taylor & Francis, London. ISBN 978-0-7484-0062-1 (1992); Reviewed in detail by M. C. Smith: Automatica **3**(2), 353–356 (1995)
5. Boulton, A.J.M., Vileikyte, L., Ragnarson, G.: The global burden of diabetic foot disease. The Lancet (2005)
6. Raming, L.O., Sapir, S., Fox, C.: Changes in vocal loudness following intensive voice treatment in individuals with Parkinson's disease: a comparison with untreated patients and normal age-matched controls (2001)
7. Manolis, A.: The diagnostic potential of breath analysis. Clin. Chem. (1983)
8. Maciocia, G.: Tongue Diagnosis in Chinese Medicine. Eastland Press (1987)
9. Kono, M., Miura, N., Fujii, T., Ohmura, K., Yoshifuji, H., etc.: Personal authentication analysis using finger-vein patterns in patients with connective tissue diseases-possible associaton with vascular disease and seasonal change. PLOS ONE (2015)
10. Liu, S., Huang, B., Yu, Y., Li, W.: Biometric identification system's performance enhancement by improving registration progress. In: The 7th Chinese Conference on Biometric Recognition (2012)
11. Huang, B., Liu, S., Li, W.: A finger posture change correction method for finger-vein recognition. In: IEEE Symposium on Computational Intelligence for Security and Defense Applications (2012)
12. Liu, L., Zhang, D., You, J.: Detecting wide line using isotropic nonlinear filtering. IEEE Trans. Image Process. (2007)
13. Strogatz, S.H.: Nonlinear Dynamics and Chaos: with Applications to Physics, Biology and Chemistry. Perseus (2001)
14. Egwald Mathematics: Linear Algebra: Systems of Linear Differential Equations: Linear Stability Analysis. Accessed 10 October 2019

A Rail Obstacle Target Detection Technology Based on the Perspective of Unmanned Aerial Vehicle

Xing Yue Du, Yue Lin Xu, Guang Zhao Song, and Jian Jun Wang

Abstract Based on the target detection results of the rail obstacles from the perspective of the UAV, it can assist in obstacle breaking and landing operations. At present, the rail obstacle target technology from the perspective of UAV has difficulties such as small target areas and dense targets. This paper adopts the YOLO5 base model structure, and then adopts transfer learning, transformer self-attention mechanism, CBAM attention module, and multi-scale hole convolution to optimize the model structure. The final model can be run on small devices such as drones with good accuracy results. The experimental results show that the final index AP@0.5 is 0.89, AP@0.5:0.95 is 0.68, which verifies the effectiveness of the proposed technology.

1 Introduction

The rail obstacles are a kind of coastal fortifications composed of steel rails and concrete bases. They are often arranged in arrays on the coast to block landings and are the key targets in landing operations. UAVs have many advantages, such as small size, light weight, and low cost. At the same time, the UAV has strong maneuverability and good military concealment. Based on the target detection results of rail obstacles from the perspective of the UAV, it can assist in obstacle breaking and landing operations. At present, the rail obstacle target detection technology from the perspective of UAV has difficulties such as small target areas and dense targets. The improved yolo5 target detection algorithm proposed in this paper can solve the above problems to a certain extent. Therefore, the method in this paper has certain military practical significance.

Recently, with the development and practice of artificial intelligence algorithm technology, good results have been achieved in object detection and other fields.

X. Y. Du · G. Z. Song · J. J. Wang
CSSC Ocean Exploration Technology Institute Co., Ltd, Wuxi, China

X. Y. Du · Y. L. Xu (✉)
Science and Technology On Near-Surface Detection Laboratory, Wuxi, China
e-mail: 1140860425@qq.com

K. Nakamatsu et al. (eds.), *Advanced Intelligent Technologies for Information and Communication*, Smart Innovation, Systems and Technologies 365,
https://doi.org/10.1007/978-981-99-5203-8_8

In the target detection algorithm in the field of artificial intelligence, the one-stage and anchor-free deep learning neural network [1–6] is the research hotspot, because this type of algorithm has the advantages of both accuracy and speed. Running deep learning object detection networks on small devices such as drones requires model optimization [7–12].

This paper adopts the baseline model based on YOLO5, and at the same time, aiming at the difficulties of small target detections and dense targets, the model structure is optimized by using transfer learning, transformer self-attention mechanism, CBAM attention mechanism module, and multi-scale hole convolution method. Models can be run on small devices such as drones with good accuracy results. The experimental results show that the method proposed in this paper has a good performance in the rail obstacle target detection from the perspective of the UAV.

2 YOLO5 Base Model Structure

As a one-stage target detection classical network, Yolo5 can quickly complete target detection and classification tasks. As shown in Fig. 1, for the input image, Yolo5, uses the anchor points generated by clustering and the convolution-based deep neural network structure to extract image features at multiple scales. Then the prediction layers predict the confidence of the target in the box, the coordinate offset of the target's bounding rectangle, and the classification confidence of the target. Finally, the soft-NMS method is used to delete the prediction redundant frame results with low confidence, and the final target detection result is obtained. Compared with the traditional method, the YOLO5 algorithm based on neural network has much faster detection speed. At the same time, YOLO5 is based on the PAN multi-scale method, which greatly enhances the detection ability of multi-scale targets. Therefore, the Yolo5 baseline model is used, which improves the comprehensive performance and robustness. However, compared with the general target detection task, the target detection task based on the UAV's perspective is difficult because most of the targets are small targets, and there may be actual situations where the target overlap density is high. Therefore, it is necessary to make some improvements over the baseline model.

The model consists of three parts: feature extraction backbone network BackBone, feature fusion Neck, and detection head Head.

The BackBone part is mainly composed of modules such as Focus, Conv, CSP, and SPP. The Focus structure integrates the width and height information of the image into the channel by slicing operation. The width and height are decreased by two times, and the number of channels is increased by four times. The main purpose of the Focus structure is to reduce the number of network layers, reduce the amount of parameters and computation, and speed up the detection while hardly affecting the detection accuracy. The Conv structure is the basic convolution unit in the detection model, including convolution, batch normalization BN, and activation function SiLU, as shown in Fig. 2.

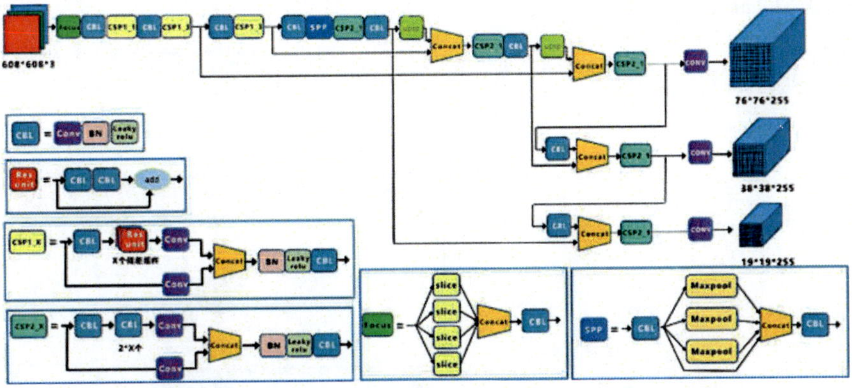

Fig. 1 YOLO5 base model structure

Fig. 2 Conv structure

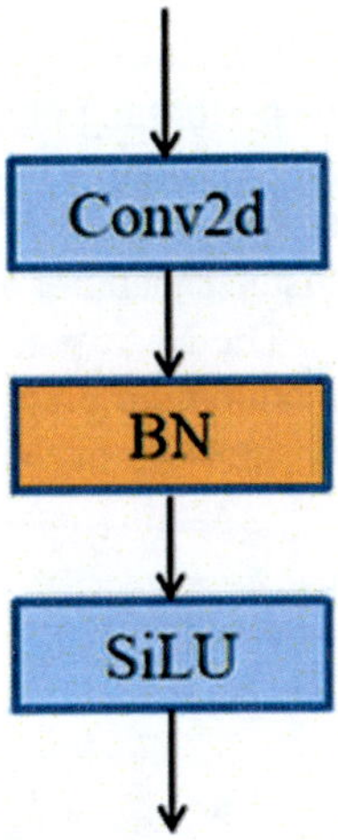

The SPP (spatial pyramid pooling) module, as shown in Fig. 3, first compresses the channel through 1×1 convolution, then performs three maximum pooling operations of different sizes (5, 9, and 13) on the feature map and splices it with the original feature map in the channel dimension, and finally uses 1×1 convolution to modulate the same number of channels as the input channels. The SPP module enriches the receptive field through max-pooling operations of different sizes, which helps to solve the alignment problem of anchors and feature maps.

The CSP (cross-stage partial) module, as shown in Fig. 4, performs two different operations on the input feature map. The left branch first reduces the channel to half of the input channel with 1×1 convolution, and then inputs a residual block, and features of different depths can be obtained by stacking different numbers of residual blocks. The right branch reduces the channel to half of the original input with a 1×1 convolution. Finally, the features of these two branches are spliced and the spliced

Fig. 3 SPP module

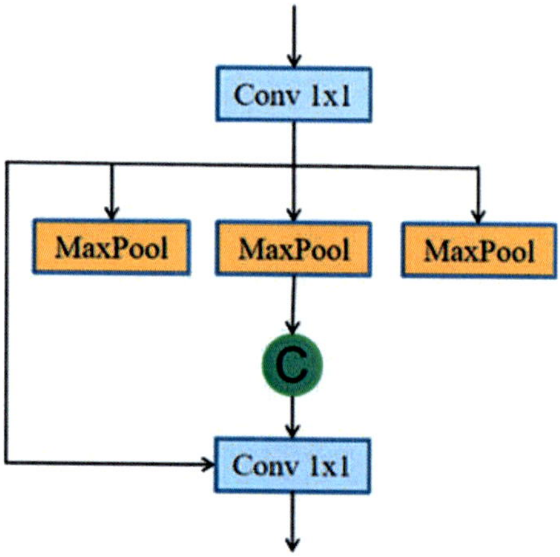

features are optimized with 1×1 convolution. The CSP module reduces redundancy by performing deeper feature extraction on some features, while supplementing the original feature information, reducing the amount of computation and improving the learning ability of the model. CSP is also used in Neck to strengthen the feature fusion capability of the network.

The Neck part is PANet (path aggregation network). The PANet module adds a bottom-up branch on the basis of FPN (feature pyramid network). The position information of the shallow layer in the FPN is gradually passed to the upper layer

Fig. 4 CSP module

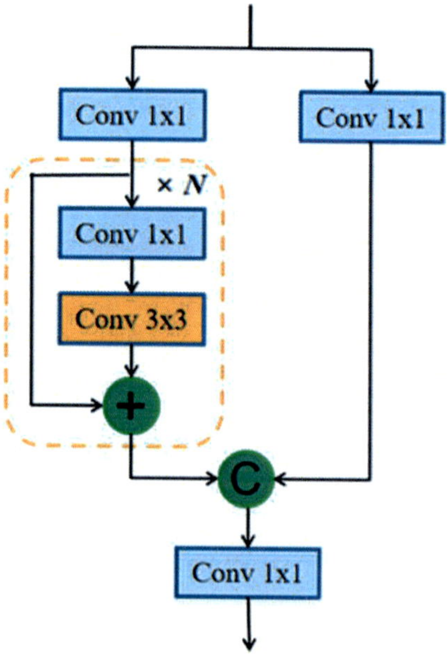

through this branch. The Neck part enhances the position information and semantic information of each output level through top-down and bottom-up branches and unilateral connections, and further strengthens the feature extraction ability of the model. The core of the Head section is prediction. Finally, the redundant detection frame is eliminated by Non-maximum Suppression (NMS) to obtain the final prediction frame.

3 The Improved Yolo5 Structure

3.1 Transfer Learning

Transfer learning refers to transferring a model trained in one task to another task. Transfer learning has received continuous attention in the field of deep learning recently. In the orbital village task based on the perspective of the drone, there are some contradictions in the training of the deep learning model, such as the scarcity of sample data. When this type of problem occurs, you can consider using the idea of transfer learning to solve it.

This type of problem can be solved with the help of fine-tuning in transfer learning. Fine-tune is a commonly used method in the field of transfer learning. The role of fine-tune is to use the existing trained network model to make adjustments on new tasks. Its advantage is that it can use pre-trained large sample network parameters, so that the model can get better training, faster convergence, and better feature expression.

3.2 Transformer Self-attention Mechanism

Transformer is previously proposed in natural language processing tasks. Recently, the research on Vision Transformer (ViT) shows that the encoder structure inherited and modified from NLP can be used for image recognition effectively. Using image patch embedding sequence as input, ViT can successfully transfer the visual features obtained from pre-training tasks to more specific and practical target image tasks from the perspective of pure sequence to sequence. Inspired by the visual transformer, we use the encoder block of the transformer to replace some convolution blocks and CSP bottleneck blocks in the original YOLO5. The transformer encoding block is able to capture rich global and contextual information, which can improve the small object detection performance.

At the backbone level, as shown in Fig. 5, by using global self-attention instead of spatial convolution in the last three bottleneck blocks of ResNet, the performance of the baseline model in object detection can be significantly improved, and a lighter model and speed can be obtained due to the reduction of the number of parameters.

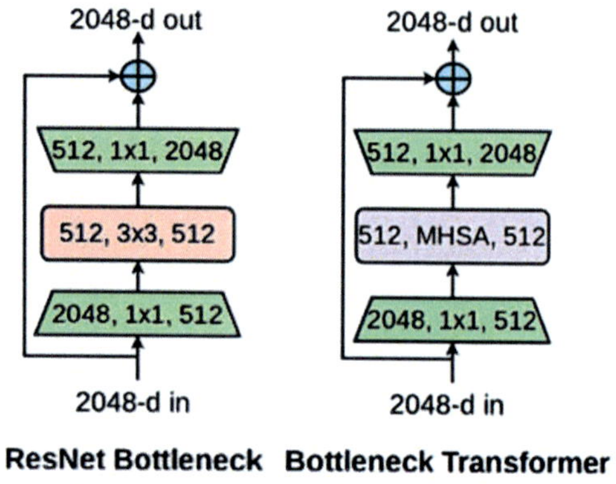

Fig. 5 Replace spatial convolution with bottleneck transformer

The MHSA layer is shown in Fig. 6. While we use four heads, we do not show them on the figure for simplicity. All2all attention is performed on a 2D FeatureMap with split relative position encodings R_h and R_w for height and width, respectively. The attention logits are $qk^T + qr^T$, where q, k, and r represent query, key, and position encodings, respectively. 1×1 represents a pointwise convolution.

At the neck level, this paper uses the transformer coding block to form the transformer prediction header (TPH), as shown in Fig. 7. The feature maps in the network backend are of very low resolution, so applying TPH on these low resolution feature maps helps reducing expensive computational costs. Furthermore, when scaling up

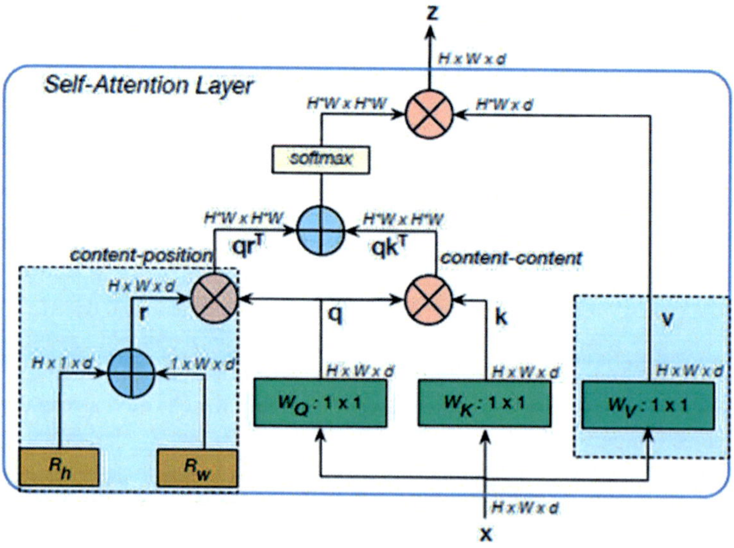

Fig. 6 MHSA layer

Fig. 7 Transformer encoding block

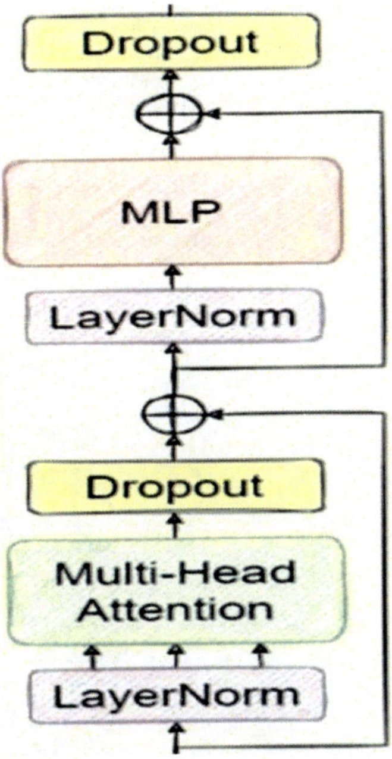

the resolution of the input image, some TPH blocks can be selectively removed in the early layers, making the training process feasible. Each transformer code contains two sublayers, the first sublayer is a multi-head attention layer and the second layer is a fully connected layer. Add residual connections in the middle of each sublayer. Through self-attention mechanism, the transformer encoding block can mine richer feature representations. In addition, the transformer encoding block has relatively better detection performance for high-density occluded objects.

3.3 *CBAM Attention Module*

CBAM is an effective attention module, as shown in Fig. 8. CBAM is a lightweight module and a convolutional attention module that can be plugged into most mainstream CNN frameworks. Given an intermediate feature map, the module sequentially inputs the attention map along the two different dimensions of channel and spatial, and then multiplies the attention map by the input feature map for adaptive feature refinement, and an adaptive refined feature can be obtained. By introducing CBAM into different models, the comprehensive performance of the model can be improved. This has been demonstrated experimentally on different datasets and models. Using CBAM can help YOLOv5 to extract attention regions, reduce confusing information, and focus on useful target information.

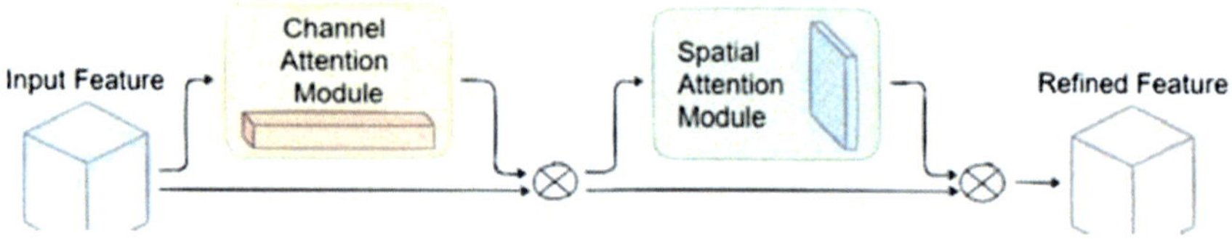

Fig. 8 CBAM structure

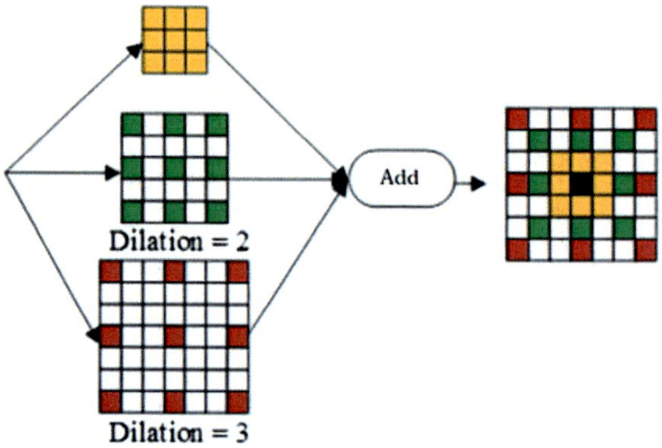

Fig. 9 Multi-scale hole convolution

3.4 Multi-scale Hole Convolution

Based on Yolo5, the scheme uses multi-scale hole convolution instead of ordinary conventional convolution kernels, as shown in Fig. 9. The neural network composed of multi-scale hole convolution combination is a network structure that can extract detailed features of different scales. The multi-scale hole convolution expands the receptive field of the convolution layer, captures multi-scale global information, reduces the loss of key information in the sampling step, and will improve the performance of small target detection.

4 Experimental Verification

Our rail obstacles dataset contains a total of 12,115 images taken by UAVs, 80% of which are used as training set and 20% as testing set. Our method uses the index AP to evaluate the detection result's effect of the test set. AP is defined as the area under the PR curve and P means Precision. The calculation formula of P is shown in Formula (1):

$$P = TP/(TP + FP) \tag{1}$$

R means Recall. The calculation formula of R is shown in Formula (2):

$$R = TP/(TP + FN) \tag{2}$$

TP means the number of detection boxes when $IOU > IOU_{threshold.}$ FP means the number of detection boxes when $IOU < IOU_{threshold}$ or the number of extra check boxes of the same ground truth found. FN means the number of ground truth boxes which is not found.

The final AP@0.5 of YOLO5 base model is 0.82 and the final AP@0.5 of the improved yolo5 structure is 0.89. The final AP@0.5:0.95 of YOLO5 base model is 0.62 and the final AP@0.5:0.95 of the improved yolo5 structure is 0.68. It can be seen that our method outperforms the Yolo5 base model.

Our method can achieve good results in the rail obstacle detection based on the perspective of unmanned aerial vehicle. Some predictive detection results are shown in Fig. 10.

5 Conclusions and Prospects

In this paper, a model based on YOLO5 is used for the rail obstacle target detection from the perspective of the UAV. For tasks such as small target detection and target density, it is necessary to optimize the model of YOLO5. Specific measures include transfer learning, transformer attention module, CBAM attention mechanism module, and multi-scale hole convolution method. The model can be run on small equipment platforms such as drones, taking into account both time and accuracy indicators. The experimental results show that the method proposed in this paper has a good performance in the rail obstacle target detection from the perspective of the UAV.

The following work is to improve the algorithm's detection accuracy of targets in bad weather conditions, such as heavy rain and fog. Subsequent works also include algorithm integration for autonomous path planning of UAVs.

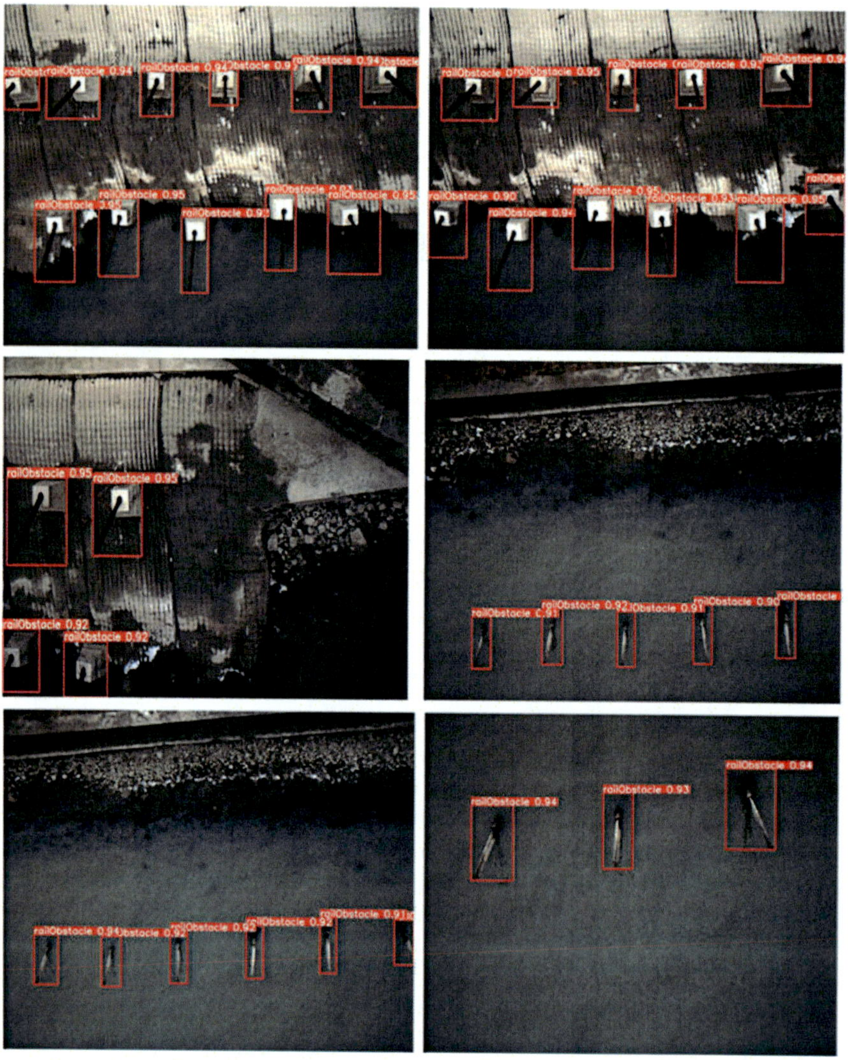

Fig. 10 Predictive detection results

Acknowledgements This project was supported by Stable Foundation Project of the Science and Technology on Near-Surface Detection Laboratory with funding number 6142414190101.

References

1. Yang, H., Cai, S., Sheng, H., et al.: Balanced and hierarchical relation learning for one-shot object detection. In: Proceedings of the IEEE/CVF Conference on Computer Vision and Pattern Recognition (2022)
2. Han, J., Ren, Y., Ding, J., et al.: Expanding low-density latent regions for open-set object detection. In: Proceedings of the IEEE/CVF Conference on Computer Vision and Pattern Recognition (2022)
3. Li, H., Pan, X., Yan, K., et al.: SIOD: single instance annotated per category per image for object detection. In: Proceedings of the IEEE/CVF Conference on Computer Vision and Pattern Recognition (2022)
4. Yu, S., Xiao, J., Zhang, B., et al.: Democracy does matter: comprehensive feature mining for co-salient object detection. In: Proceedings of the IEEE/CVF Conference on Computer Vision and Pattern Recognition (2022)
5. Chen, B., et al.: Dense learning based semi-supervised object detection. In: Proceedings of the IEEE/CVF Conference on Computer Vision and Pattern Recognition (2022)
6. Wu, J., Chen, J., He, M., et al.: Target-relevant knowledge preservation for multi-source domain adaptive object detection. In: Proceedings of the IEEE/CVF conference on computer vision and pattern recognition (2022)
7. Srinivas, A., Lin, T.Y., Parmar, N., et al.: Bottleneck transformers for visual recognition. In: Proceedings of the IEEE/CVF Conference on Computer Vision and Pattern Recognition (2021)
8. Woo, S., Park, J., Lee, J.Y., et al.: Cbam: convolutional block attention module. In: Proceedings of the European Conference on Computer Vision (2018)
9. Sayed, M., Brostow, G.: Improved handling of motion blur in online object detection. In: Proceedings of the IEEE/CVF Conference on Computer Vision and Pattern Recognition (2021)
10. Otani, M., Togashi, R., Nakashima, Y., et al.: Optimal correction cost for object detection evaluation. In: Proceedings of the IEEE/CVF Conference on Computer Vision and Pattern Recognition (2022)
11. He, L., Todorovic, S.: DESTR: object detection with split transformer. In: Proceedings of the IEEE/CVF Conference on Computer Vision and Pattern Recognition (2022).
12. Xu, D., Deng, J., Li, W.: Revisiting AP loss for dense object detection: adaptive ranking pair selection. In: Proceedings of the IEEE/CVF Conference on Computer Vision and Pattern Recognition (2022)

A Novel Global–Local Feature Extraction Method Based on Deep Learning for Football Movement Training

Ning Xu

Abstract Recently, deep neural network-based methods have been used to solve some intractable hot problems. In addition, image features are extracted by deep learning algorithms, which enables researchers to make great breakthroughs in image retrieval and detection area. The new feature extraction algorithm using deep neural networks can simultaneously extract the low- and high-level semantic information of the image. We propose a novel global—local feature extraction method for football movement training in this article. A global—local feature extraction model is constructed based on the convolutional network structure. The football images in different regions are extracted and different convolution features are obtained. Finally, the similarity of global features is measured to obtain the retrieval results. Experiments on the constructed football image dataset demonstrate that the new method can effectively improve the training accuracy.

1 Introduction

The most important two aspects of domestic education and educational reform are educational method reform and educational content strengthening. With the development of China's sports industry, China is paying more and more attention to the game of football, and also attaches great importance to the establishment of football courses in universities [1]. For college football education, introducing a combination of teaching methods can not only effectively stimulate students' interest in football learning, but also significantly improve students' overall physical quality, football skills, tactics, and other professional abilities, so as to effectively achieve this goal and improve the quality of college football education. Therefore, the in-depth study of combination training method and its application in college football teaching practice can effectively promote the development of college football education.

N. Xu (✉)
Department of Physical Education, Harbin Finance University, Harbin 150000, China
e-mail: 910675024@qq.com

K. Nakamatsu et al. (eds.), *Advanced Intelligent Technologies for Information and Communication*, Smart Innovation, Systems and Technologies 365,
https://doi.org/10.1007/978-981-99-5203-8_9

In the current situation of college football teaching, football education innovation is insufficient, teaching theory is relatively simple. From the perspective of football education teams in universities, professional teachers have few resources, and the teaching concepts of existing teachers will not change with the passage of time. In the end, they cannot fully reflect the core of football training and lack educational significance. According to a related survey, about 40 percent of college football teachers do not have the title of player coach, about 30 percent of college football teachers have the title of second-tier player coach, and only about 20 percent of college football teachers have the title of line athletic coach. At the same time, many existing college football teachers rarely participate in any form of football training after receiving education [2, 3], which slows down the updating speed of football knowledge, lack of innovation ability, and ultimately affects the level and effect of football education.

Computer vision means that the computer performs a series of analysis and processing on the obtained image or video to capture the content that the user is interested in and give feedback according to the needs of the user [4]. In order to accomplish this task, how to understand the image is a very critical part of it. In a nutshell, a computer is fed an image and it needs to figure out what objects are in it and where they are. According to the basic purpose of image understanding, it can be divided into a variety of different research directions. In the process of image understanding, how to extract feature vectors is the first problem to be solved. In other words, the data features are extracted by the model, and the data features are used to express an image. The goal is to find representative and robust features from the image. Image similarity, object recognition, and other information can be obtained by comparing the features of different images.

Traditional manual descriptors, such as HOG and SIFT, are mainly based on the manual labels related to the scene [5]. Therefore, the performance of such search engines depends on the quality and availability of human tags. The label design needs enough expert knowledge to be completed with high quality, and the labeling process is very time-consuming [6]. Moreover, manual descriptors only use the grayscale property of the image and lose other information in the image. These are the disadvantages and problems of traditional feature extraction methods.

In view of this situation, He et al. [7] proposed FashionNet and key point annotation to classify and retrieve clothing images. Gao et al. [8] used detection and tracking algorithms to extract fine-grained features of clothing in the video, conducted similarity calculation and ranking, and realized retrieval. Naka et al. [9] extracted attention maps from clothing images, obtained attention features, and fused attention feature maps and feature maps to conduct clothing image retrieval. Yin et al. [10] proposed to extract global features and local features at different scales based on the similarity pyramid graph model network, so as to carry out fine-grained retrieval of fashion clothing images. Liang et al. [11] extracted multi-level scale features and combined with traditional visual features to retrieve clothing images. Wang et al. [12] proposed the Hamming embedded hashing method for image retrieval. The above methods are all aimed at fashion clothing, and the existing methods are difficult to classify and retrieve fine-grained attribute features in images, which need to be combined with reordering to complete the fine-grained retrieval task. Aiming at the above problems, a new global—local feature extraction method is proposed.

2 The Global–Local Feature Extraction Model

Football sports image has the following characteristics: complex movement structure, rich patterns, and local texture details. The differences of fine-grained attribute features of football motion images are reflected in different local areas such as action and texture [13, 14]. The football image itself has some problems such as occlusion and deformation of body parts, which leads to inaccurate feature extraction.

Therefore, this paper proposes a global–local feature extraction method, and the specific block diagram is shown in Fig. 1. The trained faster R-CNN (faster region-CNN) is used for region detection of the input image. The component regions of the input image are identified and extracted. According to the segmentation results, the convolution features of each region are extracted as the global features of the input image and the local features of the components.

2.1 Region Detection

According to the structure of football sports images, Lableme [15] is used to manually label the human body regions on the football sports image dataset collected in this paper. The annotated categories show that football movement images have the same basic movement structure. The annotation area is divided into global and local parts. The global annotation is the whole body part of the image. The global feature is to remove the complex background, retain the complete human body area, and complete the preliminary screening of the retrieval task.

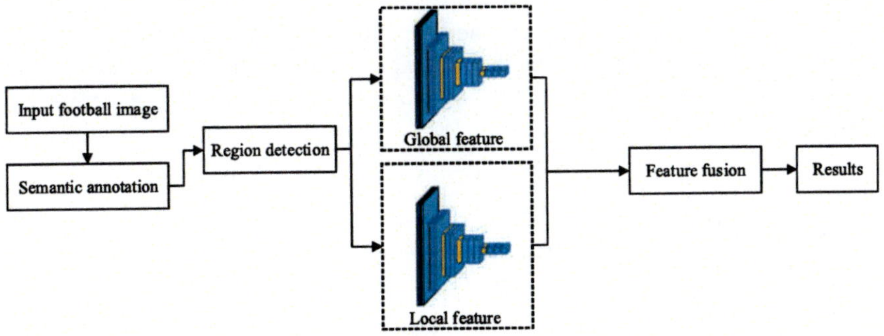

Fig. 1 Flowchart of global local feature extraction model

In order to achieve effective extraction of global and local features of football motion images and meet the requirements of retrieval task, faster R-CNN network is used for region detection in this paper. Football movement images have more fine-grained attribute features and occupy a small area of the image. The network model can be used to detect and recognize effectively, and it also has strong robustness against the occlusion problem caused by posture deformation. Firstly, the image is input into the convolution network of the network model, and the convolution feature map is obtained through CNN, and the convolution feature is input into the region recommendation network. Region recommendation network is a network added to the original method by faster R-CNN, which significantly enhances the accuracy of regional feature detection and recognition, and also has better recognition and detection effect on small targets. The results are input into the Region of Interest (ROI) pooling layer to obtain the classification detection result image of the input image, and the detection results obtained by segmentation are used for the subsequent feature extraction steps.

2.2 *Global Feature Extraction*

The global feature extraction branch uses VGG16 network as the backbone network. Compared with other convolutional neural networks [16], the entire network uses the same size of convolution kernel size and maximum pooling layer size, which reduces the number of parameters of the network model and has better performance. The VGG16 network is used as a branch to extract the backbone network, and the FC6 and FC7 layers of VGG16 are removed. In order to save time, the learning rate of the first two convolutional layers in the network is set to 0. Starting from the convolutional layer 3 of VGG16 network, the learning rate is set as 1 and the step size is 2. The convolution feature map is obtained from Conv5_3 of the convolution network, and the feature map is obtained from the input maximum pooling layer, which is used as the global feature of the input image.

The feature map with global feature $12 \times 15 \times 512$ is output. The width and height of the output feature map are 2 and 15, and the dimension of the feature map is 512. The global features of the query image are represented as L_1^1, and all the global features extracted from the football movement image dataset are represented as $\{L_1^1, L_2^1, \cdots, L_n^1\}$.

2.3 *Local Feature Extraction*

Because the global features cannot better reflect the fine-grained attributes of football motion images, the fine-grained attributes in the key areas cannot be well matched during feature matching, resulting in low retrieval accuracy, and the local features can better make up for this problem. Local features can accurately extract the areas containing fine-grained attributes, which becomes the key to fine-grained retrieval of football motion images [17]. Therefore, the local feature extraction branch is divided into three branches that can describe the fine-grained properties of football motion images, and local features such as size, shape, and texture are extracted from football motion images, respectively.

Network structure is similar to the global feature extraction branch, it uses the VGG16 network as a branch to extract the backbone network, from conv5_3 convolution of the input image feature. It inputs the ROI pooling layer; extracts the ROI features, combines the pose prediction loss function, shape prediction loss function, and texture prediction loss function defined in this paper; outputs the local features; and uses the interpolation algorithm to transform into a fixed size (3 × 3 × 512), the dimension is the same as the global features.

Three loss functions are defined in the local feature extraction branch, namely, the pose prediction loss function, the shape prediction loss function, and the labeled position prediction loss function, which are constrained when extracting pose, shape and texture features, respectively. In order to learn local feature descriptions of football movement images, Triplet Loss is used to calculate the distance constraints of training images of positive and negative samples in enhanced images, which are defined as follows:

$$L = \sum_{i=1}^{|N|} \max\{0, m + d(I_i, I_i^+) - d(I_i, I_i^-)\} \tag{1}$$

where N represents the number of training samples, d() represents the distance function, I_i^+ represents the image similar to image I_i, that is, the positive sample, I_i^- represents the image that is not similar to image I_i, and m is the parameter.

2.4 HOG Feature Extraction

Histogram of oriented gradient (HOG) is a method to calculate and collect histogram of gradient direction of local image areas to represent image features. HOG expresses the distribution of image shape information, and forms image feature descriptors by segmentation, statistics, and integration of histograms within cells [18, 19]. Compared with other feature description methods, the feature extracted by HOG is not only stable, but also has a better representation of image contour information.

The detailed steps of HOG feature extraction are as follows:

(1) Gamma correction to adjust the unevenness of image brightness and reduce the influence of shadow and illumination changes of image. A nonlinear Gamma correction method is adopted in this paper, as shown in 2.

$$g(I) = I^{\gamma} \tag{2}$$

where I refers to the pixel value of a certain point in the image, $g(I)$ is the pixel value after transformation, and γ is Gamma value. When $\gamma < 1$, the pixel contrast in the low gray value range is improved, and the image frame appears dark gray. When $\gamma > 1$ is used to enhance the contrast of high gray value interval, highlighting the overall structural features of the image.

(2) Gradient calculation. Obtaining horizontal and vertical gradient components is the prerequisite for obtaining HOG features. The horizontal template $k = [-1, 0, 1]$ and the vertical template $k = [-1, 0, 1]^T$ are used to perform convolution operation on the original image, and the horizontal and vertical gradients are obtained. As shown in Eqs. (3)–(6).

$$G_x(x, y) = pixel(x+1, y) - pixel(x-1, y) \tag{3}$$

$$G_y(x, y) = pixel(x, y+1) - pixel(x, y-1) \tag{4}$$

$$V(x, y) = \sqrt{G_x(x, y)^2 + G_y(x, y)^2} \tag{5}$$

$$\theta(x, y) = arctan(G_x(x, y)/G_y(x, y)) \tag{6}$$

where $G_x(x, y)$ and $G_y(x, y)$ represent the gradient in horizontal and vertical directions, respectively, $pixel(x, y)$ represents the pixel value of the image at the point (x, y), and $V(x, y)$ and $\theta(x, y)$ represent the gradient amplitude and the included angle at pixel point (x, y), respectively.

(3) Constructing the histogram of cell gradient and combining it to form blocks. The minimum processing unit of HOG feature extraction is cell. The image is divided into several cells, which are composed into blocks, and the local area of the image is represented by blocks. HOG divides the gradient direction into 9 bin gradients, and the gradient size is used as the weight of the projection.
(4) Normalization of interval features. Due to the overlapping feature extraction between the blocks, the variation of local illumination is not equal. Therefore, feature normalization is particularly important. L2 normal form is used as the normalization formula, as shown in Eq. (7).

$$L2 - norm : v \leftarrow \frac{v}{\sqrt{||v||_2^2 + \omega^2}} \tag{7}$$

where v represents the feature vector, $||v||_k$ is the k norm of v, and ω is a very small coefficient to keep the denominator from being zero.

In the traditional algorithm, if a 128 × 128 pixel window (window) is used, 8 pixels are scanned as the stride length, 8 × 8 pixels are used as a cell, and 16 × 16 = 256 cells are formed. 4X × 4 adjacent cells are regarded as a block. Then, an image contains 13 × 13 = 169 pixel blocks, and the gradient direction is 9. Therefore, the dimension of HOG feature vector is obtained as follows: $4 \times 4 \times 169 \times 9 = 24{,}336$ dimension. It is not difficult to find that the feature dimension of traditional HOG image feature extraction is huge.

Based on the above analysis, it is not difficult to see that the feature dimension of HOG is too large when extracting image features. The reason is that feature block overlap occurs in HOG feature extraction, resulting in an increase in the number of feature blocks. Each block contains multiple 9-gradient cells feature histograms. This results in a final feature dimension that is too large. In order to solve this problem, based on the feature extraction algorithm of grid technology [20], an improved HOG feature extraction method is proposed.

Grid-based hierarchical HOG feature extraction method, under the idea of grid division, optimizes the feature block overlap phenomenon in HOG feature extraction. The HOG algorithm is used to divide an image into several same no-cross regions, and the direction gradient histogram of each region is counted. Although this method can solve the overlapping phenomenon of feature blocks, it will lead to the grid marginalization problem, that is, the feature fracture phenomenon in the grid separated area.

Under the image with 128×128 size, the grid is divided with 32×32 pixels, and each grid contains 4×4 cell units. Let the feature of the $i - th$ cell unit be F, as shown in Fig. 2. The feature in a grid can be expressed as $F = F1 + F2 + \cdots + F16$, where S represents the feature of marginal linearity that is easily influenced by grid. Then

$$S_{upper-left} = F4 + F5 + F12 + F13 + F14 + F15 + F16 \quad (8)$$

$$S_{upper-right} = F1 + F8 + F9 + F16 + F15 + F14 + F13 \quad (9)$$

$$S_{lower-left} = F1 + F2 + F3 + F4 + F5 + F12 + F13 \quad (10)$$

$$S_{lower-right} = F4 + F3 + F2 + F1 + F8 + F9 + F16 \quad (11)$$

where $S_{upper-left}$, $S_{upper-right}$, $S_{lower-left}$, and $S_{lower-right}$ represent the easily marginalized feature set of the upper-left grid, upper-right grid, lower-left grid, and lower-right grid in the image, respectively. Similarly, $S - middle = F1 + F2 + F3 + F4 + F5 + F12 + F13 + F14 + F15 + F16 + F9 + F8$ represents the easy edge feature set in the intermediate grid.

To solve the grid marginalization problem, this paper adds a bottom HOG feature extraction layer, and divides the image into 4 cell intervals, each cell has 9 bin channels, so the bottom layer generates a 36-dimensional feature vector. In this paper, the following three mesh partitioning methods are considered, including uniform mesh partitioning for the same mesh area.

The bottom HOG feature extraction plans to adopt the form of uniform grid. Compared with the other two division methods, the features of uniform grid division

Fig. 2 Grid feature

F1	F2	F3	F4
F8	F7	F6	F5
F9	F10	F11	F12
F16	F15	F14	F13

are more in line with the structural features of the image, especially for the left and right parts of the motion posture, the division is clear, and the grid edge interference generated is minimal. Although diagonal mesh can extract features stably, it also produces more grid edge interference. The features extracted from circular mesh do not accord with the structural features of moving images, and the effect of retaining some posture edges is the worst, and more grid interference information is generated at the same time. This paper proposes a grid-based hierarchical HOG method to extract image features, which can effectively reduce the image dimension and reduce the grid marginalization phenomenon.

The steps of HOG feature extraction method proposed in this paper are as follows:

Step 1. Take the 128×128 image with a pixel neighborhood grid of 32×32 as the sampling window and traverse the entire image in a non-overlapping way. Thus, a moving image is divided into 4×4 blocks, a total of 16 grids. The introduction of grid results in the marginalization of images in the grid and brings interference information to the whole motion contour feature.

Step 2. The underlying HOG feature extraction vector is introduced, and the extraction method is similar to step 1. The image is divided into 2×2 grid cells with uniform distribution, and the feature vectors in each grid cell are counted, respectively. Then, the feature vectors of the two layers are added to form the hierarchical HOG feature. Then the horizontal template $[-1, 0, 1]$ and vertical template $[-1, 0, 1]^T$ are used to calculate the gradient direction and amplitude of hierarchical HOG.

Step 3. According to the interval of $0° \sim 180°$, it is divided into 9-gradient channels to calculate the gradient histogram of each layer grid area, and integrate the feature gradient histogram of the two layers.

Step 4. Feature normalization operation. The normalization eliminates the phenomenon of local illumination imbalance and standardizes the extracted HOG features of the two layers as much as possible. The standard formula of Eq. (7) is adopted in the experiment. The difference is that v represented the feature vector in the previous grid, $||v||_2$ is the 2-norm operation, and ω is a minimum that prevents infinite computation.

Step 5. The normalized histogram vector is formed into an $n \times m$ matrix to represent the HOG feature of the image, where n is the vector dimension of the histogram in the grid and m is the number of grids to be calculated for the whole image.

3 Experiments and Analysis

The experimental platform is as follows: Ubuntu18.04, InterCore I9-9900K, and GTX2080Ti. The training is carried out on the GPU and the TensorFlow deep learning framework. The football dataset is derived from a video recorded during the gym class of sophomore year in one university.

3.1 Analysis of Feature Classification Results

In order to accurately learn the category information of football sports images, this paper uses the accuracy rate as the index of feature classification, and selects several values with the largest probability vector from all the predicted labels. If these values have the same result as the manual labeled category, the prediction is accurate, otherwise the prediction is wrong. In this paper, VGG16 is used as the backbone network to extract image convolution features for subsequent image retrieval. Therefore, the Mean Average precision (MAP) of image feature extraction and classification results is compared with three backbone networks, namely, VGG16, RESNET-101, and inception-V4 as shown in Table 1.

According to the above data, compared with other networks,VGG16 network has a better MAP value in feature extraction and classification results for the football sports image dataset used in this paper. Therefore, VGG16 is selected as the main network of the experiment to extract features from the input images and compare the current commonly used image retrieval methods, including DARN [21], FashionNet, and Match R-CNN (match region-convolutional neural network) [22].

The comparison results are shown in Table 2. It can be concluded that, compared with Match R-CNN and DARN, the proposed method has certain advantages in accuracy. It classifies features more accurately and extracts convolution features. The proposed method is similar to MAP of FashionNet, but the proposed method does not need to manually annotate a large number of key points, and can also obtain more accurate fine-grained attribute features.

Table 1 Comparison of MAP with different models

Method	MAP
VGG16	0.594
RESNET-101	0.587
inception-V4	0.575

Table 2 Comparison of MAP with different methods

Method	MAP
DARN	0.806
FashionNet	0.884
Match RCNN	0.803
Proposed	0.881

Table 3 P, R values/%

Index	P	R
Global feature	69.8	53.4
Local feature	72.1	44.8
Global–local feature	89.6	26.7

3.2 Analysis of Ablation Experiment

In order to better illustrate the performance of the proposed method, Precision (P) and Recall (R) are used as evaluation criteria to evaluate the ablation experimental results of the proposed method.

In the results (Table 3): global features refer to the preliminary retrieval results that are output after similarity calculation between the global features of the query image and the global features in the feature database. Local features refer to the initial retrieval results returned without combining global features, which are directly used to measure the similarity between the local features of the query image and all the local features in the feature database. Global and local features are to point to after global calculated on the basis of preliminary results, take images of the search results ranking Top-50, indexed in the feature library, combined with the local characteristics of multiple local characteristics and the index calculation of similarity and global similarity weighted reordering, and output the final retrieval results.

4 Conclusion

Aiming at the problem of football motion image retrieval, this paper takes football motion image as an example, and builds a global—local feature extraction model based on the custom semantic attributes of football posture, which is used to extract the global and local features of the input image. Furthermore, a global—local feature extraction method for fine-grained football retrieval is proposed to realize fine-grained football image retrieval. In addition, on the basis of the existing football movement image dataset, the football movement image dataset is expanded, and the effectiveness of the proposed method for fine-grained image retrieval is tested on this dataset.

References

1. Todorovich, E.J.: Sport education. J. Phys. Educat. Recreat. Dance **82**(8), 12–13 (2011)
2. Siedentop, D.: Sport education: a retrospective. J. Teach. Phys. Educ. **21**(4), 409–418 (2002)
3. Hastie, P.A.: An Ecological Analysis of a Sport Education Season. J. Teach. Phys. Educ. **19**(3), 355–373 (2010)
4. Shoulin, Y., Zhang, Y., Shahid, K.: Large scale remote sensing image segmentation based on Fuzzy region competition and Gaussian mixture model. IEEE Access. **6**, 26069–26080 (2018)
5. Deena, G., Raja, K.: Keyword extraction using latent semantic analysis for question generation. J. Appl. Sci. Eng. **26**(4), 501–510 (2022)
6. Huang, J., Feris, R.S., Chen, Q., et al.: Cross-domain image retrieval with a dual attribute-aware ranking network. In: Proceedings of the IEEE international conference on computer vision. 1062–1070 (2015)
7. He, T., Hu, Y.: FashionNet: personalized outfit recommendation with deep neural network. arXiv preprint arXiv:1810.02443 (2018)
8. Gao, S.: A Two-channel Attention Mechanism-based MobileNetV2 And Bidirectional Long Short Memory Network For Multi-modal Dimension Dance Emotion Recognition. Journal of Applied Science and Engineering **26**(4), 455–464 (2022)
9. Naka, R., Katsurai, M., Yanagi, K., et al.: Fashion Style-Aware Embeddings for Clothing Image Retrieval[C]//Proceedings of the. International Conference on Multimedia Retrieval. **2022**, 49–53 (2022)
10. Yin, S., Liu, J., Li, H.: A self-Supervised Learning Method for Shadow Detection in Remote Sensing Imagery. 3D Res. **9**(4) (2018)
11. Liang, Z., Guo, Y., Feng, Y., et al.: Stereo matching using multi-level cost volume and multi-scale feature constancy. IEEE Trans. Pattern Anal. Mach. Intell. **43**(1), 300–315 (2019)
12. Wang, S., Kang, X., Liu, F., et al.: Supervised discrete hashing for hamming space retrieval. Pattern Recogn. Lett. **154**, 16–21 (2022)
13. Feng, T.: Mask RCNN-based Single Shot Multibox Detector For Gesture Recognition In Physical Education. J. Appl. Sci. Eng. **26**(3), 377–385 (2022)
14. Wang, L., Shoulin, Y., Alyami, H., et al.: A novel deep learning-based single shot multibox detector model for object detection in optical remote sensing images. Geosci. Data J. (2022). https://doi.org/10.1002/gdj3.162
15. Nemade, S.: Refinement of CNN based Multi-label image annotation. Turkish J. Comput. Math. Educat. (TURCOMAT) **12**(13), 1935–1941 (2021)
16. Shoulin, Y., Hang, L., Asif Ali Laghari, et al.: A Bagging Strategy-Based Kernel Extreme Learning Machine for Complex Network Intrusion Detection. EAI Endorsed Transactions on Scalable Information Systems. **21**(33), e8, 2021.https://doi.org/10.4108/eai.6-10-2021.171247
17. Jing, Y., Li, H., Yin, S.: New intelligent interface study based on K-means gaze tracking. Int. J. Comput. Sci. Eng. **18**(1), 12–20 (2019)
18. Yang, S., Shoulin, Y., Hang, L., Lin, T., Shahid, K.: GPOGC: Gaussian Pigeon-Oriented Graph Clustering Algorithm for Social Networks Cluster. IEEE Access. 7:99254-99262
19. Zhang, T., Zhang, X., Ke, X., et al.: HOG-ShipCLSNet: A novel deep learning network with hog feature fusion for SAR ship classification. IEEE Trans. Geosci. Remote Sens. **60**, 1–22 (2021)
20. Molderink, A., Bakker, V., Bosman, M.G.C., et al.: Management and control of domestic smart grid technology. IEEE Trans. Smart Grid **1**(2), 109–119 (2010)
21. Zhang, Y., Tian, J., Zhong, C., et al.: Darn: deep attentive refinement network for liver tumor segmentation from 3d ct volume. In: 2020 25th International Conference on Pattern Recognition (ICPR). IEEE, 7796–7803 (2021)
22. Yin, S., Li, H.: Hot region selection based on selective search and modified Fuzzy C-means in remote sensing images. IEEE J. Selected Topics Appl. Earth Observat. Remote Sensing **13**, 5862–5871 (2020). https://doi.org/10.1109/JSTARS.2020.3025582

Remaining Useful Lifetime Prediction Method of Aviation Equipment Based on Improved Particle Filter

Gao Yangjun and Wang Zezhou

Abstract Aiming at the problem that the traditional remaining useful lifetime (RUL) prediction method based on particle filter has low prediction accuracy due to particle degeneracy and impoverishment, an improved RUL prediction method based on the improved particle filter for aviation equipment is proposed. The importance sampling function of particles is randomly generated by introducing the variational auto-encoder, and a new particle reinforced mechanism is proposed in the resampling stage, which effectively overcomes the problem of particle degeneracy and impoverishment, and improves the performance of the RUL prediction method. The accuracy of the method is verified by the example analysis.

1 Introduction

Remaining Useful Lifetime (RUL) prediction is a very challenging problem in the field of Prediction and Health Management (PHM) technology, which needs to be solved urgently. RUL usually refers to the runtime of the equipment from its current state to failure, also known as usable time or uptime. This time can provide important data reference for maintenance and support personnel, and help them take targeted measures in maintenance operations to improve maintenance and support efficiency. The functional structure of aviation equipment is complex, and the maintenance cost is high. Excessive maintenance will inevitably lead to serious waste of resources, while the lack of maintenance may lead to catastrophic accidents, resulting in even greater loss of personnel and property. Therefore, from the perspective of ensuring safety or saving costs, it is of great significance to accurately predict the RUL of aviation equipment and scientifically formulate maintenance support plans.

G. Yangjun
Equipment Management and Unmanned Aircraft Engineering College, Air Force Engineering University, Xi'an, China

W. Zezhou (✉)
No. 93920 Unit of PLA, Han Zhong, China
e-mail: 350276267@qq.com

K. Nakamatsu et al. (eds.), *Advanced Intelligent Technologies for Information and Communication*, Smart Innovation, Systems and Technologies 365,
https://doi.org/10.1007/978-981-99-5203-8_10

For aviation equipment with complex nonlinear degradation characteristics, the traditional RUL prediction method based on the distribution of the first hitting time [1] has become extremely difficult to implement the algorithm, and the probability distribution of the RUL needs to be given in the maintenance decision-making research. Therefore, the method based on particle filter (PF) has begun to receive extensive attention from researchers, and has been applied and verified in products such as ball bearings [2] and lithium batteries [3].

However, traditional PF algorithms have inherent defects, namely, particle degeneracy and impoverishment, which will seriously affect the accuracy of RUL prediction. There are two main methods for overcoming particle degeneracy and impoverishment in PF. One is to select a suitable importance sampling function and the other is to improve the resampling method of PF. At present, the types of importance sampling functions are very rich, mainly including particle swarm optimization, genetic algorithm, extended Kalman filter, and unscented Kalman filter [4–8]. However, the selection of the above importance sampling functions is based on the specific problems which are specifically raised, which leads to the inapplicability of the method. For improved resampling methods, currently used methods include parallel resampling, fine resampling, distributed resampling, and residual resampling [9–11]. However, simply improving the resampling method will lead to the problem of particle impoverishment.

Aiming at the shortcomings of existing research, this paper proposes a more general improved PF algorithm. On the one hand, a Variational Auto-Encoder (VAE) is used to select the importance sampling function in the particle update stage. On the other hand, a new particle reinforced mechanism is proposed in the particle resampling stage. Thereby, the problems of particle degeneracy and impoverishment in the traditional PF algorithm are effectively overcome, and the accuracy of the RUL prediction is improved.

2 Improved PF Algorithm

2.1 VAE

VAE is a probabilistic graphical model based on neural networks [12]. In practical application, VAE is usually used to generate data and learn features. The main process of VAE is shown in Fig. 1.

In practical applications, the mini-batch gradient descent algorithm is often used to train the VAE network. This algorithm overcomes the problems of low accuracy and efficiency of traditional batch gradient descent and stochastic gradient descent algorithms by packing the training set and then solving the gradient in sub-packets. The algorithm implementation process can be found in [13].

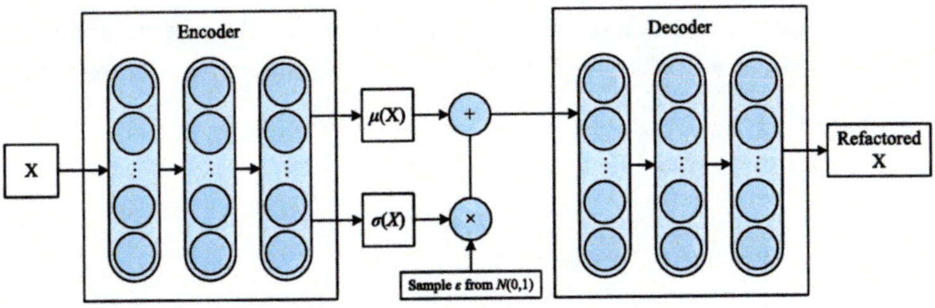

Fig. 1 VAE

2.2 *Particle Reinforced Mechanism*

The particle reinforced mechanism proposed in this paper is mainly used to reduce degeneracy and impoverishment. First, the mechanism will detect the degeneracy of particles in the posterior probability density function. When the particle degeneracy is identified, the algorithm will generate new high-weight particles to replace the existing low-weight particles, characterize the high-likelihood region, and improve the diversity of particles in the posterior probability density function.

Generally speaking, the number of efficient particles (particles with larger weights) on the posterior probability density function can be measured by the number of effective particles E_S.

$$E_S = \frac{\left(\sum_{i=1}^{N} w_k^i\right)^2}{\sum_{i=1}^{N} \left(w_k^i\right)^2} \times 100\%$$

where w_k^i is the particle weight of the i-th particle at time t_k. E_S reaches the maximum (i.e., 100%) when most of the particles are located in the high probability region (i.e., when $w_k^i \geq 1/N$). It is easy to know that if the E_S is smaller, particle degeneracy is more likely to occur.

In this paper, the effective sample size ES is used to characterize the degree of particle degeneracy on the posterior distribution. It is assumed that when $E_S \geq 60\%$, it means that most of the particles have a larger weight, and no particle enhancement is required. When $E_S < 60\%$, it means that the particle degeneracy is obvious, and an enhanced particle algorithm needs to be performed to improve the posterior distribution and reduce the sample degeneracy. The specific steps of the particle reinforced mechanism are as follows:

Step 1: Specify the upper boundary (U_b) and lower boundary (L_b) of the posterior probability density function to search.

$$U_b = \begin{cases} x_k^i + \sigma_k, \ x_k^i \geq x_B \\ x_B + \sigma_k, \ x_k^i < x_B \end{cases}, \quad L_b = \begin{cases} x_k^i - \sigma_k, \ x_k^i < x_B \\ x_B - \sigma_k, \ x_k^i \geq x_B \end{cases}$$

where σ_k represents the standard deviation of the particle; the particle weight $w_k^i <$ 1/N; x_B is the particle with the largest weight and its weight is $w_B = \max(w_k^N)$.

Step 2: Generate a new particle in the high probability region of the posterior probability density function space.

$$\overset{\frown}{x}{}_k^i = L_b + (U_b - L_b)\beta$$

where $\beta \in [0,1]$.

Step 3: Calculate the $\hat{w}_k^i$ of the generated particle $\overset{\frown}{x}{}_k^i$. If $\hat{w}_k^i \leq 1/N$, use Step 2 and Step 3 to generate a new particle, until $\hat{w}_k^i \geq 1/N$. If $\hat{w}_k^i > w_B$, let $x_B = \overset{\frown}{x}{}_k^i$, $w_B = \hat{w}_k^i$ to update the x_B and w_B.

Based on the above analysis, the process of the improved particle filter algorithm can be obtained as follows (Fig. 2).

3 RUL Prediction

In this paper, the nonlinear proportional Wiener process is used to establish the degradation model of aviation equipment. And it can be expressed as

$$Y(t) = \lambda\Lambda(t|\boldsymbol{\theta}) + \sqrt{\lambda k}B(t) + \sigma_\varepsilon\varepsilon(t) \tag{1}$$

where $Y(t)$ represents the monitoring degradation data of the aviation equipment, $X(t)$ represents the real degradation state, λ is the drift coefficient and satisfies the normal distribution $\lambda\ N(\mu_\lambda, \sigma_\lambda^2)$, σ_ε is the Gaussian observation noise, and $\varepsilon(t)$ satisfies $N(0, 1)$.

By (1) and the state-space model of the available aviation equipment is

$$\begin{cases} X_k = X_{k-1} + \lambda\tau_k + \sqrt{\lambda k}W_k \\ Y_k = X_k + \sigma_\varepsilon V_k \end{cases} \tag{2}$$

where $Y_k = Y(t_k)$, $X_k = X(t_k)$, $\tau_k = \Lambda(t_k|\boldsymbol{\theta}) - \Lambda(t_{k-1}|\boldsymbol{\theta})$, and W_k and V_k are standard normal variables.

The degradation state estimation of aviation equipment mainly includes the real degradation state $\hat{X}$ and the unknown parameter $\hat{\lambda}$, $\hat{k}$, $\hat{\sigma}_\varepsilon^2$, $\widehat{\boldsymbol{\theta}}$ in the degradation model. The parameters $\hat{k}$, $\widehat{\boldsymbol{\theta}}$, $\hat{\sigma}_\varepsilon^2$ do not change with the iterative process of particle filtering, so the maximum likelihood method can be used for estimation. The $\hat{X}$ and $\hat{\lambda}$ can be estimated by using the improved PF algorithm proposed above. On this basis, by giving the failure threshold ω, the RUL of aviation equipment can be obtained.

Let L_k^i denote the lifetime estimate obtained by the i-th particle. Then this particle obtains the RUL of the equipment as

Fig. 2 The process of the improved PF

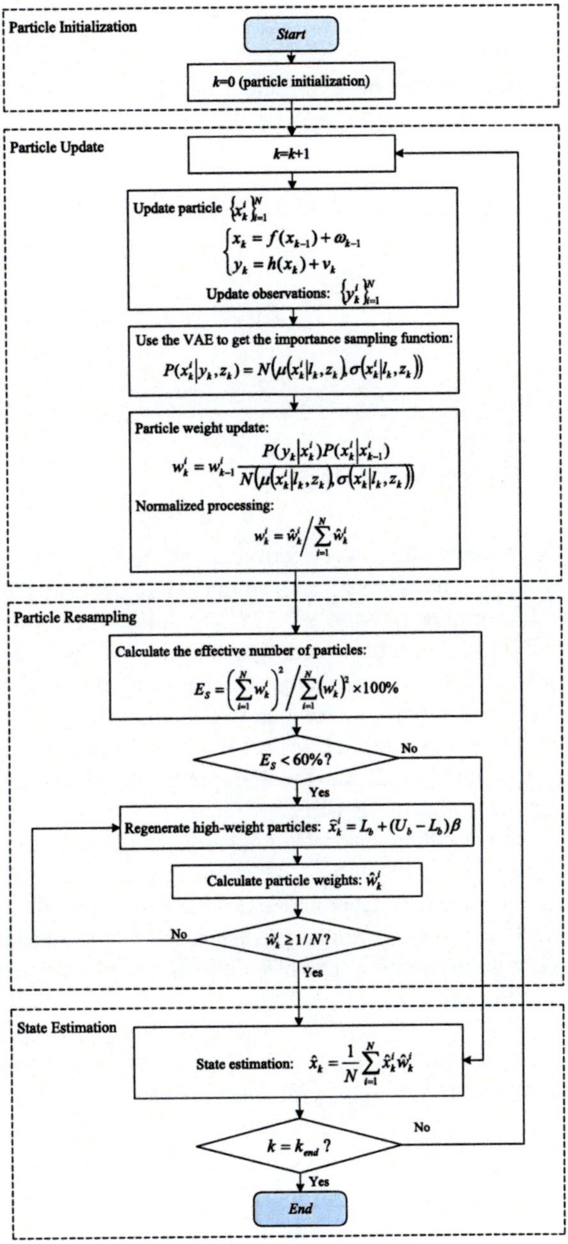

$$\mathrm{RUL}_k^i = L_k^i - t_k \tag{3}$$

Based on the basic principle of PF, the probability distribution and expected expression of the RUL can be written as follows:

$$f\left(\mathrm{RUL}_k^i|\omega\right) \approx \sum_{i=1}^{N} w_k^i \delta(\mathrm{RUL}_k - \mathrm{RUL}_k^i) \tag{4}$$

$$E\left(\mathrm{RUL}_k^i|\omega\right) = \sum_{i=1}^{N} w_k^i \mathrm{RUL}_k^i \tag{5}$$

4 Example Analysis

This paper verifies the effectiveness of the method based on the degradation data of aviation lithium batteries given in [14]. In this paper, the charge and discharge data of four groups (A1#, A2#, A3#, and A4#) of lithium batteries under the condition of 0.9Ah rated capacity are selected for analysis, and the failure threshold is given as 80% of the rated capacity. The degradation data of lithium batteries are shown in Fig. 3. To facilitate the comparative analysis, the RUL prediction method proposed in this paper is denoted as VAEe-PF. The RUL prediction method based on traditional PF is denoted as PF. And the RUL prediction method based on the unscented Kalman particle is denoted as UPF.

For the lithium battery degradation data, this paper sets the generative model and recognition model of the VAE network as a three-layer network model, and sets the number of hidden layer nodes to 14 and the mini-batch size to 48. Thus, the RUL of the lithium battery can be predicted. First of all, analyze the A1 lithium battery as the target equipment. The life prediction results corresponding to different methods are shown in Table I.

In Table 1, AE and CIW are representing the RUL prediction Absolute Error (AE) and 95% Confidence Interval Width (CIW), respectively. The calculation methods of AE and CIW can be written as follows:

$$\mathrm{AE} = |RUL_t - RUL_e|$$

$$\mathrm{WCI} = H\_CI - L\_CI$$

where RUL_t and RUL_e are representing the real RUL and predicted RUL of the lithium battery, respectively; H_CI and L_CI are representing the upper and lower boundaries of the 95% confidence interval of the RUL, respectively.

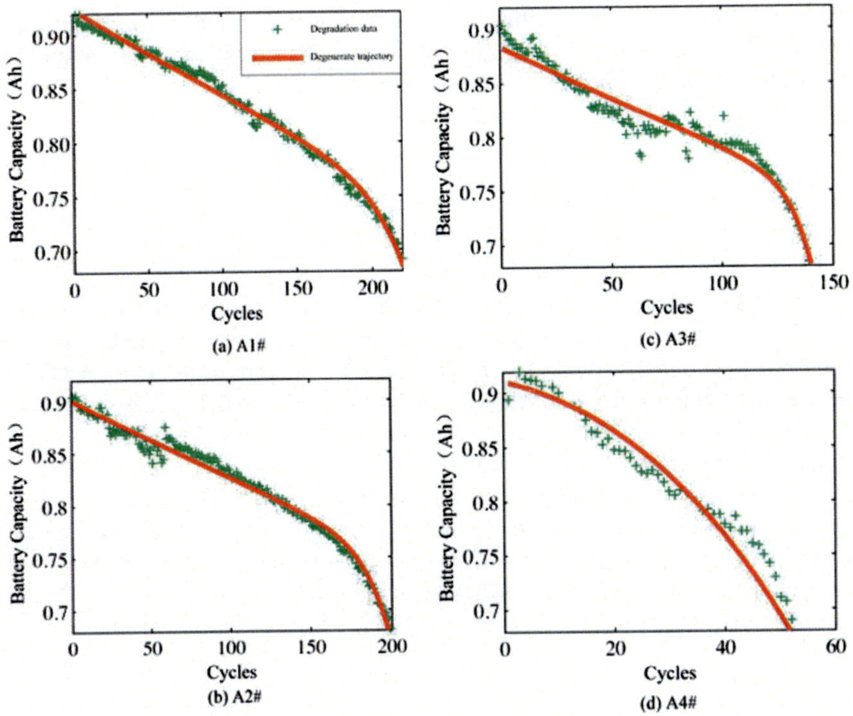

Fig. 3 The degradation data of lithium batteries

Table 1 RUL prediction results

Method	t_k	True lifetime	Predicted lifetime	True RUL	Predicted RUL	AE	95% WCI
PF	125	212	153	87	28	59	21
UPF	125	212	201	87	76	11	13
VAEe-PF	125	212	209	87	84	3	11
PF	167	212	187	45	20	25	20
UPF	167	212	202	45	35	10	11
VAEe-PF	167	212	210	45	43	2	10

It can be seen from Table 1 that compared with UPF and VAEe-PF, PF has the largest AE and the widest WCI, indicating that the PF method has no advantages in the accuracy and precision of RUL prediction. Compared with PF, the prediction performance of UPF and VAEe-PF is improved, indicating that the improved particle filtering algorithm can better overcome the particle degeneracy and impoverishment, and improve the prediction accuracy. Further analysis shows that VAEe-PF not only improves the RUL prediction accuracy but also achieves higher precision than UPF, which also verifies that the method of simultaneously improving the importance sampling function and resampling is better than the method of simply improving the importance sampling function, which can improve the prediction accuracy of RUL and reduce the uncertainty of the prediction.

5 Conclusion

Aiming at the problem that the traditional RUL prediction method based on PF has low accuracy due to the problem of particle degeneracy and impoverishment, this paper proposes an RUL prediction method based on improved PF. Different from the traditional improved algorithm that only improves the particle update process or the resampling process, this paper introduces a VAE in the particle update process to generate the importance sampling function, and introduces a new particle reinforced mechanism in the resampling process, so as to effectively overcome particle degeneracy and impoverishment. Based on the lithium battery degradation example, it is verified that the method proposed in this paper can effectively improve the accuracy and reduce the uncertainty of the RUL prediction.

References

1. Wang, X.: Wiener processes with random effects for degradation data. J. Multivar. Anal. **101**(2), 340–351 (2010)
2. György, K., Kelemen, A., Dávid, L.: Unscented Kalman filters and particle filter methods for nonlinear state estimation. Proc. Technol. **12**, 65–74 (2014)
3. Zhang, Y., Xiong, R., He, H.: Lithium-ion battery remaining useful life prediction with box-cox transformation and Monte Carlo simulation. IEEE Trans. Industr. Electron. **66**(2), 1585–1597 (2018)
4. Joaquin, M.: Analysis of parallelizable resampling algorithms for particle filtering. Signal Process. **87**(12), 3155–3174 (2007)
5. Andrieu, C., Doucet, A., Holenstein, R.: Particle markov chain monte carlo methods. J. R. Stat. Soc. **72**(3), 269–342 (2010)
6. Zhang, Q., Wang, P., Chen, Z.: An improved particle f lter for mobile robot localization based on particle swarm optimization. Expert Syst. Appl. **135**, 181–193 (2019)
7. Yu, M., Li, H., Jiang, W.: Fault diagnosis and rul prediction of nonlinear mechatronic system via adaptive genetic algorithm-particle filter. IEEE Access **7**, 11140–11151 (2019)
8. Doucet, A.: On sequential Monte Carlo methods for Bayesian filtering. Stat. Comput. **10**(3), 197–208 (1998)
9. Bolic, M., Duric, M., Hong, S.: Resampling algorithms and architectures for distributed particle filters. IEEE Trans. Signal Process. **53**(7), 2442–2450 (2005)
10. Fu, X., Jia, Y.: An improvement on resampling algorithm of particle filters. IEEE Trans. Signal Process. **58**(10), 5416–5422 (2010)
11. Qian, Y., Yan, R.: Remaining useful life prediction of rolling bearings using an enhanced particle filter. IEEE Trans. Instrum. Meas. **64**(10), 2696–2707 (2015)
12. Liu, S., Huang, Y., Hu, J.: Learning local responses of facial landmarks with conditional variational auto-encoder for face alignment. In: IEEE International Conference on Automatic Face and Gesture Recognition. IEEE (2017)

13. Khirirat, S., Feyzmahdavian, R., Johansson, M.: Mini-batch gradient descent: Faster convergence under data sparsity. In: 2017 IEEE 56th Annual Conference on Decision and Control (CDC). IEEE (2017)
14. He, W., Williard, N., Osterman, M.: Prognostics of lithium-ion batteries based on dempsterC-shafer theory and the bayesian monte carlo method. J. Power Sources **196**(23), 10314–10321 (2011)

Research on CBRN Practical Assessment Technology Based on Artificial Intelligence Technology

Junhua Wang, Hongyu Yang, Wenbin Dong, Minghu Zhang, He Zhang, Yunke Jing, and Xin Zhao

Abstract Combined with the common practical assessment problems in the field of CBRN, which have the problems of high subjective factors and inconsistent assessment standards, this paper proposes the feasibility design scheme of the assessment system based on intelligent identification technology, and analyzes the key technologies such as target recognition technology and expert system involved in the system. While verifying the feasibility of the assessment system design scheme based on intelligent identification technology, it provides ideas for the construction of the assessment system based on intelligent identification technology. It also provides a good solution to improve the quality of assessment.

1 Introduction

In recent years, although the automated detection and monitoring technology in the field of CBRN has been greatly developed, many specific tasks still need to be completed manually, and the equipment operation ability of the staff will have a great impact on the measurement results. Taking nuclear radiation monitoring as an example, the level of staff's practical operation ability is related to whether the measurement result of nuclear radiation level is accurate, which will affect the safety status monitoring of nuclear facilities, and even affect the emergency rescue operations related to nuclear radiation accidents.

The practical operation ability of CBRN equipment is usually tested by practical operation assessment. However, due to the inconsistency of assessment standards, the difficulty of manual assessment organization and implementation, and the easy influence of assessment results by subjective factors, the practical operation assessment cannot fully achieve the purpose of testing, which is not conducive to staff to improve the practical operation ability of equipment. At present, with the development of artificial intelligence technology such as deep learning, especially in the

J. Wang (✉) · H. Yang · W. Dong · M. Zhang · H. Zhang · Y. Jing · X. Zhao
Dalian Naval Academy, Dlian 116018, China
e-mail: wangjunhua19901991@163.com

K. Nakamatsu et al. (eds.), *Advanced Intelligent Technologies for Information and Communication*, Smart Innovation, Systems and Technologies 365,
https://doi.org/10.1007/978-981-99-5203-8_11

field of car driving [1], face recognition [2], medical diagnosis, fitness methods [3], and other fields, object recognition technology and expert system construction technology are becoming increasingly mature, which provides a good way for the field of equipment intelligent operation assessment.

2 Problems Faced by Equipment Practical Assessment

In the field of CBRN such as nuclear radiation monitoring, it is highly dependent on efficient and accurate equipment operation ability. Therefore, professional practical operation assessment [3, 4], as the basic test method of operation ability, has attracted much attention.

(1) The assessment standards are not uniform

For the practical assessment of the same equipment, there is a problem of inconsistent assessment standards. The main reason is that the people who formulate and implement the assessment standards in each unit are different, and the direction they grasp is also different, which objectively affects the quality of the assessment and leads to the lack of effectiveness of the practical assessment.

(2) Subjective factors affect the effectiveness of assessment.

The current assessment method is carried out by manual score evaluation, but the assessment results are greatly affected by subjective factors, which is not conducive to reflecting the real ability of reference personnel, and is not conducive to the later improvement of the ability of these personnel.

(3) Difficulties in organization and implementation

Because the practical assessment is mostly implemented by manual method, the assessment efficiency is low for the units with many reference personnel and a wide distribution area, and it is difficult to complete a large-scale assessment in a short time. At the same time, due to the impact of the epidemic situation and the limitation of personnel flow, the implementation of the assessment by manual method becomes more difficult.

(4) The applicability of traditional unmanned assessment method is not strong

The traditional unmanned assessment method generally uses simulation equipment or modification on the basis of existing equipment, but it has long research and development cycle, high modification cost, insufficient applicability, and is difficult to be applied on a large scale.

3 Feasibility Design of Assessment System Based on Intelligent Recognition Technology

In recent years, artificial intelligence technologies such as object detection have been developed rapidly. The deep learning model [5, 6] of YOLO (you only look once) [7] based on CNN (convolutional neural network) [8]. It has shown great advantages in object detection rate and accuracy, so the acquisition of practical operation information based on images has become feasible, and it also provides a method to solve the problems of organization and implementation difficulties, high economic cost, and long development cycle existing in the current assessment method. In addition, the technical advantages of strong rules and high objectivity of expert system also provide a good solution to solve the problems such as the existing assessment methods and standards are not uniform and easy to be affected by subjective factors.

Figure 1 shows the structure of the assessment system based on intelligent recognition technology. According to the assessment background extracted by the system, the reference personnel selects the appropriate equipment and equipment mode to measure and submit the results. The camera set at the operation site is used for operation image acquisition. The image intelligent recognition system with object detection as the key technology performs object detection according to the acquired image, and analyzes and generates data such as equipment selection, mode selection, and measurement results. The expert system analyzes the assessment background, the measurement results of the reference personnel, and the data of the image intelligent recognition system to give the assessment results.

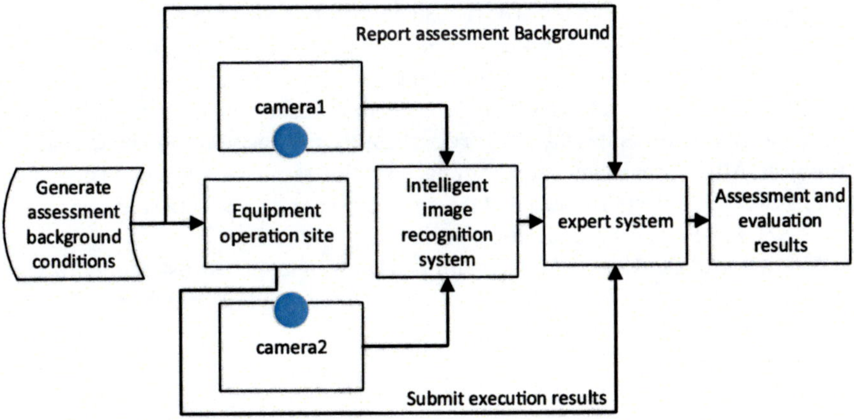

Fig. 1 Intelligent assessment system structure

4 Verification of Target Detection Technology and Design of Expert System

4.1 Object Detection Technology

As one of the core problems in the field of machine vision, the task of object detection is to find out all the objects in the image, and determine their position and size. Object detection has always been the most challenging problem in the field of machine vision due to the different appearance, shape, and pose of various objects, as well as the interference of illumination and occlusion during imaging. Therefore, in essence, object detection consists of two main tasks: object image recognition and object localization in the image. At present, object detection is mainly applied in pedestrian detection, vehicle detection, face recognition, medical image detection, and so on.

The core problems of object detection can be summarized into three points:

(1) The classification of objects in the image.
(2) Determination of target position in the image.
(3) Consideration of changes in the size and shape of targets.

They can be roughly divided into two categories, they are two-stage and one-stage, according to the proposals whether regions need to be generated in the middle of the algorithm. There were also multi-stage algorithms, but they were rarely used because of gaps in speed and accuracy.

The two-stage detection algorithm divides the detection problem into two stages. Firstly, the candidate regions are generated, and then the candidate regions are classified after the location is refined. The two-stage detection algorithm has a low recognition error rate and a low missing recognition rate, but it is slow and cannot meet the requirements of real-time detection scenarios, such as video object detection.

The one-stage detection algorithm does not need to generate the candidate region stage, but directly generates the category probability and position coordinate value of the object. After a single detection, the final detection result can be directly obtained, so it has a faster detection speed, but the general recognition accuracy and accuracy of the two-stage algorithm are a little worse.

According to the above classification of algorithms, the mainstream two-stage and one-stage target detection algorithms are shown in Table 1, and the mAP values of the main algorithms are shown in Table 2.

Through analysis of Table 2, it can be found that the one-stage YOLO has stronger adaptability in general. YOLO becomes more accurate in iterations, and its one-stage-based calculation speed makes it more practical and effective in practical applications, especially in situations where real-time factors need to be considered. Therefore, for real-time assessment methods based on artificial intelligence, YOLO method is of more practical significance.

The key of the image intelligent recognition system is the object detection technology. Here, the YOLO model is used to conduct deep learning and verification of the surface contamination detection device Equipment A. Due to space limitations,

Table 1 The mainstream two-stage and one-stage target detection algorithms

One-stage algorithm	Two-stage algorithm
YOLO V1, YOLO V2, YOLO V3	Fast R-CNN, Faster R-CNN
G-CNN	HyperNet, MS-CNN, PVANet
R-SSD, DSSD, DSOD, FSSD	MR-CNN, FPN, CRAFT
RON	CoupleNet, Mask R-CNN
–	OHEM, Soft-NMS

Table 2 Network model evaluation indicators % [9]

Network model	mAP
Faster R-CNN	80.8
YOLO v3	90.7
YOLO v4	87.6
YOLO v5	91.7
YOLO v5-CB	96.8

the technical verification of the identification of the selected equipment in the intelligent assessment system is mainly carried out. The data in the model were divided into training set, validation set, and test set according to 7:2:1.

According to the training iteration results (Fig. 2) and target detection results (Fig. 3), the loss rate will rapidly converge during training, and good training results can be achieved in a very short time, with high training rate and short training period. The recognition rate of target detection results is the recognition result under the condition of small samples, and the accuracy will also improve rapidly with the expansion of sample data. Therefore, for the target recognition technology based on YOLO model, both its training rate and detection results can meet the needs of short development cycle of intelligent assessment system and high target recognition rate.

4.2 *Design of Expert System*

The expert system in the structure of the intelligent assessment system contains a large amount of knowledge and experience of experts in the field of CBRN practice, such as precautions during operation, background condition discrimination, and measurement parameter analysis., and the system design is mainly based on fault tree. Take device A as an example. Figure 4 shows the operation flow.

Combined with the operation process in Fig. 4, and through the data acquisition method and process shown in Fig. 1, a fault tree model taking device A as an example is designed (Fig. 5), and the model is used to output whether the results are valid.

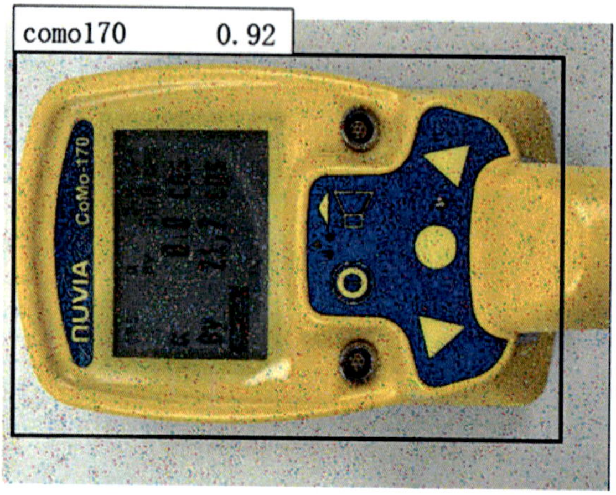

Fig. 2 Object detection results

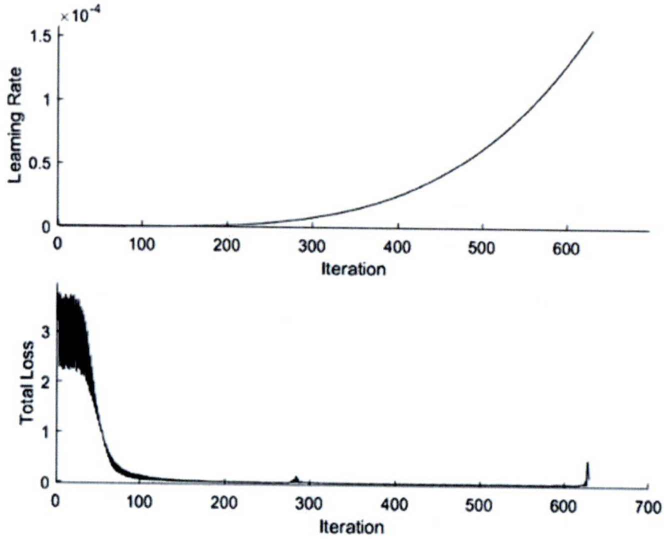

Fig. 3 Results of training iterations

In fault tree expert system with Equipment A, according to the real operation process, in turn, the selection of equipment type, monitoring mode, monitoring position, data such as step, step by step through the analysis of the various steps to realize the accurate judgment of the process of operation, and then according to the range of allowable error of the measurement results and the actual data comparison results are given. In the steps of device type, monitoring mode, monitoring posture, and data reading, image information such as device image, device interface mode,

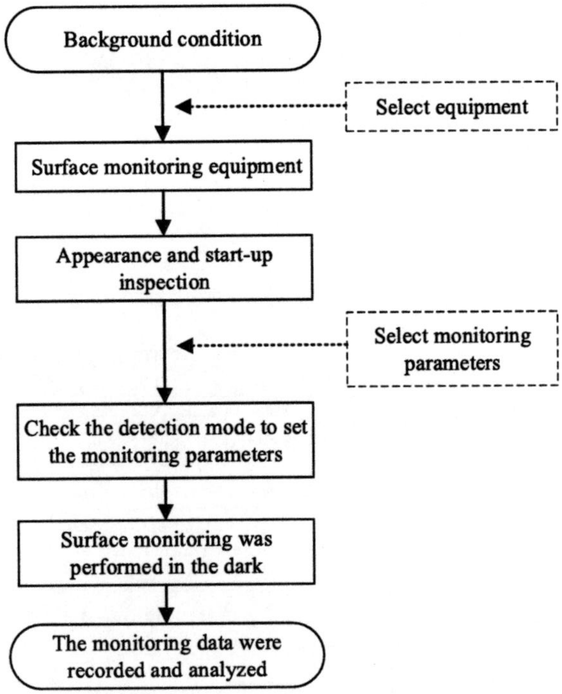

Fig. 4 Operation process(equipment A)

body posture, and data interface are analyzed for identification, and the identification results are used for specific judgment process.

The assessment background conditions in the expert system are randomly selected from several types of situations set in advance. Because the conditions to be monitored may be γ dose, surface contamination, or nuclide types, the selected equipment is different, and the corresponding operation mode is also different. The example studied in this paper is an example of surface contamination inspection. During surface contamination inspection, it is necessary to determine whether the distance is too close or too far. At the same time, it is necessary to consider whether the detector of the device is aligned with the measuring surface. If the object surface is not aligned, the actual parameter values will change, as shown in Fig. 6.

To reduce the error in the measurement, therefore, S and S' should to be as parallel as possible, if through the recognition, computer vision technology due to problems such as shade image information acquisition, which can lead to fail to judge the S and S', therefore in the device was installed on A probe in the plane four detectors, as shown in Fig. 7. Detection range is one of the inputs of expert system, its data is obtained from the sensors (1–4), it was used to comparing the value of the four sensors whether discriminant detection range is appropriate within the scope of the permit; moreover, the difference between the maximum (Lmax) and the minimum value (Lmin) delta L is within the scope of the permit which can distinguish whether measuring surface detector approximately parallel.

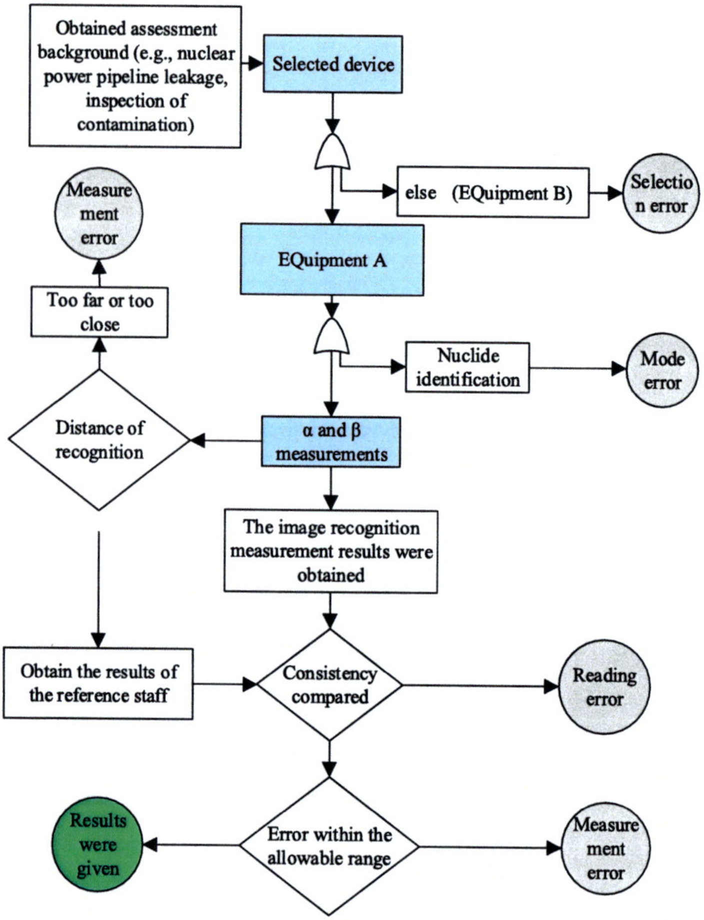

Fig. 5 Fault tree with equipment A as an example

From the above content, the construction of expert system is based on the real operation process, and uses the equipment information, body posture, measurements, and other data as input of the system, so the idea also applies to other equipment designs of expert system, in practice, as long as the equipment operation standard, the construction of expert system will be very easy to implement.

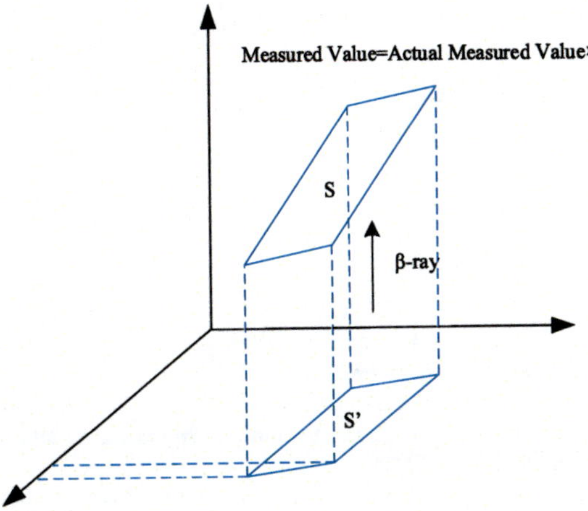

Fig. 6 Relationship between measured values and actual measured values

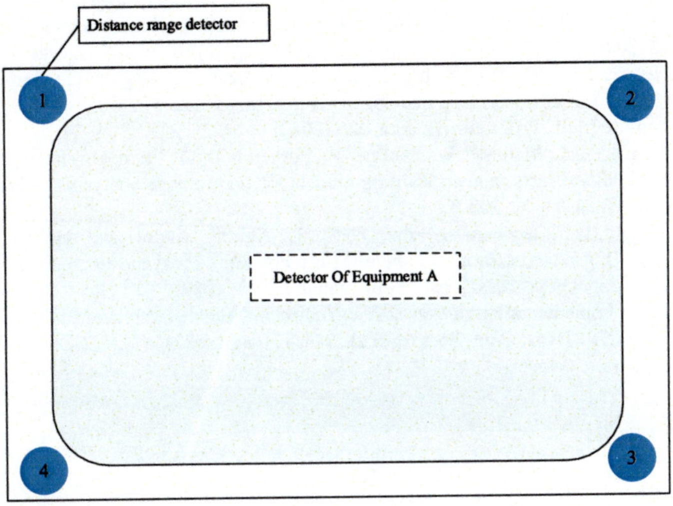

Fig. 7 Distribution of detectors

5 Conclusion

Combined with the actual needs of equipment practical operation assessment in the field of CBRN, this paper analyzes the problems existing in the current assessment methods and relies on the advantages of artificial intelligence technology to propose an assessment method based on intelligent recognition technology. The feasibility of the assessment method based on intelligent recognition technology is demonstrated through the verification of the target recognition technology based on YOLO model and the advantages of the construction of the expert system based on fault tree, which provides good guidance for the further development of the assessment system based on intelligent recognition technology.

References

1. Chan Jennifer S.K., Choy S.T. Boris, Udi, M., Ariel, S., Vered, S.: Variable selection algorithm for a mixture of Poisson regression for handling overdispersion in claims frequency modeling using telematics car driving data. Risks **10**(4)
2. Yaozhe, S., Hongying, T., Fangzhou, M., Chaoyi, W., Mengmeng, W., Ziting, S., Guanjun, T.: A transformer-based low-resolution face recognition method via on-and-offline knowledge distillation. Neurocomputing 509 (2022)
3. Yunlong, T.: Assessment of physical fitness and health status of athletes based on intelligent medical treatment under machine learning. Comput. Intell. Neurosci. 2022 (2022)
4. Stanisław, D., Jacek, P., Aneta, H., Radosław, D., Arkadiusz, O., Marek, W.: Assessment of the reliability of wind farm devices in the operation process. Energies **15**(11) (2022)
5. Athirah, Z.N., Mokhzaini, A.M., Khairunnisa, H., Salwa, M.K.A., Uswah, K.: An automated solid waste detection using the optimized YOLO model for riverine management. Front. Public Health **10** (2022)
6. Marcel, F., Thomas, K., Marcel, N., Daniela, T., Andreas, S., Sergios, G.: Self-supervised learning for automated anatomical tracking in medical image data with minimal human labeling effort. Comput. Meth. Programs Biomed. **225** (2022)
7. Stanley Emma, A.M., Matthias, W., Pauline, M., Forkert Nils, D.: Fairness-related performance and explainability effects in deep learning models for brain image analysis. J. Med. Imaging (Bellingham, Wash.) **9**(6) (2022)
8. Xin, S., Gaolu, H., Xiaochen, H., Yue, Y., Ze, L.: Sliding window and dual-channel CNN (SWDC-CNN): A novel method for synchronous prediction of coal and electricity consumption in cement calcination process. Appl. Soft Comput. J. **129** (2022)
9. Research and Implementation of Locust Recognition in Ningxia Grassland Based on Improved YOLO v5 [J/OL]. Trans. Chin. Soc. Agricult. Mach. 1–10 (2022)

Research on the Application of Voice Acoustics for Speech Intelligence in Folk Songs

Shiliang Lyu, Shuang Chen, and Hanyun Qi

Abstract Folk songs take language as the carrier, and the singing tune and cultural connotation of singing are the important research contents of many music scholars, ethnologists, and linguists. In the process of folk song singing, the vocalization of the singing voice has an important influence on the singing voice and artistic performance. The state of the singing pronunciation can be carefully observed through the acoustic analysis of the voice. This paper will introduce the specific application of voice acoustic analysis method in the research of folk song singing. Combined with the singing characteristics of folk songs, the vocal type of singing voice can be obtained through the parameters of voice acoustic analysis. The significance and the feasibility of the application of voice acoustic analysis in folk song research will be further explored. The research parameters and conclusions are helpful to speech recognition, speech synthesis, and artificial intelligence.

1 Introduction

In recent years, with people's attention to the protection of the world's intangible cultural heritage, the study of folk songs has made great progress in both form research and social and cultural connotation research. However, many scholars' research methods on folk songs tend to be traditional. The research on Chinese folk songs tends to be in the category of morphological introduction, morphological description, or research, and there is less research on various activities related to folk songs and their social and cultural connotations [1]. Many valuable conclusions

S. Lyu (✉)
Key Laboratory of Linguistic and Cultural Computing Ministry of Education, Northwest Minzu University, Lanzhou 730000, Gansu, China
e-mail: shilianglv@126.com

S. Chen · H. Qi
Gansu Provincial Key Laboratory of Intelligent Processing of National Languages, Northwest Minzu University, Lanzhou 730000, Gansu, China

K. Nakamatsu et al. (eds.), *Advanced Intelligent Technologies for Information and Communication*, Smart Innovation, Systems and Technologies 365,
https://doi.org/10.1007/978-981-99-5203-8_12

have been drawn in the study of singing voice by using acoustic technology, but there are few achievements in the research on the vocalization of folk songs from the perspective of voice.

Voice acoustic analysis is a method of comprehensive research on vocalization by collecting voice and speech signals through laryngeal instruments and voice acoustic equipment [2]. Singing is an artistic expression of language, and singing pronunciation can also perform statistical analysis on related voice parameters by collecting voices, so as to obtain the characteristics of voice and speech in the singing process. This paper will introduce the application of voice acoustic analysis method in folk song singing, and comprehensively analyze the vocal types of singing voice through specific experimental cases, so as to realize the visual analysis of folk songs' singing voice, which can be used for singing teaching, creation, inheritance, and protection of folk songs, it provides methodological reference and data reference [3].

2 Introduction of Voice Acoustics Analysis

Folk song is a singing art with language as the carrier, and the vocal information of its singing voice is an important content of singing acoustic research. Different folk songs have their own unique singing styles. In previous researches, we have focused on the analysis of the characteristics of the voice from the aspects of singing style and sense of hearing, on resonance characteristics, etc. [4]. Using voice acoustic analysis is possible to carry out statistical analysis of voice parameters on the characteristics of various arias, and to realize the classification of arias and the determination of vocal types [5]. The quantitative analysis of singing voice and voice parameters can also provide examples and guidance for the teaching of folk song singing skills and the intelligent protection and inheritance of folk songs [6].

The main steps of voice acoustic analysis include recording, speech signal processing, parameter extraction, data analysis, and summarization. Commonly used voice acoustic analysis parameters include: fundamental frequency (F0), which represents the number of times the vocal cords vibrate per second, in Hz [7]. The fundamental frequency can roughly reflect the pitch information, and the magnitude of the fundamental frequency can reflect the vibration of the vocal cords, and is related to the subglottic pressure and the tension of the vocal cords [8]. The faster the vocal cords vibrate, the greater subglottic pressure, or the greater vocal cord tension means the higher fundamental frequency. Open quotient (Open Quotient) and speed quotient (Speed Quotient) are important voice vocalization type parameters. The voice signal is collected by the electronic glottal instrument, and the open quotient and speed quotient parameters are extracted from the voice signal [9]. A cycle of the voice signal includes two parts, the contact phase and the non-contact phase. The quotient is the ratio of the non-contact phase time to the cycle time, which can reflect the spectral slope characteristics of the voice and further reflect the change of high-frequency energy. The speed quotient is the ratio of the time between the opening of the glottis and the closing of the glottis in the contact phase [10]. The speed quotient can be understood as the ratio of the opening speed and closing speed of the glottis, which can reflect the high-frequency energy characteristics of the voice. It can also reflect changes in the glottis airflow. In the specific research process, combining these three parameters can comprehensively reflect the mode of vocal cord vibration and the situation of the opening and closing of the glottis. Then the vocal type of the voice can be determined [11].

3 Equipment and Software for Voice Acoustic Analysis

In the process of applying voice acoustic analysis methods to folk song research, techniques from musicology, linguistics, acoustics, and other disciplines are integrated. It is necessary not only to master the theoretical knowledge of voice acoustics, but also to master the use of recording equipment, to use voice analysis software proficiently, and to have the ability of processing data and analyzing parameters. In addition to Real-Time EGG Analysis Model 5138 from KAY Company [12], the commonly used voice analysis software also includes VoiceSauce software [10], which can extract and analyze various voice parameters. VoceVista software is a speech visualization analysis software developed by Sygyt speech software company in the Netherlands, which can display voice and speech waveforms in detail. The multi-dimensional voice analysis software MDVP (multi-dimensional voice program) of KAY Company in the United States can extract 33 voice parameters, but it is still difficult to study the characteristics of voice. Researchers can choose appropriate software according to specific research needs. The study of voice is inseparable from the comparison and reference with the speech signal, so it is also essential for the application of speech analysis software. In the research of phonetics, the commonly used speech analysis software is the Praat software developed by the Speech Science Laboratory of the University of Amsterdam in the Netherlands, the software can perform preprocessing of voice and speech signals, extract multiple speech acoustic parameters, and some voice parameters [13].

4 Acoustic Analysis of Folk Singing Voice

The research mainly takes the eastern folk song "Precious Horse on the Prairie" of the Yugur ethnic group in China as an example, and introduces the whole process of voice acoustic analysis. During the experiment, the fundamental frequency, opening quotient, and velocity quotient parameters were extracted, and the vocal parameters of the whole song and the vocal parameters of the dragging method were systematically analyzed.

4.1 Selection and Recording of Speakers

In the experiment, an electronic glottal instrument was used to collect voice signals, and a microphone was used to collect speech signals. The speaker of the folk song is a female singer, aged 52 years old, who has been singing original ecological folk songs for a long time and has good singing skills and level. During the signal acquisition process, Adobe Audit CC software is mainly used to simultaneously collect voice and glottis signals. The hardware equipment mainly includes Lavalier microphone, EGG for Singers 7050A electronic glottal instrument, external sound card, mixer, and laptop. The speech signal is the first channel, the voice signal is the second channel, the sampling rate is 44 kHz, and the file is saved in Windows PCM Wav format.

4.2 Voice Parameter Extraction

Real-Time EGG Analysis Model 5138 software is used for voice parameter extraction. In the process of extracting parameters, the singing signal is segmented and marked by referring to the speech signal. The voice signal marking is mainly based on the research needs, referring to the speech signal to mark the part of the voice segment, which mainly includes the part of the musical segment that needs to extract parameters. The specific operation steps are as follows.

Figure 1 shows the voice parameter extraction software interface. The entire signal analysis interface is divided into three parts, namely, speech fundamental frequency, voice speed quotient, and sound pressure signal.

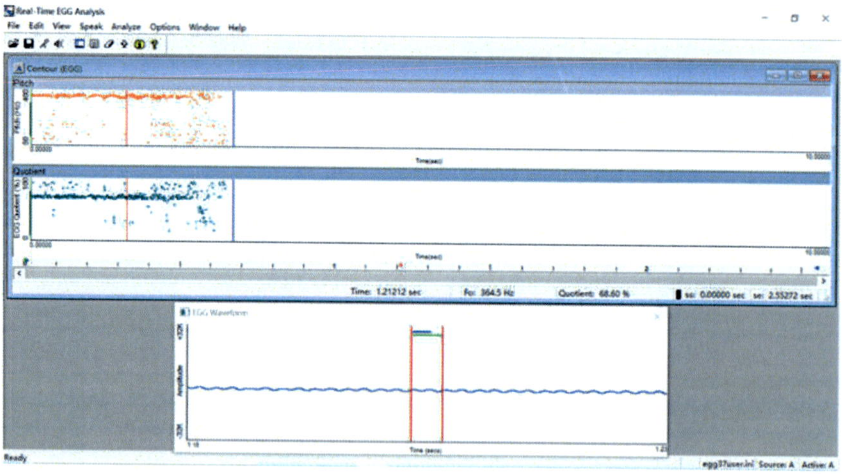

Fig. 1 Voice signal marking interface diagram

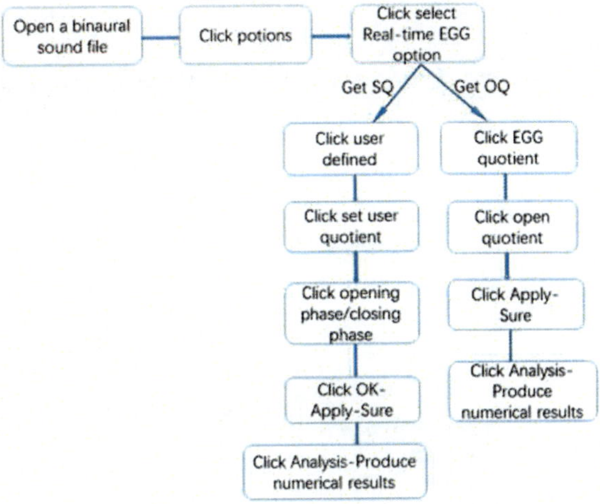

Fig. 2 Flowchart of voice parameter extraction

In order to ensure the accuracy of the signal marking, the voice signal is segmented according to the played sound during the marking process. Figure 1 shows the glottal signal of the singing drag section. After opening the voice file, select the position where the parameters need to be extracted, and select the voice signal at this position to save it separately. After the signal is segmented, the parameters are extracted and analyzed according to different files. The red vertical line in Fig. 1 is the position pointed by the cursor, and the EGG signal graph at this position is the waveform between the two red vertical lines in the bottom window. According to the marked signal, the fundamental frequency, open quotient, and speed quotient parameters can be directly extracted using the software. The parameter extraction process is shown in Fig. 2.

First open the two-channel file, the two-channel files are speech and voice, respectively. Then select and mark the voice segment whose parameters need to be extracted, and set the opening quotient and speed quotient parameters, respectively. Finally, the software extracts and calculates the open quotient and speed quotient parameters according to the signal marker position, as shown in Fig. 3. The parameter data is arranged by sampling point and by row.

The extracted voice parameters can be saved as text files in txt format. During the analysis process, the data can be imported into Excel software for summary analysis. Before data analysis, it is also necessary to systematically screen the extracted data, remove invalid data, and then perform statistical analysis. The parametric statistical software used in this experiment are Microsoft Excel and IBM SPSS Statistics 19.0. According to the data type and research needs, choose appropriate software for data statistical analysis.

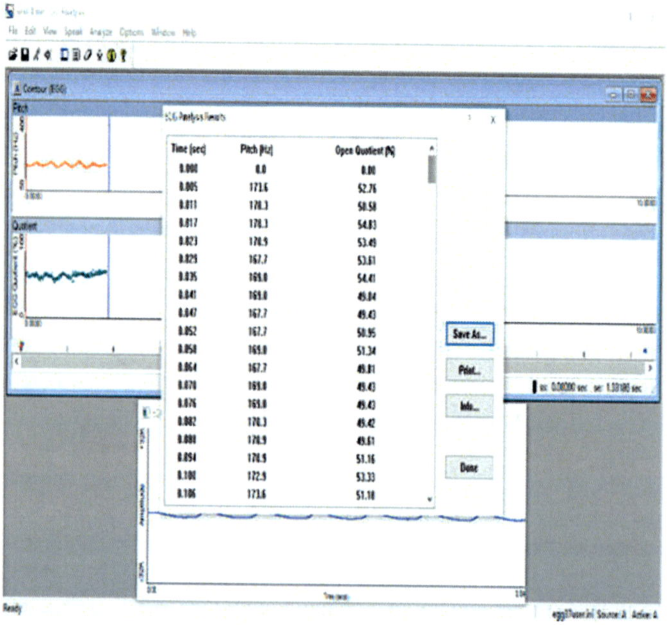

Fig. 3 Schematic diagram of voice parameter storage

5 Voice Parameter Analysis

5.1 Analysis of the Vocal Type of Singing Voice

During the experiment, in order to analyze the vocal type of singing voice, the voice signal of the whole folk song was extracted, and the vocal part included the vowel and voiced consonant part of the singing voice. The extracted parameters include fundamental frequency (F0), open quotient (OQ), and speed quotient (SQ).

Figure 4 shows the distribution of voice parameters in the "Precious Horse on the Prairie" song of the speaker: the X-axis is the fundamental frequency, the Y-axis is the opening quotient, and the Z-axis is the speed quotient. From the parameters in Fig. 4, the range of fundamental frequency parameters is between 70.4 Hz and 396 Hz, the opening quotient is in the range of 7.16–84.43%, and the speed quotient is in the range of 100–526.47%. The vocal parameters of the whole song have a large distribution span, which reflects the diverse characteristics of the singing voice of this folk song.

From the perspective of the parameter distribution, with the increase of the fundamental frequency parameter, the speed quotient and the opening quotient parameter both show an increasing feature. The distribution of the fundamental frequency shows the adjustment mode of the pitch change during the singing process, and the rhythm of the singing pitch is represented by the way of up and down, and it also shows

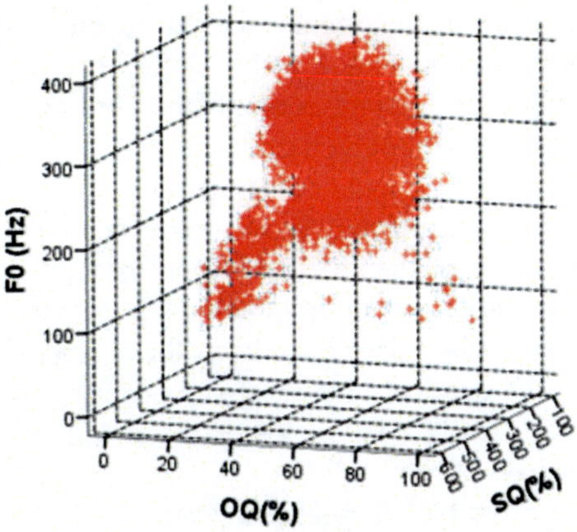

Fig. 4 Distribution map of vocal parameters of the folk song

the vocal range distribution of the singer during the singing process. Combining the opening quotient and speed quotient parameters, the opening quotient parameters are relatively concentrated, but the speed quotient shows a large distribution range, showing that the vocal cord opening changes less during the singing process, but the vocal cords open and close faster.

In order to analyze the distribution of voice parameters and the type of singing voice in detail, the voice parameters of the entire folk song were averaged, and the standard deviation analysis was performed, and two decimal places were reserved. Table 1 shows the average value and standard deviation of the voice parameters of singing voice. When the parameters are extracted, only the vowels and voiced consonants in the singing process are also extracted, and the voices with non-vibrated tones are not considered.

From the view of voice parameters, the average value of fundamental frequency parameters is higher than the fundamental frequency of speech when speaking, and the standard deviation of parameters shows a larger degree of dispersion, which is greatly affected by the variation of singing range and singing pitch. The open quotient is a key parameter variable of the frequency characteristics of the voice, and has a close relationship with the spectral characteristics of the voice. The increase of the open quotient is manifested in the large difference between the low-frequency energy and the high-frequency energy in the spectrum, which belongs to the characteristics of the air voice. If the quotient parameter is small, it means that the difference between high-frequency energy and low-frequency energy is small, and it can show the characteristics of squeezing and tight voice. The quotient parameters are relatively concentrated, and the average value is about 30.12%. According to the opening

Table 1 Data sheet of vocal parameters of the folk song

Parameters	F0 Hz)	OQ (%)	SQ (%)
Mean	289.96	30.12	304.51
SD	53.75	15.78	160.99

parameters of normal voice, the opening parameters of normal voice are about 55%, 45% for tight voice, and 58% for air voice [7]. Therefore, the singing voice of this folk song is more concentrated, and the distribution range is consistent with the characteristics of a tight voice.

The speed quotient is the ratio of the opening and closing speed of the glottis, which can reflect the high-frequency energy characteristics of the voice. With the increase of the speed quotient, the high-frequency energy of the voice gradually increases. The speed quotient parameters of the whole folk song vary widely, and the standard deviation shows the characteristics of large parameter dispersion. The speed quotient of normal voice is roughly in the range of 250%, and the average value of 304.51% is close to the distribution range of tight voice (350%) and bubble voice (420%) [7].

From the comprehensive voice parameters, it can be concluded that the overall pitch of the folk song singing voice is high, and the parameters of opening quotient and speed quotient show the characteristics of tight voice, indicating that the pronunciation type of the folk song singing voice is tight voice. The characteristics of the tight voice are mainly influenced by the pronunciation of the Mongolian language group of the Altaic language family. The folk song is sung in the eastern language of the Yugur ethnic group, which belongs to the Mongolian language group. The singing tune and pronunciation are close to Mongolian folk songs. In the singing process, in order to enhance the loudness and achieve the characteristics of loud timbre, a method with high-frequency energy is adopted, combined with the characteristics of the overall pitch increase, so as to achieve the artistic expression effect of singing.

5.2 *Voice Analysis of Drag Cavity in Folk Songs*

In the folk songs of this experiment, there are many drag cavities in the singing process. Influenced by Mongolian long-tuned folk songs, most of the drag cavities are expressed as the vowel extension at the end of the passage. Drag cavity can make the timbre of folk songs tactful and long, and under the control of different pitches and lengths, it can play a broad and loud singing sound. The drag tune in the folk song "Precious Horse on the Prairie" can be divided into two types in terms of timbre: one has a relatively low pitch and sounds relatively low, slow, and stable, which can be classified as low drag cavity. The other is high pitched and loud, and it is classified as high drag cavity. In this experiment, we selected low drag cavities with better signal as the research object, and analyzed the voice characteristics of the drag cavity part of folk songs from the aspect of voice parameters.

A total of 19 voice signals with a low-pitched tone were selected in the experiment, which have an average duration of 2.47 s. The extracted voice parameters are shown in Table 2. The fundamental frequency variation range in the drag cavity is 79.5 Hz–392.2 Hz, and the range of the sound range is 312.7 Hz. In addition, the average fundamental frequency in the overall drag cavity signal is high, which shows a treble effect in singing voice. Although the pitch of the low drag cavity is lower than that

Table 2 Data sheet of voice parameters of drag cavity

Parameters	F0 (Hz)	OQ (%)	SQ (%)
Maximum	392.20	94.20	543.75
Minimum	79.50	19.63	100.00
Mean	321.24	41.88	242.61
SD	73.98	21.00	124.60

of the high drag cavity, the single low drag cavity still has a bright and melodious timbre. The large range of opening quotient indicates that the opening degree of the glottis varies greatly during the singing of the drag cavity. Combined with the performance of the speed quotient, it can be shown that the opening speed and degree of the glottis in the drag cavity are overall larger. In the speed quotient parameters, the variation range is 100–543.75%, the variation range and the standard deviation are large, which indicates that the opening and closing process of the glottis is unstable, and the vocalization exhibits the characteristics of high-pitched voices.

Common types of vocalizations are bubble voice, air voice, tight voice, normal voice, and high-pitched voice [14]. Among them, the main physiological mechanism of high-pitched voice is the increase of vocal cord tension and the relaxation of vocal cord muscles. This means that the effectiveness of the opening and closing of the glottis during vocalization is reduced, with only a small portion of the vocal folds at the glottal boundary participating in the vibration. When the vocal cords are always open, the fundamental frequency is high and the opening quotient is high. The parameters of the drag cavity in Table 2 show the same performance. At the same time, the EGG signal also shows a high-pitched voice waveform. The high-pitched voice waveform has a short period and a relatively regular periodicity, is symmetrical to the left and right, approximately conical, and has a short contact phase.

From the perspective of comprehensive vocal parameters, in the process of singing the drag cavity of the folk song, the singer increases the pitch and loudness of the voice by increasing the frequency of vocal cord vibration, and expresses the high-pitched tonal characteristics of the voice by adjusting the opening and closing of the glottis, making the whole drag cavity full of emotions and distant and melodious timbre, which shows the unique artistic style of Yugur folk songs and long-tune folk songs.

6 Conclusion and Discussion

When using an electronic glottal instrument to collect voice signals, male speakers can obtain better voice signals than female speakers, which is due to the difference in the physiological structure of the larynx between males and females. The folk singers in this study are female. Among the collected voice signals, some high-pitched voice signals have poor effect. Therefore, the voice parameter data of the whole song can only roughly describe the range of the folk song and the overall characteristics of the voice. Judging from the collected data, the eastern Yugur folk song "Precious Horse on the Prairie" covers a wide range of sounds, the overall pitch is high, and it is characterized by high-pitched and loud singing. The whole song has a rich variety of voices when it is sung.

Judging from the voice signal of the drag cavity part of the folk song, the overall pitch of the low drag cavity part is also high. The change of the speed quotient indicates that the high-frequency energy is larger during the vocalization process, the glottal opening time is longer, and the vocal cords vibrate faster. It shows the characteristic of a high-pitched voice. Due to the large signal noise in the high drag cavity, the analysis of it was not carried out, and it was impossible to compare and summarize the two drag cavities.

At present, in the research of folk songs, the application of voice acoustic analysis is less, and the main difficulty is the mastery of voice acoustic knowledge and the use of voice analysis software. The application of voice acoustic analysis can research different vocal types in folk songs, and finally get the singing characteristics and artistic styles of different folk songs. It can also realize the quantitative comparison of signal parameters between different folk songs, analyze the differences between different folk songs, and make the ontological research of folk songs more detailed and in depth.

Under the trend of multi-disciplinary research, using the method of speech acoustics experiment to realize folk song singing research will be an important field of future singing research. The method of voice acoustics is used to quantify the singing voice of folk songs, which can realize the datamation of various singing styles and special pronunciations of folk songs from the perspective of singing and vocalization, and provide new ideas and methods for the construction of the acoustic database of folk songs, protection, and inheritance. In addition, the large amount of folk song data obtained through experiments can establish vocal standards for different types of folk song singing, and provide methods and references for the teaching and inheritance of folk songs.

Acknowledgements This research was financially supported by the Fundamental Research Funds for the Central Universities (Grant No. 31920220064).

References

1. Mu, Y.: Thinking about the problems in the study of Chinese folk songs from the current situation of Hua'er research. Music Res **4**, 13 (2004)
2. Lyu, S., Lin, N., Fan, B.: The application of phonetic acoustic analysis in Ewenki traditional folk songs. In: CIPAE 2021
3. Qichao, H.: Analysis on the difference of vibrato between Jinsheng and Guansheng in Kunqu Opera. J. Nanjing Acad. Arts: Music Perform. Ed. **2**, 6 (2022)
4. Yang, L., Jinrang, L.: Research on voice classification based on breath sound. Chin. J. Hearing Lang. Rehabil. Sci. **20**(1), 4 (2022)

5. Lyu, S..: The research on the singing voice Timbre of the eastern Yugur traditional folk songs. In: 7th International Conference on Education, Management, Information and Computer Science (2017)
6. Permission, Wei, H., Yuzhu, L., et al.: Fundamental frequency perturbation and amplitude perturbation in Chinese dialects. J. Chin. Phonet. **2,** 11 (2021)
7. Kong, J.: On Language Phonation. Minzu University of China (2001)
8. Sawashima, M., Hirose, H.: Laryngeal gestures in speech production. In: MacNeilage, P.F. (ed.) The Production of Speech, pp.11–38. Springer-Verlag, New York (1983)
9. Fujisaki, H.: A note on the physiological and physical basis for the phrase and accent components in the voice fundamental frequency contour. In: Vocal Fold Physiology: Voice Production, Mechanisms and Functions (1988)
10. Shue, Y.L., Keating, P., Vicenik, C., et al.: VoiceSauce: a program for voice analysis. J. Acoust. Soc. Am. **126**(4), 2221–2221 (2009)
11. Du, X., Zhu, B., Kong, Q., et al.: Singing melody extraction from polyphonic music based on spectral correlation modeling. In: ICASSP 2021—2021 IEEE International Conference on Acoustics, Speech and Signal Processing (ICASSP). IEEE (2021)
12. Kay PENTAX. MDVP Instruction Manual. USA: IEEE **5**, 11–12 (2007)
13. Praat, N.A.: Doing phonetics by computer. Ear and Hearing **32**(2), 266 (2011)
14. Kong, J.: Phonation patterns of Tone and Diatone in Mandarin, from traditional phonology to modern speech processing. In: Fant, G., et al. (ed.) Foreign Language Teaching and Research Press (2004)

Acoustic Parameters Analysis of Special Singing Method of Folk Songs for Singing Speech Synthesis

Shuang Chen, Shiliang Lyu, and Hanyun Qi

Abstract Singing speech synthesis is a method to convert the information of lyrics and music score into singing speech. It usually uses statistical parametric synthesis, hidden Markov model-based method, and deep learning method. The naturalness of the synthesized singing voice has been significantly improved, but the naturalness of the special singing method for folk songs is low. Special singing methods are important methods for expressing the artistic characteristics of various folk songs, and it is also an important content in the study of folk song's ontology. At present, in the field of artificial intelligence, singing speech recognition and synthesis has gradually become a hot research field. In the research of folk song speech recognition and synthesis, the research of special singing method is the key and difficult point. Through the method of phonetic and acoustic analysis, we can quantify and refine the study of special singing methods in folk songs. Taking the folk song "The Precious Horse on the Prairie", which is from the eastern of Yugur nationality in China as an example. Based on the speech signal acquisition, this paper performs an acoustic analysis by using Praat software to extract the fundamental frequency, intensity, jitter, shimmer, the first formant, and the third formant from the segments of drag cavity in this song. What's more it also summarizes the acoustic characteristics of the drag cavity, in order to provide some method and data reference for the study of special singing methods of folk songs and singing speech synthesis.

S. Chen · H. Qi
Gansu Provincial Key Laboratory of Intelligent Processing of National Languages, Northwest Minzu University, Lanzhou 730000, Gansu, China

S. Lyu (✉)
Key Laboratory of Linguistic and Cultural Computing Ministry of Education, Northwest Minzu University, Lanzhou 730000, Gansu, China
e-mail: shilianglv@126.com

K. Nakamatsu et al. (eds.), *Advanced Intelligent Technologies for Information and Communication*, Smart Innovation, Systems and Technologies 365,
https://doi.org/10.1007/978-981-99-5203-8_13

1 Introduction

Speech synthesis is an important part of human–computer interaction. Singing speech synthesis allows the computer to realize singing. After using the lyric text input, the singing rhythm is analyzed according to the music score information, so as to obtain the pitch, resonance, and rhythm characteristics. After obtaining the characteristics of speech and music, the singing speech waveform is finally synthesized by statistical parameters or waveform splicing [1]. From the current research on singing voice synthesis, the singing voice synthesis of folk songs is a difficult point, especially in the special singing part, which has low naturalness and is difficult to restore the artistic expression form in the singing process. In the process of singing synthesis, for special singing methods such as singing vibrato, the method of generating dynamic characteristics of fundamental frequency parameters is usually adopted, without comprehensive consideration of voice intensity and formant information, and the natural degree of the synthesized folk song vibrato is low [2].

Usually, the study of folk song's ontology is an important part in the field of folk song research, which has a long history of research and background. Among the study of folk song ontology, the research about special singing methods of folk song is a significant section. However, looking at the research content of the special singing methods of folk songs in the past, it is mainly biased toward the induction of theory and the description of vocal techniques, and the methods used are mainly investigation methods and analysis methods. The traditional method is difficult to restore the singing characteristics of folk songs, which is not conducive to the protection and learning of folk songs.

With the development of modern information technology, the method of speech acoustic analysis has gradually been applied to the study of folk songs. Speech acoustic analysis takes the voice signal of folk songs as the analysis object, which can realize the digitization of various types of folk songs, so that the related research of folk songs is more evidence based. Singing speech synthesis is one of the digital technologies to protect folk songs.

This paper is mainly aimed at the demand of singing voice synthesis, taking the voice signal of the special singing part of folk songs as the research object, and extracting the fundamental frequency, energy, and formant parameters after signal marking. Combined with acoustic analysis, voice experiment, and statistical methods, the parametric characteristics are analyzed to obtain the parametric characteristics of folk song singing. The folk song selected for the study is "The Precious Horse on the Prairie", a folk song of the eastern Yugur, which contains a special singing method of trill and drawl. The research results can provide reference for the parameters of singing speech synthesis. Acoustic analysis of the special singing methods of folk songs can not only verify the theories studied by predecessors, but also make vocal skills manifest as specific data, so as to establish data standards for various special singing methods, classify the special singing methods of folk songs, and provide a basis for the teaching and inheritance protection of folk songs.

2 Experimental Process

2.1 Pronunciation Informant and Recording Methods

The pronunciation informant of this experiment is a Yugur female singer, aged 52, who has been singing Yugur eastern folk songs for a long time and has a good singing level. Recording equipment is Lavalier microphones, external sound cards, mixers, and laptops, and the recording software is Adobe Audition CC. Recording in a room with ambient noise less than 45 dB SPL, taking a sample rate of 44 kHz and files saved in Windows PCM.wav format.

2.2 Voice Signal Processing

The speech analysis software used in this experiment is Praat developed by the Speech Science Laboratory of the University of Amsterdam in the Netherlands [3]. In the experiment, Praat is used for speech signal processing and parameter extraction. At first, open the recording file in Praat, and the voice signal is displayed as shown in Fig. 1. The signal shown in the upper part of the figure is the voice signal of the whole song of "The Precious Horse on the Prairie", and the lower half shows the spectrum map corresponding to the speech signal.

In the song "The Precious Horse on the Prairie", the drag cavity is a very obvious and regular special way of singing. The drag cavity mentioned here refers to the singing method in which the singer drags the vowel of the lyrics until the next lyric or interlude during the singing process of the song [4]. What's more, the stabilization

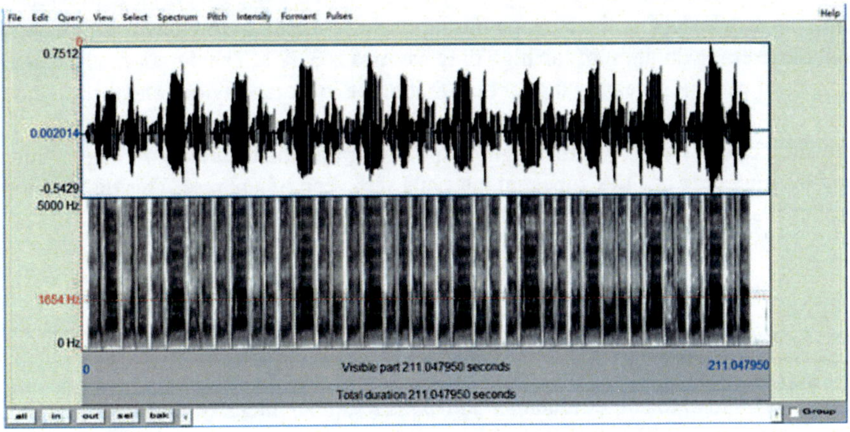

Fig. 1 Speech signal diagram of the song "The Fine Horse on the Grassland"

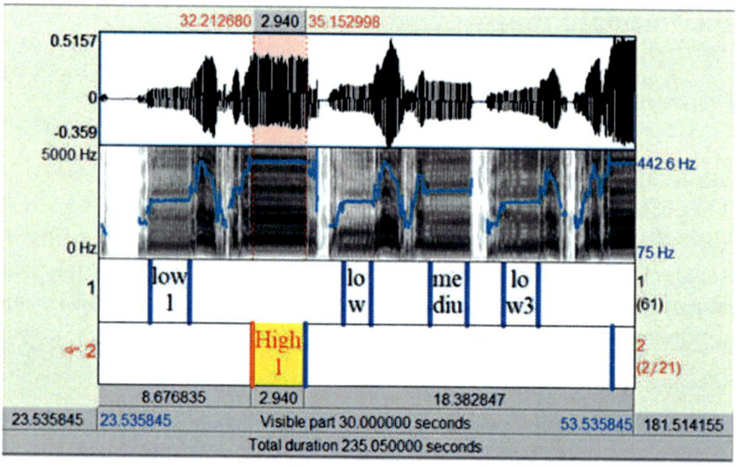

Fig. 2 Speech signal annotation diagram of the song

sections inside the drag cavities are intercepted and labeled to analyze the data regularity of the drag cavity stabilization segment. Combined with the characteristics of speech hearing and the spectrum map, the drag cavities of the song are obviously manifested in three situations: The first one has low pitch and calm sound, which is named low drag cavity; the second one has medium pitch and flat and open sound, named medium drag cavity; and the third one has high pitch and loud sound, named high drag cavity. The speech annotations are shown in Fig. 2.

The blue line in the spectrogram is the pitch line, the first labeling layer is the low and medium drag cavity layers, and the second labeling layer is the high drag cavity layer. The whole song is marked with 20 low drag cavity segments, 10 medium drag cavity segments, and 10 high drag cavity segments. The average duration of the low drag cavities is 1.64 s, the average duration of the medium drag cavities is 2.06 s, and the average duration of the high drag cavities is 3.30 s. The duration of the high drag cavities is significantly higher than that of the low drag cavities and the medium drag cavities. The distribution of the three kinds of drag cavities shows regularity, the low drag cavities appear at the beginning of a phrase, and the medium drag cavities and the high drag cavities appear at intervals at the end of a phrase. This distribution makes the melody ups and downs, the sound melodious and powerful.

2.3 *Acoustic Parameter Extraction*

Use a script to extract the fundamental frequency (F0), intensity, first formant (F1), and third formant (F3) of the labeled speech segments, and extract the fundamental frequency perturbation (Jitter) and amplitude perturbation (Shimmer) at the same time. The fundamental frequency can reflect the pitch and range of singing melodies, intensity is related to the rhythm and emotion of singing, and every annotation segment will extract 30 F0 quantities and 30 intensity quantities used for analysis. The formant parameter reflects the characteristics of pronunciation resonance during singing, and studies have shown that the first formant and the third formant are the objective reference for measuring the singer's singing level [5]. Therefore, this experiment extracts the first formant and the third formant of drag cavity for data statistics. Jitter and shimmer are important parameters to measure the characteristics of the voice, and jitter refers to the rate of change of the frequency of sound waves during adjacent cycles, reflecting the irregularity of vocal cord vibration. Shimmer refers to the rate of change in amplitude during adjacent cycles, reflecting the instability of vocal cord vibrations. In this experiment, the extracted data is saved and imported into an Excel table for collation, and the organized data is imported into IBM SPSS Statistics 19.0 for further analysis.

2.4 Data Processing

Calculate the mean value, maximum value, and minimum value of the data, and make standard deviation analysis, keep two digits after the decimal point, and make a statistical table to analyze the data variation range and characteristics of various drag cavities. Statistical graphs and spectrograms of fundamental frequency, intensity, the first formant, and the third formant are made to visually compare the parameter distributions between various drag cavities.

3 Analysis of Results

3.1 Fundamental Frequency Analysis

Figure 3 shows the fundamental frequency distribution of various types of drag cavity, and the fundamental frequency distribution is relatively concentrated, the fundamental frequency of low drag cavity is concentrated at about 300 Hz, the medium drag cavity is concentrated at about 350 Hz, and the high drag cavity is concentrated at about 450 Hz.

Table 1 is a statistical table of the fundamental frequency data of various types of drag cavity; from the table, we can find that the change of fundamental frequency is small, the minimum changing amplitude of low drag cavities is 6 Hz, the maximum changing amplitude is up to 40 Hz, and the average changing amplitude is 16.90 Hz. The minimum changing amplitude of medium drag cavities is 5 Hz, the maximum changing amplitude is 10 Hz, and the average changing amplitude is 7.30 Hz. The minimum changing amplitude of high drag cavities is 4 Hz, the maximum changing amplitude is 16 Hz, and the average changing amplitude is 9.40 Hz.

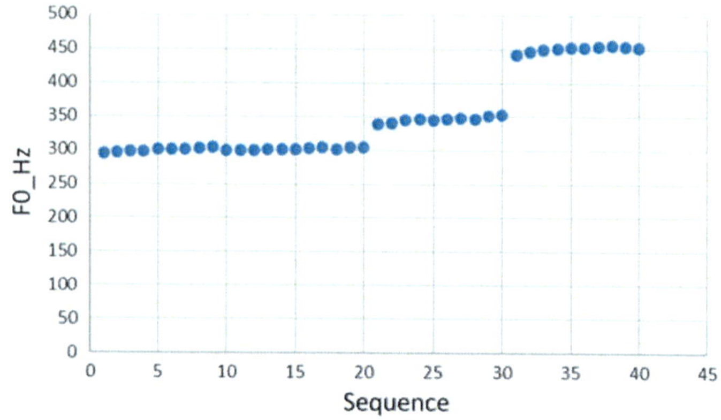

Fig. 3 Drag cavity fundamental frequency distribution plot

Table 1 Drag cavity fundamental frequency data statistical table

Drag cavity type	Minimum	Maximum	Mean	Standard deviation
Low drag cavity	272	310	300.38	3.31
Medium drag cavity	335	355	345.68	1.69
High drag cavity	439	465	451.60	1.96

On the whole, the fundamental frequency of three types of drag cavity is on the overall high side, which means the vocal cords vibrate fast when the drag cavity is sung, so that the entire drag cavity tune shows bright and high characteristics. However, the fundamental frequency change of each type of drag cavity is small, which shows that in the singing process of the drag cavity, the vibration frequency of the vocal cords changes little, it means when the drag cavity happened, the position of the larynx is relatively stable. This makes the space of the pharyngeal cavity larger, providing enough space for the resonance of the basal tone [6]. Therefore, the sound of drag cavity singing is natural and beautiful.

3.2 *Intensity Analysis*

Figure 4 shows a line chart of the intensity parameter distribution of each drag cavity, the intensity distribution is relatively concentrated. The intensity fluctuation of the low drag cavity is about 70 dB, the intensity fluctuation of the medium drag cavity is about 72 dB, the intensity fluctuation of the high drag cavity is about 82 dB, and the intensity of the high drag cavity is significantly greater than that of the medium drag cavity and the low drag cavity.

Table 2 shows a statistical table of intensity data of each type of drag cavities, and the intensity fluctuations of each drag cavity are small. The maximum fluctuation

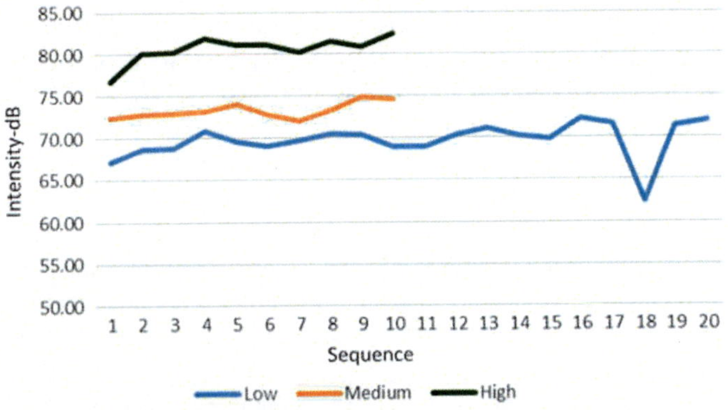

Fig. 4 Drag cavity intensity distribution line chart

Table 2 Statistical table of drag cavity intensity data

Drag cavity type	Minimum	Maximum	Mean	Standard deviation
Low drag cavity	62	74	69.52	0.84
Medium drag cavity	71	76	73.30	0.64
High drag cavity	75	85	80.63	0.93

amplitude of the low drag cavity is 6 dB, the minimum is 1 dB, and the average fluctuation amplitude is 3.25 dB; the maximum fluctuation amplitude of the medium drag cavity is 3 dB, the minimum is 1 dB, and the average fluctuation amplitude is 2.10 dB; the maximum fluctuation amplitude of the high drag cavity is 5 dB, the minimum is 1 dB, and the average fluctuation amplitude is 3.30 dB.

Overall, the intensity of the three types of drag cavities is high, which indicates that during the singing process the vibration amplitude of the vocal cords is large and the air flow through the glottis is high. The small range of intensity fluctuations means that the singer has a stronger ability to control the breath air, making the air flow through the glottis more stable. The intensity's change of the three types of drag cavity from low to high reflects the uniform and stable singing characteristics of the song and the emotional changes.

3.3 *Formant Analysis*

In the research of the formant of songs, the singing formant is an essential content. Singing formant refers to the apparent peak of spectrum that appears at a frequency of around 3 kHz in the singing voice emitted by a traditional opera singer, which can make the singer's voice more full and enhance the penetration of the singing voice [7]. Figures 5, 6, and 7 selected the vowel length average spectrogram of one drag cavity segment in each type of drag cavities to make a representative analysis of the formant. Singing formant theory believes that there are relatively dense high-intensity formants in the frequency domain of 2500 Hz–3200 Hz, and relatively dense low-intensity formants near 500 Hz [8]. From the spectrogram, it can be seen that there are thick black bars or intensity concentration areas including F1 in various drag cavities between 500 and 1000 Hz, and thick black bars or intensity concentration areas including F3 appear around 3000 Hz. Low drag cavities appear at 1000 Hz–1500 Hz, the thick black bar including F2. The intensity concentration area including F2 appears between 1500 Hz and 2000 Hz in the middle drag cavity and the high drag cavity. It can be seen that the female singers in this experiment have singing formants when singing.

The formant parameter reflects the voice resonance and tone quality. The low formant or frequency distribution of the voice forms the depth or resonance of the voice, and the high formant forms the brilliance of the voice, resulting in a beautiful singing voice [9, 10]. In the research of singing voice, the first and third formants are often used to evaluate the sound quality of singers. Therefore, based on previous research, the experiment extracts the first and third formants for analysis. Table 3

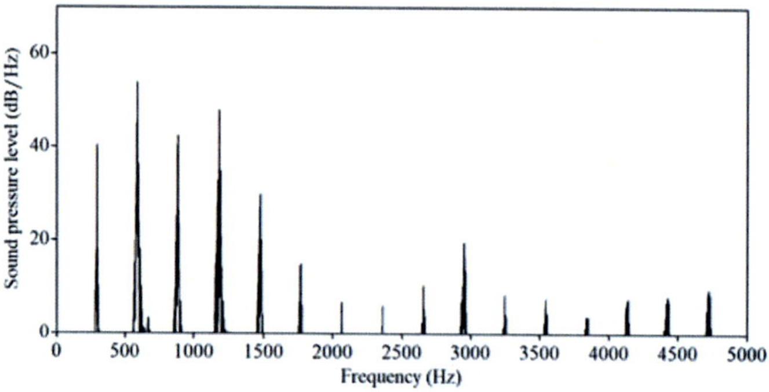

Fig. 5 Long-time average spectrum of low drag cavity

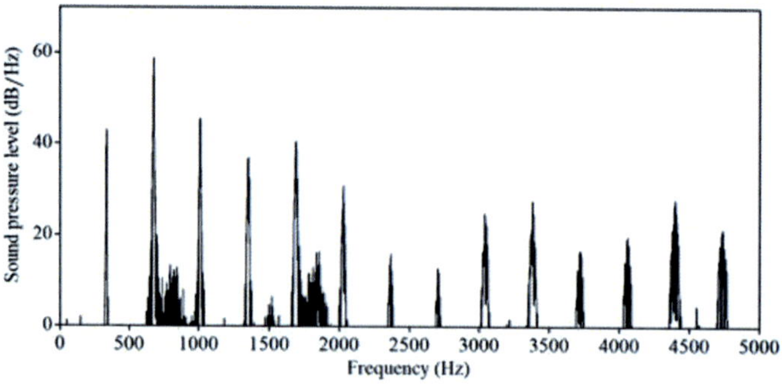

Fig. 6 Long-time average spectrum of mid-drag cavity

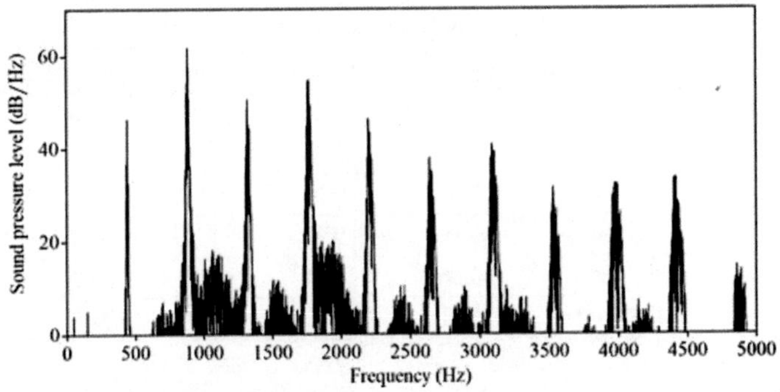

Fig. 7 Long-time average spectrum of high drag cavity

Table 3 Statistical table of drag cavity formant data

Drag cavity type	Formant	Minimum	Maximum	Mean	Standard deviation
Low drag cavity	F1	555	860	655.85	70.98
	F3	2847	3263	3053.60	91.04
Medium drag cavity	F1	713	758	736.40	12.95
	F3	3004	3270	3133.10	92.78
High drag cavity	F1	910	924	920.10	3.75
	F3	2621	3099	2888.60	186.39

shows the first formant, the third formant data statistics table. It can be seen from the table that the F1 distribution of the three types of drag cavities is in the range of 500–1000 Hz, and the F3 distribution is in the range of 2600–3200 Hz. These two distribution ranges are exactly in line with the formant interval pointed out by the singer formant theory. That means the singer's singing level is relatively high, and the existence of the singing formant enhances the sound quality of drag cavity, making the singing loud and open.

3.4 Perturbation Analysis

In this experiment, the jitter and shimmer of each drag cavity segment are extracted and analyzed, in order to explore the characteristics of vocal cord vibration in drag cavity singing. Table 4 shows a statistical table of perturbation data for each type of drag cavity segments, in which the average value of jitter of the high drag cavity segments is 0.08%, and the fluctuation range is 0.06–0.11%. The mean value of the shimmer is 1.18% and the fluctuation range is 0.67–2.13%. Jitter and shimmer of the high drag cavity segments are generally smaller than those of the low drag

Table 4 Statistical table of dragging cavity perturbation data

Drag cavity type	Perturbation	Minimum	Maximum	Mean	Standard deviation
Low drag cavity	Jitter	0.06	0.26	0.14	0.06
	Shimmer	0.72	5.64	1.71	1.36
Medium drag cavity	Jitter	0.08	0.29	0.15	0.07
	Shimmer	0.86	3.10	1.60	0.64
High drag cavity	Jitter	0.06	0.11	0.08	0.02
	Shimmer	0.67	2.13	1.18	0.48

cavity segments and the medium drag cavity segments, which indicates that during the singing process, the vocal cord vibration is more stable and the frequency change is smaller when singing the high drag cavity segment, which is also reflected in the above fundamental frequency and intensity. In summary, the voice parameters of the three types of drawstring are relatively small, which means that better voice control is required to sing this song. Mainly manifested in the control of breathing during the singing process, as well as the control of the throat and vocal cavity, to ensure stable singing pronunciation and full timbre.

4 Conclusion and Discussion

To sum up, in the acoustic parameters of the drag cavity singing of the eastern Yugur folk song "The Precious Horse on the Prairie", the fundamental frequency and intensity parameters are generally high, which are expressed as high pitched and bright voice, conveying full emotions. In the three types of drag cavity segments, the distribution of fundamental frequency is: high drag cavity segments > medium drag cavity segments > low drag cavity segments. The intensity distribution is as follows: high drag cavity segments > medium drag cavity segments > low drag cavity segments. In addition, the variation range of fundamental frequency and intensity is relatively small of all drag cavities, which is reflected in the smooth and melodious singing, and in line with the characteristics of long and straight tones in the original singing method of folk songs [11]. In the formant parameters of the drag cavities, the first formant appears in the intensity concentration area of 500–1000 Hz, and the third formant appears in the intensity concentration area of 2600–3200 Hz, which is consistent with the female singer formant interval in the singing formant theory. It shows that the sound of the singer has achieved a very well resonance effect in the process of drag cavity singing, which is manifested as the high-pitched sound. Besides, the statistics of perturbation parameters further illustrate that strong breath control ability and vocal cord vibration control ability are required in the singing process, so that the singing sound is more stable and far reaching.

The language of the eastern Yugur nationality belongs to the Mongolian language of the Altaic language family, which makes the pronunciation of the eastern Yugur folk songs close to the pronunciation of Mongolian folk songs [12]. Therefore, the singing of the eastern Yugur folk songs shows strong loudness and loud timbre characteristics, which are reflected in the acoustic parameters that the fundamental frequency is high and the intensity is strong. This shows unique national art features.

In this experiment, a total of six acoustic parameters were extracted to describe the procrastination singing in the eastern Yugur folk song "The Precious Horse on the Prairie", and helped to realize the detailed classification on drag cavities in this song. At the same time, the analysis of the drag cavity sound reflected by each parameter is carried out. The characteristics of all parameters reflect the vocal characteristics of the singer in the process of procrastination singing. Due to the limitations of the experimental conditions, this experiment also has some shortcomings. First, there is only one female speaker, and the collected data has individual differences, which is not enough to reflect the detailed characteristics of drag cavity singing. Second, it fails to collect the speech signal of male speakers to make a comparative analysis.

The fundamental frequency, energy, and formant parameters of the singing voice of folk songs can not only be used in the research of singing voice synthesis, which is conducive to improving the naturalness of voice synthesis, but also to establishing a certain singing method or a certain type of folk song data standard or database to provide guidance for the singing teaching of folk songs. At the same time, it can also make contributions to the speech synthesis of folk songs and the research on the inheritance and protection of folk songs. In the acoustic research of folk songs, the main difficulty lies in the integration of interdisciplinary knowledge and technology. Researchers should not only have rich theoretical knowledge about folk songs, but also be proficient in the use of acoustic analysis equipment and related software, and also need well acoustic-related knowledge. At present, under the trend of interdisciplinary research, more and more scholars have the above-mentioned conditions, and use acoustic analysis in the study of folk songs, which will be an important field of folk song singing research.

Acknowledgements This research was financially supported by the Fundamental Research Funds for the Central Universities (Grant NO. 31920220064).

References

1. Yuanhao, Y.: Research on singing speech synthesis method based on deep learning. University of Science and Technology of China (2020)
2. Cai, L., Hou, J., Liu, R., et al.: Pitch-guided HMM parametric singing synthesis. In: 5th National Conference on Human-Computer Interaction (CHCI2009) (2009)
3. Boersma, P: Praat: Doing phonetics by computer. Ear Hear **32**(2):266 (2011)
4. Yimeng, X.: Exploration of learning from opera singing techniques in national vocal music singing. Art Educat **21**, 212–213 (2017)
5. Lan'e, L., Jianming, X.: Application of acoustic parameters in singing art voice. Shanxi Electronic Technology (02):32–33+55 (2009)
6. Zeng, S.: The important role of the relative stability of the throat in singing. Northern Music **39**(12):56+58 (2019)

7. Ke, L.: Mechanism characteristics of Bel canto resonance. Music Creation **10**, 162–163 (2018)
8. Written, F.: Singing resonance peak theory and singing training. J Sichuan Univ Educat **25**(11), 40–42 (2009)
9. Shilin, Y.: Acoustic detection of voice. J. Audiol. Speech Diseases **9**(4), 255–258 (2001)
10. Daniel, R.B.: The three ages of voice: the singing acting voice in the mature adult. J. Voice **11**, 161–164 (1997)
11. Wenhao, Q.: On several special vocaling techniques in Folk songs. Art. Rev. **16**, 44–45 (2018)
12. Shiliang, L.: The research on the singing voice timbre of the Eastern Yugur Traditional Folk Songs. In: 7th International Conference on Education, Management, Information and Computer Science (ICEMC 2017) (2017)

Simulation of Crop Planting Decision System Based on "U + CSA" Public Welfare Agriculture and Machine Learning Algorithm

Ying Zhang

Abstract "U + CSA" University + CSA (Community Supported Agriculture)) public welfare assistance to agriculture is an alternative agriculture gradually emerging on the basis of re-examining industrialized agriculture. The "U + CSA" public welfare assistance to agricultural development has economic rationality. It is an institutional choice made by consumers and producers with "self-interest" as the starting point, and has produced side benefits such as protecting ecological environment, promoting agricultural culture and rural community development in practice. In order to promote the development of "U + CSA" agriculture, this paper will study the crop planting decision system through machine learning algorithm, and study the influence of machine learning algorithm on the accuracy and precision of the crop planting decision system. The research results show that the system has no wrong response and handling behavior, and the test accuracy and precision reach 98.57%. Therefore, it is found that the test indexes of the crop planting decision system based on the machine learning algorithm can reach the expected indexes of the normal system, thus promoting the development and progress of "U + CSA" agriculture.

1 Introduction

U + CSA mode is the most equal and mutually beneficial mode among Chinese agricultural U + CSA modes, and there is still a lot of room for future development. However, the U + CSA model is still in its infancy in China, and it is a very young agricultural model [1]. Moreover, the U + CSA model has greatly promoted the development of China's agricultural economy, and China's agricultural economy is an important part of the national economy, which closely links the U + CSA model with China's economic development [2]. By comparing with other agricultural models and U + CSA models in other countries, we can clearly realize the achievements and relevant measures taken by these countries, and make the development of U +

Y. Zhang (✉)
Yunnan Normal University, Kunming 650500, China
e-mail: zhangying27@tju.edu.cn

K. Nakamatsu et al. (eds.), *Advanced Intelligent Technologies for Information and Communication*, Smart Innovation, Systems and Technologies 365,
https://doi.org/10.1007/978-981-99-5203-8_14

CSA model in China take its essence and discard its dross [3]. At present, China's CSA industry as a whole presents the situation of a small farm scale, few kinds of agricultural products and low profit level [4]. How to achieve the goals of making full use of land resources, diversified agricultural products and steady increase of economic income in a small space poses a big challenge to CSA producers' scientific planting decision-making ability [5].

In order to promote the development of China's "U + CSA" public welfare agriculture, this paper will study the crop planting decision system through the machine learning algorithm. The so-called machine learning means that robots can simulate real people to carry out agricultural-related operations through a series of predetermined instructions, and they can learn from real people's behaviors in their daily operations to help them improve themselves, so that they can better engage in agricultural activities [6]. Robot learning algorithm can pre-compile the robot in advance, so that the robot can initially complete an agricultural operation. Machine learning is a data analysis technology, which allows the computer to perform the innate activities of humans and animals and learn from experience [7]. Machine learning algorithms use computational methods to "learn" information directly from data, without relying on a predetermined equation model [8]. When the number of samples available for learning increases, these algorithms can adaptively improve performance [9]. Based on the current CSA industry tense in China, this paper studies the crop planting decision system through machine learning algorithm, in order to provide reference for exploring machine learning algorithm more suitable for China's agricultural development. Therefore, it is necessary to carry out research on the crop planting decision system based on the machine learning algorithm, and put forward methods and schemes to solve the current problems in China's agricultural system, so as to realize low-risk decision-making and more accurate intelligent control.

2 Analysis of the Existing Mode of "U + CSA" Public Welfare Assistance to Agriculture in China

The situation varies greatly in different parts of China, and the U + CSA model is different. According to different classification standards, different types of classifications can be made [10]. According to the different identities of CSA sponsors, China's CSA models can be divided into the following types: citizen partnership organizations, small farmers cooperatives, and Industry-University-Research models.

(1) Citizen partnership organization mode. As the initiator of CSA, the citizen organization. The basic characteristics are that CSA is initiated by representatives or institutions of urban citizens. First, members are recruited. Members are mainly citizens who can participate in farming through various forms; The second is to establish direct contact with farmers, such as renting farmers' land for a long time. The vast majority of CSA projects in China belong to this category,

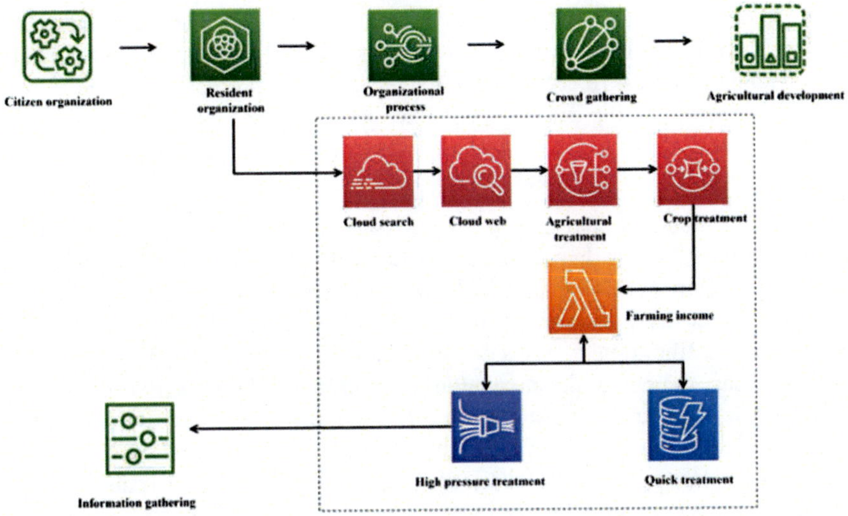

Fig. 1 Citizen partnership organization mode

concentrated in the first-and second-tier cities and other economically developed cities. Summarizing its development, we can see that its mode is summarized in Fig. 1.

The characteristic of this kind of CSA is that the citizen partnership organization, as an advanced element, can lead the development of CSA, which is conducive to the formation of a long-term cooperation mechanism between both parties. As a sponsor or management, citizens do not or rarely participate in specific labor, but employ farmers for production. With its abundant capital and social resources, citizens can quickly build a CSA relationship, and have a thorough understanding of the concept of CSA, which is conducive to operating in strict accordance with CSA standards. The sponsor can rent farmers' land for a long time and provide technical guidance for farmers' production, which is conducive to the scientific production of organic agriculture. Citizens personally participate in it, monitor production, and farmers produce according to residents' needs, which will enhance mutual trust and help establish long-term cooperation.

(2) Small farmers' cooperative mode. CSA (Small Farmers Cooperative) is a type with a large number in China. Its basic feature is that a number of small farmers with the same idea of CSA form a cooperative, and each small farmer is responsible for the production of different agricultural products. This kind of small farmers can supervise each other, and at the same time, different farmers can make use of their own advantages to avoid collective risks.

(3) Industry-University-Research combination mode. The basic feature of this mode is mainly Industry-University-Research base, which is implemented in a small area. The mode is easy to change, can adapt to the environment, can win the

support and trust of the citizens, and is conducive to development and growth. To a certain extent, this kind of CSA project is supported by the state. With the help of social resources of colleges and universities, it is promoted through media, conferences, training and education, etc., which is conducive to the formation of a wide range of social influences, so as to attract consumers to join the club and promote CSA to various places.

"U + CSA" public welfare to help farmers is the inevitable choice to practice social responsibility. Public welfare is an important attribute of the media, and fulfilling social responsibility is also an inevitable choice of the media [11]. As far as China's agricultural industry is concerned, in the face of the economic challenges in the epidemic era and the arduous task of decisive victory in getting rid of poverty, it is even more necessary to practice the instruction of putting social benefits first and actively participate in the cause of public welfare and helping agriculture. Public welfare helping agriculture is an effective way to fully embody public welfare. CSA, as an alternative model emerging on the basis of re-examining industrialized public welfare agriculture, fully embodies the development characteristics of resource conservation and environmental friendliness. Firstly, CSA adopts a large number of sustainable farming methods and technologies such as organic or nearly organic and recycling in production, which is conducive to the organic unity of agricultural production cycle and ecological cycle, and then realizes agricultural sustainable development with a new reproduction path. Secondly, by establishing direct contact with local consumers, CSA realized the re-localization of food production, greatly shortened the distance between food producers and consumers, and achieved remarkable benefits in saving energy consumption and reducing carbon emissions.

3 Machine Learning Algorithm Model

Machine learning algorithm can improve the accuracy of classification, detection and recognition. Secondly, machine learning algorithm has good generalization and universality. The machine learning algorithm can be used to collect the growth state of crops and some necessary data, thus facilitating the management of dependent crops. The use of this high-tech can effectively help agricultural development. There are some image acquisition technologies such as cameras in smart agriculture, which can analyze the maturity of crops, automatically calculate the maturity date, remind farms to harvest, and quickly identify image technologies. These technologies are very helpful for agricultural development, and conform to China's "U + CSA" public welfare to help farmers. Machine learning is also a part of smart agriculture. Using camera image acquisition technology and fast image recognition technology to machine learning system can effectively increase the efficiency of machine work and help agricultural management. The current machine learning algorithm has been able to monitor crops well and help analyze the growth state of crops, so as to better manage them. It is difficult to solve the problem directly by machine learning.

Usually, iterative optimization is used to solve the problem. Firstly, k random values in the same dimensional space as the sample are randomly selected as the initial clustering centers. The formula is as follows:

$$\arg s \min \sum_{i=1}^{k} \sum_{x \notin s} x - ui \tag{1}$$

$$(x + a)^n = \sum_{k=0}^{n} \binom{n}{k} x^k a^{n-k} \tag{2}$$

$$F1 = \frac{1}{P} + \frac{1}{R} = \frac{2TP}{2TP + FP + FN} \tag{3}$$

$$AP + 2TP + FP + FN = 1 \tag{4}$$

For most application scenarios, the four basic evaluation indicators, accuracy, precision, recall and $F1$ value, are enough to meet the demand. The "number of samples" here refers to the number of times a certain training point collects data for each AP. Due to the influence of small-scale fading and Gaussian noise, the signal strength of a certain AP received at the same training point is different at different times. Theoretically, the more the number of samples, the more the random change of the received signal strength can be removed by the averaging or filtering algorithm, so as to obtain more accurate signal strength data as shown in Table 1 below.

As shown in Table 1, after machine learning algorithm processing, the random change of received signal strength is removed, and more accurate signal strength data is obtained. It fully solves the problem of signal receiving intensity in CSA agricultural development, and the processing method is effective and energy-saving, which fully meets the standards of public welfare agriculture in China.

Table 1 Signal processing data table

Number of samples	Positioning time	Minimum positioning error	Maximum positioning error	Average positioning error
12	205.04858	0.0247	0.4856	0.1556
42	244.422746	0.0246	0.4728	0.1528
72	318.848586	0.0586	0.3868	0.1328
99	369.503854	0.0369	0.3899	0.1367

4 Machine Learning Algorithm Flow

Essentially, machine learning is a method of automatically analyzing its laws from data, and then using the obtained laws to predict and estimate unknown data. There are many ways to classify machine learning. The most common one is to divide machine learning into supervised learning, unsupervised learning and reinforcement learning. Supervised learning is the most widely used one at present, which can be used for most classified forecasting problems, such as abnormal detection and house price forecasting. Unsupervised learning is mostly used to automatically discover potential patterns in data, such as classification of consumer groups; Reinforcement learning is mostly used in robot control and other fields. The representative algorithms among these three types of machine learning will be introduced and analyzed below. The specific steps of the algorithm are shown in Fig. 2.

The machine learning algorithm used in this study, because of the EM algorithm, harmonizes and averages the point estimation value of the sample with the historical agricultural production data, so that even if the average value is the same, the variance of the sample is obviously smaller than that of point estimation, least squares and weighted least squares methods. Because point estimation is to calculate the expectation and variance directly by arithmetic average of real-time data, historical data is not considered; The least squares and weighted least squares are all historical data, and the error is the introduction of white Gaussian noise. Because the historical estimated values are not considered in the characteristic parameters, the accuracy of estimation and the risk of decision-making are not as good as those of EM algorithm as shown in Fig. 3.

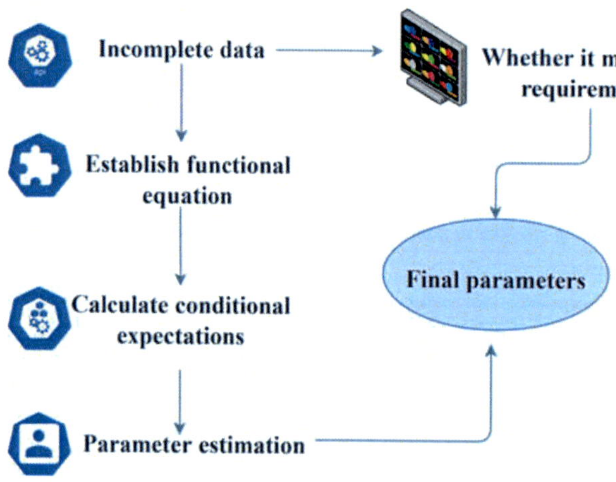

Fig. 2 Flow of machine learning algorithm

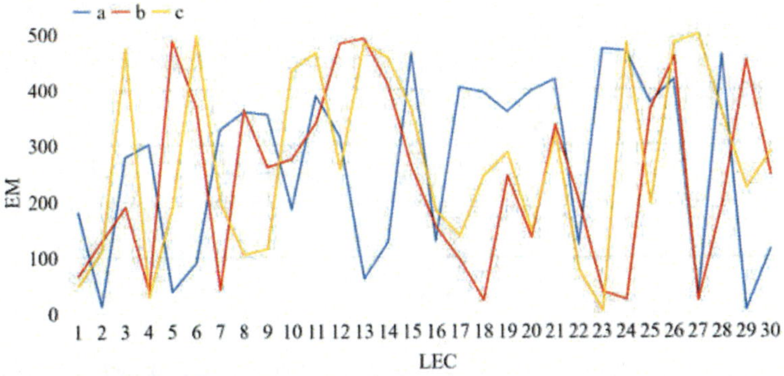

Fig. 3 Function comparison

The EM algorithm adds the defective data to the data obtained by the sensor network to form complete data, and constructs a posterior probability density likelihood function about unknown characteristic parameters, thus obtaining a simplified statistical model, and then takes the logarithmic likelihood estimation of the posterior probability density function.

5 Crop Planting Decision System

The decision-making model of crop multi-variety planting based on the mathematical method is an effective way to deal with the spatial optimization problem, because it can accurately describe the quantitative optimization in the macro-state. With the increasing demand, this mathematical method can't deal with the complex land system with dynamic characteristics. However, the model of the crop planting decision system based on simulation method provides an effective method to solve the complex land system with dynamic characteristics. Compared with the former, it has the following advantages: (1) It can accurately deal with the complexity and dynamics of land use system; (2) It can be well coupled with GIS; (3) Advanced algorithm; (4) Based on the optimization results of the simulation model, the planning objectives of the model can be achieved to some extent.

This system can solve the management of all kinds of basic data in the raw material department. By inputting, modifying, deleting, querying, counting and printing these basic data, the raw material department can free the data entry clerk from boring manual input, reduce repeated data input, reduce repetitive labor, and help to summarize information and reduce the repetition and inconsistency of information as shown in Table 2.

Through the management of basic production data, this system can solve the information sharing among various departments and improve work efficiency. Reduce

Table 2 Analysis of decision system

Sample point	Actual value	Manual estimation	Relative error	Model value	Relative error of model value
1	6.88	5.16	−21.868	6.96	5.6698
2	6.45	7.2	−2.8899	7.80	9.7749
3	5.66	6.3	−3.5547	8.80	6.4595
4	6.99	8.0	−6.9958	6.98	2.3598
5	7.66	8.55	−6.1258	6.23	8.9635

error situations. Through the communication of information and services, strengthening the communication and cooperation between business departments or enterprises within an enterprise will get more and more information and services, and a virtuous circle will make everyone gain benefits. According to the basic data, automatically generate various statistical reports, provide decision support for leaders in various ways, such as graphics, sound, documents, etc., and make annual planting plans of enterprises conveniently. Through this system, we can make a plan for all kinds of operations and results in the whole planting process, including when to raise seedlings, when to plant, when to irrigate, when to fertilize, when to pick and so on. Based on the output and production period, and taking the economic benefits into account, the production process and expected input are formulated by the backward method.

6 Analysis of Experimental Results

Because of the complex relationship between data indicators in "U + CSA" agriculture, the threshold of machine learning will be optimized. The key parameters that affect the threshold of decision optimization are expectation, variance and superparameter. Through analyzing the historical data of environmental parameters in agricultural production sites, the theoretical distribution of various parameters (indoor temperature, light intensity, indoor CO_2 concentration) in greenhouse is obtained, as shown in Fig. 4.

As can be seen from Fig. 4, the data of Week 2 fluctuated greatly at the beginning, while the data of Week 5 to Week 11 tended to change steadily. At the same time, it can be seen that with the increase of iteration times of the machine learning algorithm, when the iteration times exceed 19 times, each parameter changes obviously, and then tends to be stable, with a total of 20 iterations ending. It can be seen that the threshold optimization method of machine learning has good robustness, which makes the threshold optimization process in the complex "U + CSA" agricultural problem simple and easy to realize, and it has good accuracy, saves resources, and accords with the idea of public welfare to help agriculture. The basic idea is: to obtain a posterior probability density likelihood function of unknown parameters

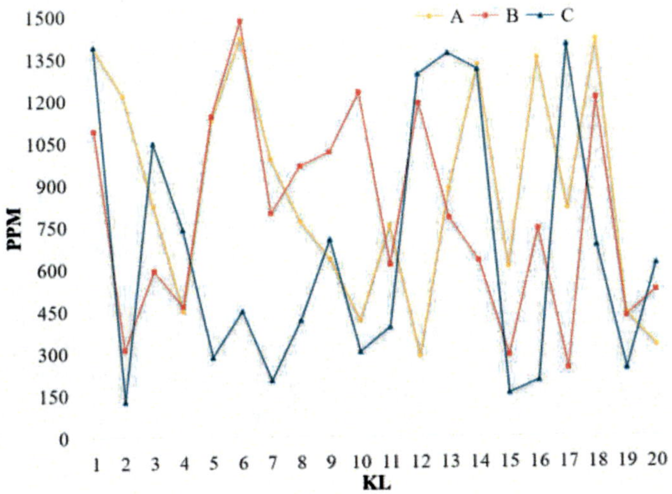

Fig. 4 Greenhouse parameter data

based on threshold by relating all environmental parameters in the database and adding defective data, then to obtain the full conditional distribution function of each unknown threshold parameter respectively, and finally to obtain the optimized control threshold by means of the machine learning algorithm, so as to realize the control of electrical equipment by means of this threshold.

Based on the machine learning algorithm, the design case and process structure of the crop planting decision system are tested to determine whether the development of the system has achieved the expected purpose. In the process of testing, most of the statements and structures of the system should be covered, so that all the codes in the system can be run again. Therefore, for the system test, the design should meet the software coverage, and the existing problems and defects should be found as much as possible in theory, so as to provide the basis for the improvement of the system and ensure the development quality. The performance test of planting a decision support system mainly refers to whether the performance index reflected by the system can meet the design requirements when it is reused in normal environment, that is, the performance of the software when processing a large amount of data. The test results are shown in Fig. 5.

After the deployment of the crop decision-making system, the online simultaneous access test of multiple users is carried out. The test shows that the system allows up to 1500 users to access at the same time. The open development environment and object-oriented programming language provide convenience for the later maintenance and expansion of the planting design system. When the system is running at full capacity, the maximum response time of the system is 30 s. During this time, there is no wrong response and handling behavior in the system, and the test accuracy and precision reach 98.57%. Therefore, it is found that the test indexes of crop planting decision system based on the machine learning algorithm all reach the expected indexes of the normal system.

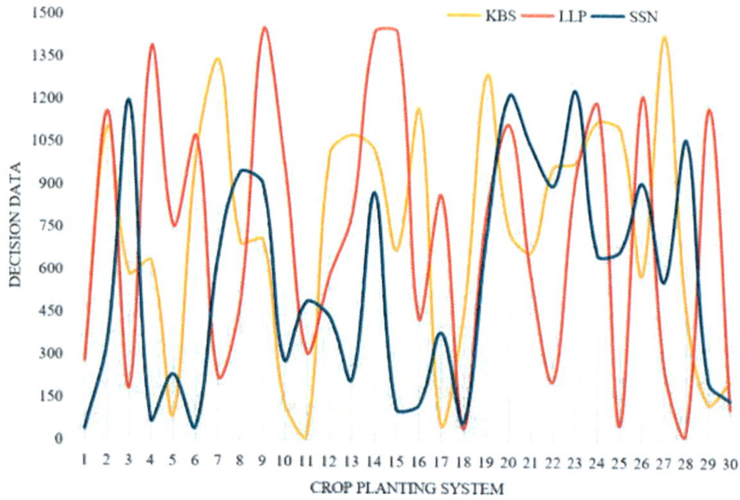

Fig. 5 Test results of crop planting decision system

7 Conclusion

"U + CSA" agriculture is a model that can benefit many parties, but its development is still in the initial stage, and there are still many problems to be discussed. As a new agricultural model which is very different from the general production or consumption process, "U + CSA" agriculture has aroused the attention of all sectors of society. However, there is very little research on "U + CSA" agriculture in China, and the development of its industry is facing great challenges at present. Based on the concept of "U + CSA" public welfare to help agriculture, this paper studies the crop planting decision system through the machine learning algorithm. CSA agricultural decision-making system based on the machine learning algorithm has basically completed algorithm simulation and system development. The whole software platform of the system is designed and developed by using key technologies such as machine learning. Through cloud database technology and data simulation tools, the threshold optimization method based on machine learning is realized. By this method, the optimization threshold of each environmental parameter can be obtained, and then the equipment control module in the software system can be guided to drive the corresponding electrical equipment to enable, and the environmental state in the greenhouse can be adjusted adaptively, thus achieving low-risk crop planting decision and more accurate intelligent control.

However, with the accumulation of time, the amount of agricultural production data entered by this crop planting decision-making system will continue to increase. Therefore, it is necessary to consider how the system should store and process data when it comes to big data, so as to achieve higher requirements for the performance optimization of the decision-making system, and thus promote the development of China's "U + CSA" public welfare agriculture industry.

References

1. Yuan, L., Zejiang, C., Qiangyi, Y., et al.: Evaluation of spatial differences of cultivated land quality in typical latosolic red soil agricultural areas in South China from the perspective of crop rotation. J. Agric. Resour. Environ. **38**(6), 1–13 (2021)
2. Kuo, L., Yinlong, X.: Research on adjustment of China's agricultural planting structure to adapt to climate change. China Agric. Sci. Technol. Herald 19(1), 10–15 (2017)
3. Chunqin, Z., Guochun, Z., Aixia, C.: Study on fuzzy random optimal allocation of water-soil-crop system. Hubei Agric. Sci. **4**, 18–24 (2019)
4. Xiaohua, H.: New technology of crop planting. Farmhouse Sci. Technol. (next xunkan)(11), 96–98 (2017)
5. Jiaru, C., Yan, X., Dongyue, D., et al.: Dynamic monitoring and decision-making system of intelligent agricultural greenhouse based on Internet of Things technology. Henan Sci. Technol. 40(36–40), 5 (2021)
6. Qing, L., Ganran, D., Fengguang, H., et al.: Research progress of monitoring system for crop planting machinery. Modern Agric. Equip. **41**(1), 1–6 (2020)
7. Jing, W., Gang, G., Zongwen, Z., et al.: Crop germplasm resources management: present situation and prospect. J. Plant Genetic Res. **23**(3), 9–15 (2022)
8. Xiaoxia, L.: Research on crop planting management technology. Farmhouse Sci. Technol.: Zhongxunkan **3**, 1–5 (2020)
9. Ming, G. S., Jinyan, G., Weizhi, W., et al.: Local optimal granularity selection of incomplete multi-granularity decision system. Comput. Res. Develop. 54(7), 10–16 (2017)
10. Xue, C., Wei, H., Linhao, Y., et al.: Research on the auxiliary decision system of distribution network planning based on multi-source data. Guangdong Electric Power **30**(1), 6–12 (2017)
11. Yinghai, W.: Application of soil moisture content (soil moisture) data in intelligent irrigation decision system. Water-saving irrigation **4**, 3–10 (2017)

Evaluation Method of Armored Soldier Simulated Training Based on BP Neural Network

Chao Song, Hua Li, Hong-Tian Liu, Dong-Jun Wang, and Yang Cao

Abstract Aiming at the assessment and evaluation of simulated training of armored forces, the evaluation index system for simulated training of armored forces is studied, and the evaluation method using BP neural network is used, and the BP neural network evaluation model is constructed, and the feasibility of the model is verified with actual data. The simulation training assessment and evaluation of the nuclear force unit provides a scientific basis and a new method, and it also replaces a certain foundation for the design of the simulation training assessment and evaluation software platform.

1 Introduction

The armored force simulation training assessment and shooting content covers a wide range of factors. It is believed that the factors have an obvious influence, a multi-level and multi-target comprehensive evaluation problem. Making a correct, scientific, and comprehensive evaluation of the simulation training of the armored forces, deepening the reform training, and improving the training level has an important role in promoting [1]. At present, the theoretical research on simulation training assessment and evaluation of armored forces is still immature. There is no special simulation training assessment software platform. In addition, there are many positions in the armored forces and the training items are complicated. The assessment and evaluation process is time-consuming and laborious, and the assessment and evaluation results cannot be reflected. The actual combat effectiveness level [2], the lack of a system simulation training platform, and the increasingly prominent contradictions with traditional combat readiness on-duty training have become important obstacles restricting the development of armored simulation training. Literature 1 proposes an assessment and evaluation method for armored soldiers simulation training based on fuzzy theory, and the evaluation results are more in line with reality. However,

C. Song (✉) · H. Li · H.-T. Liu · D.-J. Wang · Y. Cao
Department of Weapons and Control, Army Academy of Armored Forces, Beijing, China
e-mail: andysong475@163.com

K. Nakamatsu et al. (eds.), *Advanced Intelligent Technologies for Information and Communication*, Smart Innovation, Systems and Technologies 365,
https://doi.org/10.1007/978-981-99-5203-8_15

giving weight preferences based on experience is subjective and easily leads to distortion of evaluation results. At present, there is extensive theoretical research on neural network terminology evaluation, and it can reduce subjective factors in the evaluation process. To this end, this paper proposes a simulation training assessment model for armored forces based on BP neural network, and conducts multi-level assessments according to personnel categories [3]. First, determine the evaluation index system for simulated training of armored forces; secondly, establish a BP neural network evaluation model based on the index system and expert group evaluation; third, train the neural network model so that it can learn to evaluate and adjust the index weights. Make the output result and the expert group's evaluation error lower than the error accuracy; finally, through data input and expert evaluation, the performance of the evaluation method is tested.

2 Establishment of Assessment and Evaluation Index System

The purpose of the assessment and evaluation of the simulation training of the armored forces is to improve the quality of simulated training of the armored forces, to closely combine the quality of training with the improvement of combat effectiveness, and to improve the combat capability of the armored forces [4]. Therefore, the evaluation index system for simulation training of armored forces should follow the six principles of objectivity, purpose, comprehensiveness, a combination of qualitative analysis and quantitative analysis, feasibility, and scientificity.

In accordance with the current requirements for training and assessment of armored forces, combined with the six principles established by the indicators, a three-level evaluation index layer has been established. The first level: the total target factor set $C = (C_1, C_2, C_3)$; the second level: the sub-target factor set $C_1 = (C_{11}, C_{12}, C_{13})$, the sub-target factor set $C_2 = (C_{21}, C_{22}, C_{23})$ and the sub-target factor set $C_2 = (C_{21}, C_{22}, C_{23})$ The target factor set $C_3 = (C_{31}, C_{32}, C_{33})$, and so on, constitute the indicator system framework. The specific frame diagram is shown in Fig. 1.

2.1 *First-Level Indicators*

The first-level indicators of the armored force simulation training assessment system are composed of three parts: individual, unit-level, and comprehensive ability. Personal simulation training is the subject of simulation training for armored forces. It determines the training quality of the trainees of each unit of the armored force and the basic combat capabilities of the armored force; unit-level simulation training is an indispensable and important part of the armored force simulation training. In this system, the unit-level simulation determines the level of intelligence in the area of

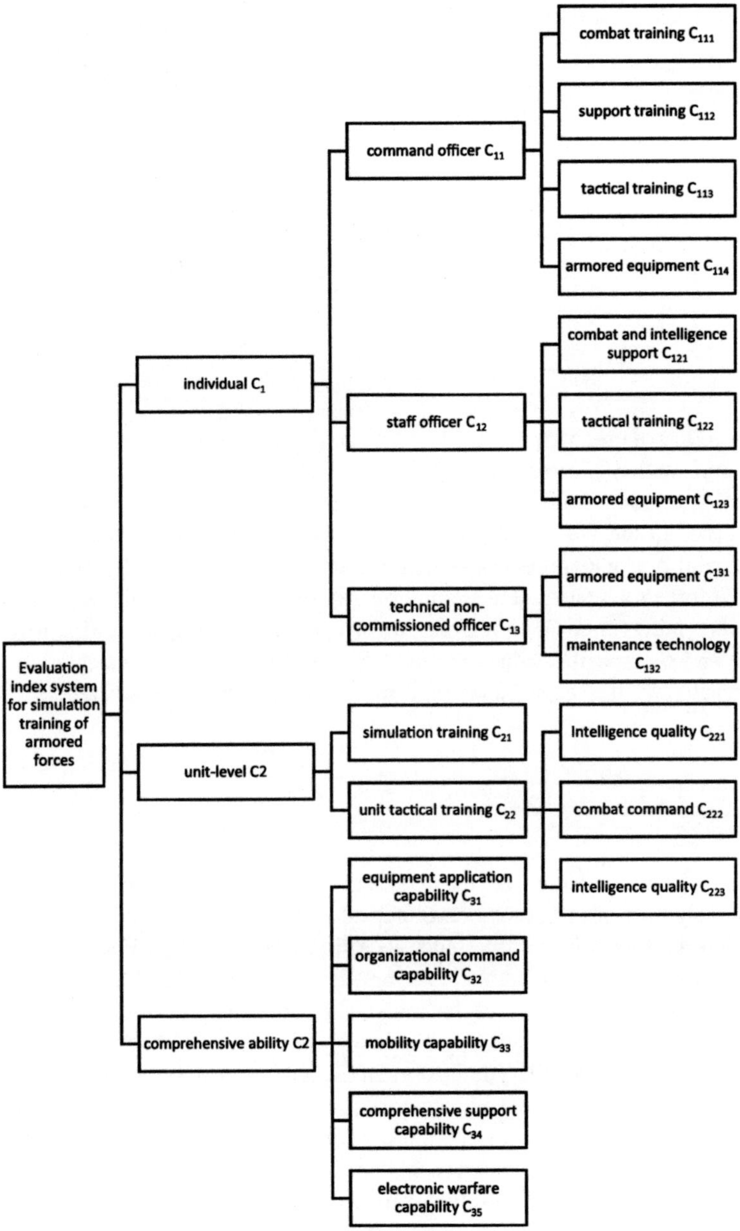

Fig. 1 Evaluation index system for simulation training of armored forces

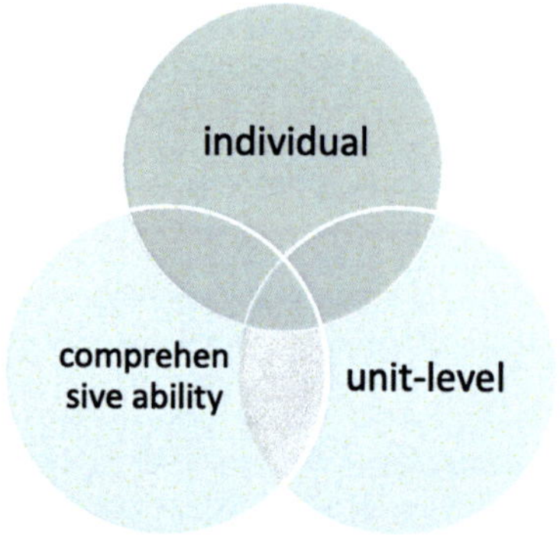

Fig. 2 Diagram of first-level indicators

responsibility of the entire armored force; comprehensive capability is a comprehensive manifestation of the combat effectiveness of the armored force's actual combat simulation training. In this system, the unit-level simulation determines the level of intelligence in the area of responsibility of the entire armored force; comprehensive capability is a comprehensive manifestation of the combat effectiveness of the armored force's actual combat simulation training, as shown in Fig. 2.

A percentile system is implemented for the performance of the first-level indicators, and the specific values are calculated by the summary calculation of the performances of the second-level indicators and the third-level indicators.

2.2 *Secondary Indicators*

The second-level indicators are the extension and refinement of the first-level indicators, and are the basis for the evaluation results of the first-level indicators. There are several second-level indicators under each first-level indicator. For example, there are three second-level indicators in the first-level indicator individual C_1: command officer C_{11}, staff officer C_{12}, and technical non-commissioned officer C_{13}. There are 2 secondary indicators in unit C_2: simulation training C_{21} and unit tactical training C_{22}. There are five secondary indicators in the comprehensive capability C_3: equipment application capability C_{31}, organizational command capability C_{32}, mobility capability C_{33}, comprehensive support capability C_{34}, and electronic warfare capability C_{35}. Some indicators in the secondary indicators are no longer refined and set as separate assessment subjects. For example, the summer simulation training C_{21} in unit C_2 and the five secondary indicators in comprehensive ability C_3 are all separate assessment subjects.

The second-level indicators are the same as the same-level indicators, and there is a considerable degree of correlation between the indicators. For example, comprehensive capability C_3 assesses the comprehensive combat effectiveness of armored forces. It does not include position deployment capability C_{31}, organizational command capability C_{32}, mobile combat capability C_{33}, as well as unit comprehensive support capability C_{34}, and battlefield survivability in a complex electromagnetic environment. C_{35} and so on. The commander of the armored force is the organizer and leader of the combat operations of the armored force and is the core figure of the combat command of the armored force. In the comprehensive ability assessment, the commander's ability to command in battle is a big uncertain factor. Different performance experiences and different knowledge systems may differ greatly in handling and commanding in different environments. Therefore, the organization and command ability evaluates the commander's theoretical knowledge, psychological quality analysis and judgment, and the ability to coordinate command. Maneuverability is an important indicator to measure the combat assessment of armored equipment. Maneuvering is not only an important means to improve its survivability, but also an important way to adjust and deploy. Comprehensive support capability is the general term for various support measures and corresponding activities undertaken by armored forces to smoothly carry out combat missions. It mainly includes communications support, equipment support, and logistical support. Anti-electronic jamming is the top priority of armored forces for electronic defense. Commanders and operators must have a deep understanding of the methods and tactics of electronic jamming under high-tech conditions, the types of armored interference and its impact on armor, and carefully study armored anti-electronics. Methods of jamming and basic measures for mastering air targets under jamming conditions were studied. At the same time, the layout of the position deployment C_{31} in the comprehensive capability C_3 will affect the mobility capability C_{33}, the comprehensive support capability C_{34} and the electronic warfare capability C_{35}. The professional level of position deployment, then the organization and command ability, mobile ability, and comprehensive support ability will be affected by the deduction of points, as shown in Fig. 3.

Fig. 3 Diagram of comprehensive capabilities

2.3 Three-Level Indicators

The third-level index is the extension and refinement of the second-level index, which involves specific simulation training assessment subjects and content, and is the cornerstone of the entire simulation training. The three-level index data directly affects the data of the second-level index and the first-level index, and the third-level index is described in detail here.

The commanding officer of the armored force is responsible for coordinating operations with friendly neighbors, organizing the training of the unit, commanding subordinate units, and providing intelligence support for joint operations. Therefore, the indicators for the simulation training of the commander of the armored force are mainly combat training C_{111}, support training C_{112}, tactical training C_{113}, and armored equipment C_{114}.

The staff officer of the armored force must possess five kinds of abilities: analysis and judgment ability, operation planning ability, organization and coordination ability, inspection and guidance ability, and research and innovation ability. Therefore, in the simulation training of staff officers of the armored forces, the staff's analysis and judgment ability, operational planning ability, and organization and coordination ability are emphasized. The evaluation indicators are composed of C_{121} for combat armor support and intelligence support, C_{122} for tactical training, and C_{123} for armored equipment.

The simulation training of technical officers of the armored force is mainly to familiarize themselves with their own business, master the methods of equipment repair, maintenance, and improvement, and continuously improve the equipment repair support and management capabilities. Therefore, the training evaluation of the technical corps officer of the armored force mainly includes two aspects: armored equipment C_{131} and maintenance technology C_{132}.

The tactical training of the unit level is the comprehensive training of a set of armored forces, and it is an important indicator for the assessment and evaluation of the armored force simulation training. Intelligence quality, combat command and intelligence quality have corresponding assessment content. The combat style is difficult to judge. Therefore, the performance of combat style in this article is based on the performance of armored troops in the annual style discipline inspection.

3 Appraisal Method of Armored Force Simulation Training Using BP Neural Network

Based on the complexity of the armored force simulation training evaluation index system, the BP neural network system model is constructed, which basically solves the problem of unscientific index weight allocation in the evaluation. It can not only take into account the expert experience but also reduce the uncertainty in the evaluation process. Factors can not only ensure the standardization of the evaluation but also maintain a high solution efficiency.

3.1 Building a BP Neural Network Model

The BP neural network is a three-layer forward network composed of an input layer, an output layer, and a hidden layer, as shown in Fig. 4. The number m of the network input nerves is the subject of simulation training assessment, which is composed of 6 secondary indicators and 12 secondary indicators as separate assessment subjects, so m = 18, n = 1 is the number of network output neurons, which is the unit The final result of simulated training is assessed, and the number of hidden layer neurons is based on the formula:

$$s = \sqrt{0.43mn + 0.12n^2 + 2.54m + 0.77n + 0.35} + 0.51$$

Get s = 8.

ω_{ik} is the connection weight from the i-th node in the input layer to the k-th node in the hidden layer. The connection weight is generally between 0 and 1, and is the output of the k-th node in the hidden layer of the sample pattern P, ω_k is the connection weight from the k-th node of the hidden layer to the output layer, and the connection weight is between 0 and 1.

The nonlinear relationship between the output and input of each node is described by the Sigmoid function:

$$f(x) = (1 + \exp(-x))^{-1}$$

The output formula of the hidden layer sample mode P is as follows:

$$y_{pk} = f(\sum_{j=1}^{s} \omega_{kj}\mu_{pj} - \theta_k)$$

In this formula: represents the bias value of the hidden layer node k.

The actual output of the output layer sample mode P is:

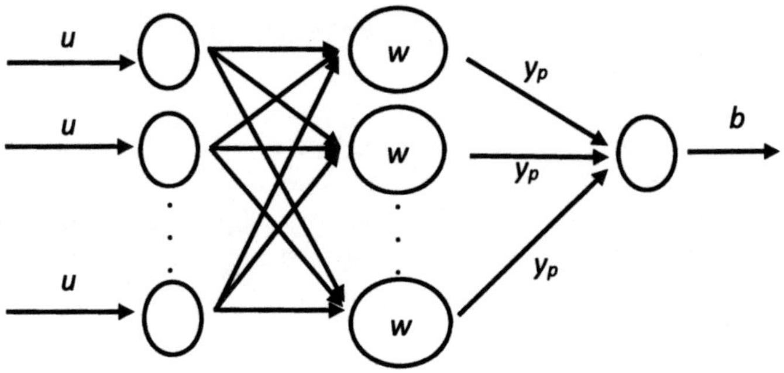

Fig. 4 Neural network model

$$b'_p = f(\sum_{k=1}^{m} \omega_k y_{pk} - \theta)$$

In this formula: represents the bias value of the output node of the output layer.

After the model is learned on Matlab and compared with the verification sample, there will be two results: the error between the result and the expert evaluation is gradually stable and the approach is lower than the fixed value as a success, and the result is not close to the expert evaluation or oscillates near the extreme point. It is considered a failure. The reason for the failure may be the unreasonable setting of the learning step length. If the learning step length is too large, the possibility of oscillation near the extreme point is increased, and repeated oscillations are difficult to converge. Here, as long as the step length is adjusted, the neural network can be stabilized by trial and error, and a satisfactory generalization result can be obtained.

3.2 Instance Verification

Collect 8 sets of samples from the armored force simulation training experts. The first 6 sets of samples are learning samples, and the last 2 sets of samples are test samples. In the samples, all indicators are sorted in the order of 1–18. For example, combat training is 1 and support is 2 etc. and so on. In addition, the expert group also evaluated each group of samples as expert evaluation to modify the model as shown in Table 1.

The error accuracy of the neural network is set to 10^{-6}, After inputting the sample data and evaluation value into the BP neural network for training, the following results are obtained: 71.000 0,81.000 0,74.000 0,90.999 8, 83.000 0,69.000 2.

After the neural network training is completed, the weight of the network is determined. The input of the above BP neural network is replaced with test value 1 and test value 2, and the results are 81.0000 and 57.0002 respectively. The output evaluation score is basically the same as the expert evaluation score.

3.3 Results and Analysis

After the model is trained and learned through 6 sets of learning samples, it has initially possessed the evaluation ability. In the latter two groups of test samples, the model's input and output evaluation scores are basically consistent with the expert evaluation scores. The error value gradually approaches and is lower than the error accuracy during continuous learning, indicating that the establishment of the model is basically successful.

After the neural network training is completed, input the performance of each subject in the simulation training of the armored forces to obtain a relatively objective

Table 1 Matlab training data

Serial number	Sample 1	Sample 2	Sample 3	Sample 4	Sample 5	Sample 6	Test 1	Test 2
1	61	69	65	67	69	75	71	97
2	83	72	75	96	66	86	69	93
3	71	96	98	71	52	82	69	92
4	96	95	98	95	83	81	85	96
5	86	68	79	85	92	86	79	89
6	91	97	91	92	31	82	71	71
7	72	69	68	69	53	72	69	77
8	65	54	65	62	55	48	59	34
9	93	76	89	73	79	65	68	59
10	91	95	97	97	88	91	93	95
11	71	76	91	72	86	77	83	82
12	61	69	65	67	69	75	61	43
13	86	68	79	85	92	86	73	99
14	93	76	89	73	79	65	79	90
15	72	77	92	73	87	78	88	97
16	65	73	69	73	79	80	72	73
17	92	91	94	91	79	77	72	60
18	67	56	67	64	57	50	85	84
Expert evaluation	71	81	74	91	83	69	83	62
BPNN output	71.0000	81.0000	74.0000	90.9998	83.0000	69.0002	81.0000	57.0002

and fair score. The operation is simple, quick, and easy to adjust. It can be applied in the simulation training evaluation and evaluation of the armored forces.

4 Concluding Remarks

The armored force simulation training system is a complex system. There are many factors to be considered in the comprehensive evaluation. Different factors are not only divided into levels, but also related to each other. If you use commonly used evaluation methods to evaluate, it will have certain limitations. For example, the determination of weights has obvious human factors, and it is difficult to objectively and effectively evaluate simulation training results. Artificial neural network, as an evaluation method for simulation training evaluation of armored forces, can produce results through continuous learning of sample data and self-calculation of the network. It is no longer necessary to manually set weights and reduce the impact of human factors on the evaluation results. It improves the objectivity, reliability and validity of the evaluation results, provides a scientific basis and new method for the simulation training evaluation and evaluation of the armored units, and also lays a certain foundation for the design of the simulation training evaluation software platform. Of course, the BP neural network also has certain limitations in simulation training evaluation. For example, when correcting the weights of different neuron connections, it is easy to make the system fall into a minimum value, which needs to be continuously improved in future research.

References

1. Soares, C.G., Teixeira. A.P.: Risk assessment in maritime transportation. Reliab. Eng. Syst. Safety **74**(3), 299–309 (2001)
2. Cai, w.: Extension set and non-compatible Problems. Advanc. Appl. Math. Mechanics in China. Peking: International Academic Publishers (1990)
3. Kaplan, S.: The general theory of quantitative risk assessment. Risk-Based Decision Making in Water Resources V. ASCE, 1991: 11–39 (1991)
4. Ayyub, B.M., Beach, J.E., Sarkani, S., et al.: Risk analysis and management for marine system. Naval Eng. J. **2**, 181–206 (2002)
5. Zhang, R.X., Mahadevan, S.: Bayesian methodology for reliability model acceptance. Reliab. Eng. Syst. Safety **80**, 95–103 (2003)
6. Farrell, J.P., Fineberg, M.L.: Specialized training versus experience in helicopter navigation at extremely low altitudes. Human Factors **18**, 305–308 (1976)
7. Yifei, H., Yingying, L.: DDoS Attack Detection System Based on Information Entropy and Naive Bayes Formula Hybrid Model. Network Security Technology and Application (08): 32–33 (2019)
8. Jimmie, H.: Analysis of Fatalities Record By OSHA. J. Construct. Eng. Manage. :26–28 (1995)
9. Yanli, M., Luyang, F., Tianling, L., et al.: Bayesian network quantitative grading method for vehicle operation risk. J. Harbin Institute Technol. 1–5 [2020–03–07]
10. Yuntao, S., Xiang, Z., Hui, D., et al.: Dynamic comprehensive assessment method of community distribution network risk based on AHP-Bayesian network. Safety Environ. Eng. 1–7 [2020–03–07]

Research on the Shape of Hair Dryer Handle Using Fuzzy Theory

Xingmin Lin, Zhen Ge, and Luting Xia

Abstract This research evaluates the visual features of different hair dryer handles. 12 home-use hair dryers of top brands in market were selected as samples, our study collected 40 most suitable adjectives out of 146 by 10 professional consumers. Then, the Semantic Differential Method (SD method) was used to design the questionnaire, whose results were analyzed with the factor analysis method, and factors were renamed afterwards. The statistical data from the market questionnaires were carried out in the triangular fuzzy numerical calculation, and the evaluation values of the 12 hair dryers were generated in 6 visual images respectively. The image evaluation values for different handle styles were trivial in terms of "robustness", "simplicity", "lightness" and "interesting", but very different in terms of "novel", "safe", "healthy" and "elegant qualities". The overall shape of these handles has several similarities between each other, consumers have more than one choice for each characteristic, such as NEUEO, PHENIX, Sima or Feike, are often deemed to be safe, healthy and rugged. Rowenta, Superpower and Panasonic tend to be interesting, personal, simple and lightweight. Povos, Haier, Sassoon were favored for their elegant quality. The study helps designers to set up the product image cognitive model of home-use hair dryers, in order to clarify the relation between customers' needs and product form.

X. Lin
Xiamen Academy of Arts and Design, Fuzhou University, Xiamen, China

Z. Ge (✉)
Xiamen Maken Tech Co., Ltd., Fujian, China
e-mail: zhengeamoy@hotmail.com

L. Xia
Zhejiang Gongshang University Hangzhou College of Commerce, Hangzhou, China

K. Nakamatsu et al. (eds.), *Advanced Intelligent Technologies for Information and Communication*, Smart Innovation, Systems and Technologies 365,
https://doi.org/10.1007/978-981-99-5203-8_16

1 Introduction

Shopping decisions can be affected by a multitude of product shapes and visual stimuli [1]. With the worldwide rise of e-commerce and online shopping, visual features play a more and more crucial role in people's shopping behaviors under a digital context [2], where e-consumers select products virtually most of the time. The pandemic has also dramatically accelerated this shift to a great extent. Therefore, visual features of products can be a differentiation strategy in keen competition and make an impression on consumers.

In addition, the culture of Generation-Z becomes trending and encourages people to self-express, show individuality and embrace diversity. People tends to choose the products that match their style and expectation, rather than just meet functional needs. Norman put forward his insight of good design, which should emphasize characteristics recognition from its users [3]. Usually, the visual features that meet consumers' preferences and expectation can generate great appeal when shopping [4]. In order to stand out from competitors, designers need to enhance their understanding of customers as well as their perception of different forms and features [5].

Since style and form has a critical impact on the consumer economy, designers need to target their users and make the right decision on visual features during the product development process. The consumers' cognition can be explored through the physiological test and market research. User-centered methods that incorporate the user's behavioral decisions, life experiences and operational perceptions can help to generate a simple and comprehensible design outcome. In consumers' cognition, styling characteristics are the most effective and direct cues to provide for users to manipulate [6].

This article will take home-use hair dryers as a case to study how the handle shape affects the perception of customers. Nowadays, types of hair dryers are varied in the market. It's not easy for consumers to make the shopping choice. Possession of aesthetic value and comfortable handle has a vital effect on consumers' decision-making, the key of product success being the excellent service and experience. In the process of design, mastering molding image of the product is the key of phase of conception development. Due to the wide-spreading of home hair dryer, meanings are given in options of molding, color, materials by designer, yet different consumers have different values of feeling in mind, which is possible to affect consumers' standards of judgement in product fondness. Confined in selection of only one objective information for designer and consumers, it will be conducive to communications between the two, also consumers will be able to get products that are more in line with their psychological feelings. The motive of this research is to explore into the review of visual image of varied types of hair dryer handles, collecting vocabulary of adjectives relevant with modeling image of the handle, and through phase 1 consumer questionnaire for the selection of vocabularies of adjectives by experts then through questionnaire of general consumers with semantics analysis method, and statistics from the elements analysis method, group and rename relevant vocabulary of adjectives and generate the conclusion finally according to the value table of image review by triangular fuzzy calculation. Therefore, for a research scheme that is in an objective manner, analysis into visual and psychological feelings about different handles of home hair dryer and evaluations on each molding image, will bring in the distance of designer to consumers in gaps of cognitions about home-use hair dryer handles, offering help to staffs of related industries, designers and consumers.

2 Research Purpose

The aim of this study is to explore consumers' cognition about the shape image of home-use hair dryers, with the following objectives:

(1) Based on the adjective vocabulary we collected, a questionnaire was designed to understand consumers' image evaluation of 12 hair dryer handles through factor analysis and triangular fuzzy number calculation;
(2) Through the integrated use of adjective vocabulary, factor analysis and triangular fuzzy calculation, inductive result was produced to enable designers to get a better understanding of consumers' needs and preference. Also, it is meaningful to provide consumers with experience reference in the selection of home-use hair dryers.

3 Literature Review

Scholars believe that imagery can take two different paths: one is mainly based on sensation (i.e., sensation-imagination), and the other is more objective [7]. In 1972, Mckellar conducted a survey of an image combining with 5 senses, among which the visual image has the most proportion, demonstrating its colossal influence. Scholars have concluded from the study of the multi-modal nature of sensory imagery in psychology that there is a direct correlation between consumer behavior and product imagery [8]. When scholars study the current Chinese Internet celebrity economy, they found that the image of the product is positively correlated with the consumers' attitude and attractiveness to them [9].

3.1 Factor Analysis

The factor analysis method can determine quantitatively of the influence direction and influence degree of each factor, according to the relationship between the analysis index and its influencing factors. The method can help researchers reduce groups of variables that reflect the nature, status, characteristics of things to a few intrinsic factors that determine the intrinsic connection of things. Factor analysis is particularly suitable to extract few factors from the large number of related variables to a more manageable number [10]. Fu examined the perceived comfort and fit using factor analysis and established a linkage between anthropometry and human perception for design uses [11]. The design management model adopted in the research provides a platform for designers to create design methods, quickly capture consumer preferences, and employ factor analysis to realize the relationship between product shape and color vocabulary in product development, and promote innovative design [12]. Some scholars use the literature analysis method, combined with the current situation of the industry, to discuss the problems and needs encountered by designers in design research, and to collect and classify the information that needs to be obtained in the research process. Through the method of factor analysis, it aims to help designers narrow the cognitive gap with consumers and effectively improve the quality of product design decisions [13].

3.2 Fuzzy Logic

Fuzzy logic is a law of science that researches the ambiguity of thinking, language and its rhythmicity by using fuzzy sets, based on the multi-valued logic. American mathematician L. Zadeh first proposed the concept of fuzzy set, marked the birth of Fuzzy Mathematics. Zadeh believed that human subjective thinking, reasoning and perception of peripheral things were vague in nature. Human must use the concept of fuzzy logic to describe the superiority and inferiority of things to make up the shortcomings of using two-valued logic, and the logic expand the relationship between elements and collections in the traditional set theory. The fuzzy set theory quantifies the fuzzy concept, so that the degree of membership of the set can be extended to any value in (0, 1) [14]. Scholars pass the four criteria of modeling creativity, cultural image, customer demand and city charm. It is applied to evaluate three cases, a multifunctional scale, a toy robot, and an agricultural specialty product and also to optimize product development suitability with Fuzzy logic quantitative computing capabilities [15].

4 Research Process and Methods

The first step was to sort out 146 adjectives describing Shaping image of the home hair dryer handles as samples in the first phase of the experiment, and then the first 40 scores of adjectives were chosen by 10 experienced users for the follow-up study. This study selected the top 12 home hair dryer brands in market share as the research samples (Table 1). The network questionnaire was conducted using the Likert Scale, and 139 valid questionnaires (65 males and 74 females) were obtained. The data were analyzed by SPSS. The 40 adjectives were obtained after the first T-test, which were analyzed through factor analysis. The principal component analysis showed that 20 adjectives that the absolute values of the factor loading are greater than 0.6, and then 20 adjective vocabularies were extracted by the second and third factor analysis. Six groups of adjective vocabularies were obtained relating to the adjective vocabulary characteristics and the renaming of factors (Fig. 1).

Semantic variable was commonly used as the evaluation method in order to help consumers to choose based on the importance of consumer needs and the ordering of the product selection. The questionnaire was designed using the seven-level semantic variables in the triangular fuzzy scale in Table 2. The subjects aged 18 to 55 with rich experience of using the products were chosen. The total number of questionnaires were 129 (65 for men and 64 for women). The consumers can rate based on their subjective image. The corresponding triangular fuzzy values of semantic variables can be looked up in Fig. 2.

The triangular fuzzy number could be used to describe the relationship between the potential characteristics of the home hair dryer handles and the degree of membership of semantic meaning in terms of the whole semantic scale. The particularity of the triangular fuzzy number lies in its membership function, and its distribution of its possibilities is to form a triangle.

In order to obtain a detailed analysis data, the fuzzy numbers in the membership function can be converted to crisp values by defuzzification. The maximum set and the minimum set were the most commonly used. In this paper, we use the method

Table 1 The numbers and pictures of 12 types of hair dryer handle

No	Name	Code	No	Name	Code
1	Meimanda	MA	7	Povos	BT
2	Neueo	NO	8	Feike	FO
3	Phenix	PX	9	Philips	PS
4	Rowenta	RA	10	Haier	HR
5	Sima	SA	11	Sassoon	VS
6	Superpower	SR	12	Panasonic	PC

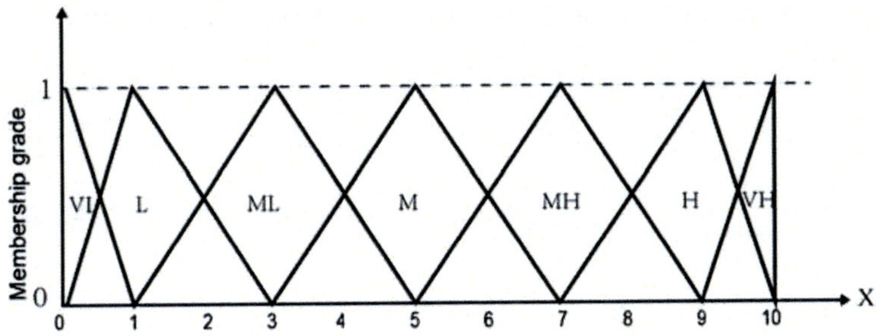

Fig. 1 Membership functions of triangular fuzzy numbers

Table 2 Linguistic variables for importance and rating

Linguistic variables	Triangular fuzzy numbers
Very low (VL)	(0, 0, 1)
Low (L)	(0, 1, 3)
Medium low (ML)	(1, 3, 5)
Medium (M)	(3, 5, 7)
Medium high (MH)	(5, 7, 9)
High (H)	(7, 9, 10)
Very high (VH)	(9, 10, 10)

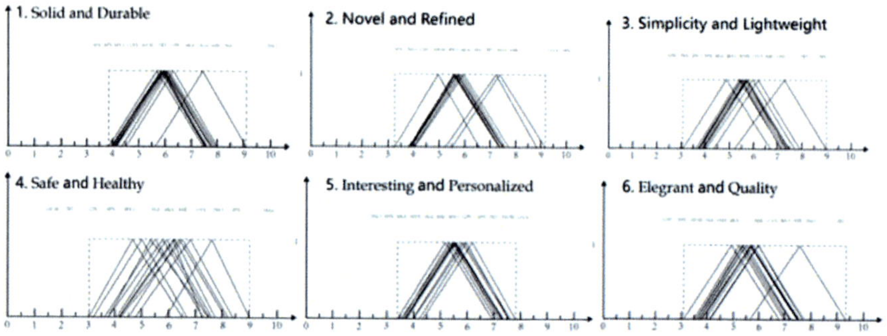

Fig. 2 The triangular fuzzy numbers of 12 hair dryer handle in each visual evaluation

of calculating the weights of the two trigonometric functions and deduce the triangular fuzzy function to the absolute utility values. The absolute value Eq. (1) of the triangular fuzzy number $\tilde{t}i$:

$$\begin{aligned}\mathrm{UT}(t_i) &= [(t_i3-\mathrm{Xmin})/(\mathrm{Xmax}-\mathrm{Xmin}) + (t_i3-t_i2)) + 1 \\ &-(\mathrm{Xmax}-t_i1)/((\mathrm{Xmax}-\mathrm{Xmin}) + (t_i2-t_i1))]/2, \mathrm{i} = 1, 2, \ldots, \mathrm{n}\end{aligned} \tag{1}$$

The result of the further comparative analysis of the 12 hair dryer handles can provide designers and consumers with reference when designing and consuming home hair dryers.

In this paper, we use the method of calculating the weights of the two trigonometric functions and deduce the triangular fuzzy function to the absolute utility values.

5 Investigation and Analysis

After factor analysis, the top 20 values adjective vocabularies were Secure (0.675), Convenient (0.745), Interesting (0.813), Personalized (0.697), Simple (0.703), Eco-friendly (0.730), Refined (0.758), Novel (0.707), Elegant (0.701), Classic (0.741), Concise (0.699), Solid (0.757), Luxurious (0.761), Durable (0.678), Delicate (0.758), Laborsaving (0.695), Unique (0.726), Lightweight (0.686), Healthy (0.710) and Precise (0.706).

After KMO and Bartlett test, the KMO value was 0.889, indicating that the data was suitable to analyze. The spherical test value was 2773.290, which was very significant.

There are six factors that the initial eigenvalues were greater than 1, and the total explanatory variance is 67.922% (Table 3).

Table 3 Transformed component matrix

Product code	Solid and durable	Novel and refined	Simplicity and lightweight	Secure and healthy	Interesting and personalized	Elegant and quality
Solid	0.847	0.167	0.170	0.218	0.101	0.036
Durable	0.767	0.258	0.210	0.157	−0.005	0.002
Classic	0.767	0.209	0.262	0.016	0.169	−0.168
Refined	0.308	0.839	0.131	0.332	0.173	−0.168
Precise	0.226	0.812	0.152	0.053	0.203	−0.116
Novel	0.131	0.762	0.329	0.290	0.147	0.082
Delicate	0.151	0.733	0.310	0.163	0.089	0.069
Lightweight	0.411	0.610	0.803	0.225	0.041	0.005
Simplicity	0.451	0.250	0.799	0.138	0.273	0.015
Convenient	0.021	0.164	0.744	0.167	0.028	0.131
Effort-saving	0.053	0.117	0.701	0.031	0.037	0.009
Secure	0.269	0.153	0.109	0.813	0.263	0.195
Healthy	0.314	0.136	0.155	0.728	0.242	0.213
Eco-friendly	0.061	0.215	0.101	0.640	0.114	0.059
Interesting	0.228	0.234	0.040	0.140	0.810	0.005
Personalized	0.478	0.109	0.231	0.252	0.795	0.106
Unique	0.086	0.301	0.199	0.127	0.762	0.218
Elegant	−0.021	−0.005	−0.029	0.065	0.192	0.826
Quality	0.478	0.280	0.214	0.031	0.046	0.743
Luxurious	0.163	0.190	0.089	0.182	0.123	0.728

The statistics in Table 4 shows that the handles of group 1 are shaped with the beauty of harmony and a sense of high quality. The handle is of a homogeneous radian, with abrupt variation on PX and SA. Handles in group 2, having a simple and sleek design is of a sense of technology. In group 3, the shape of handles is mellowed, with a high curved radian, grip comfort, and high aesthetic value of BT. Handles in group have a slim and elegant design that is fit for women consumer. Handles in group 5, is beautifully bent and is flowing with a sense of elegance and offers great friction with the bulge button Table 5.

Table 4 12 types of hair dryer handles image evaluation Values

Product Code	Solid and durable	Novel and refined	Simplicity and lightweight	Secure and healthy	Interesting and personalized	Elegant and quality
MA	0.4454	0.4172	0.4755	0.4845	0.4768	0.4094
NO	**0.6320***	0.4625	0.4205	0.5591	0.4313	0.4701
PX	0.3625	0.3391	0.4285	0.5848	0.4562	0.3766
RA	0.4454	0.4394	0.4394	0.4948	0.4712	0.4094
SA	0.4849	0.4415	0.3982	0.5183	0.4857	0.4001
SR	0.4736	0.4755	0.4647	**0.7122***	0.4895	0.4354
BT	0.4193	0.4517	0.5040	0.3717	0.5415	**0.6701***
FO	0.4084	0.4495	0.4415	0.4593	0.4901	0.4479
PS	0.4084	**0.6573***	**0.6459***	0.4338	0.5302	0.4727
HR	0.4313	0.4285	0.4287	0.3267	0.5563	0.3892
VS	0.4172	0.6394	0.6105	0.5463	**0.5892***	0.4354
PC	0.4344	0.4285	0.5404	0.4356	0.4969	0.3408

Table 5 The image of five groups hair dryer handle features had similar comprehensive visual evaluation

	Radar chart	Hair Dryer Handles Types and Visual Characteristics
1	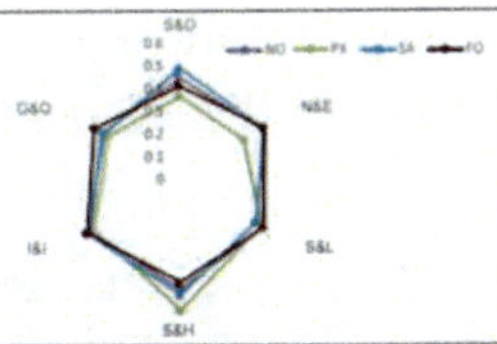	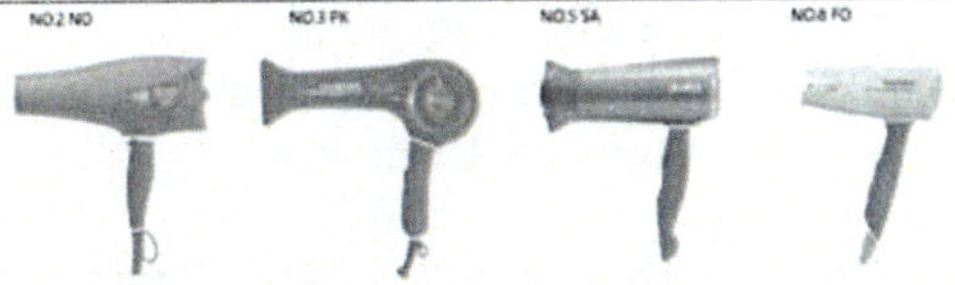
2	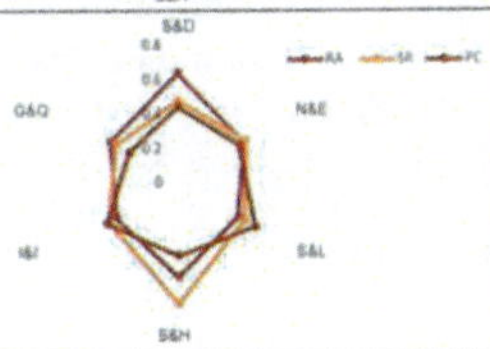	
3	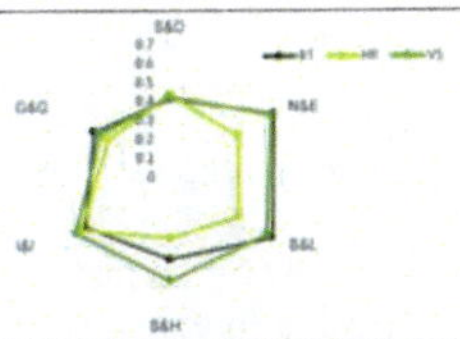	
4	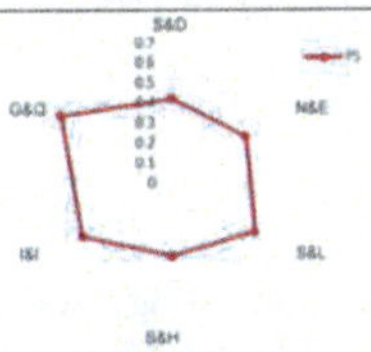	
5	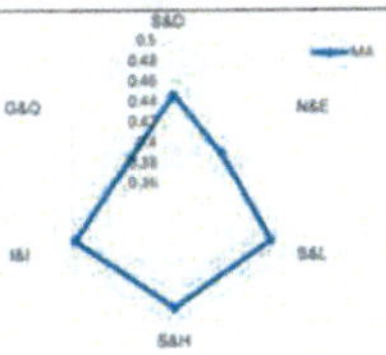	

6 Conclusion

The study aimed at the image reviews of consumers on various hair dryer handles, the analysis of the vocabulary of adjectives, the statistical analysis of varied factors, and the unique characteristic image of each handle sample. The image reviews of different handles vary little on the aspects of durability, simplicity, and fun. However, it varies largely on novelty, safety and quality. The research also shows the similarity in the overall images of handles, and the replaceability in the selection and application of consumers, as in group 1 NEUEO (NO), PHENIX(PX), Sima (SA) FO, and in group 2 Rowenta (RA), Superpower (SR), PC, and group 3 BT, Haier, Vs. From the point of the application of the above research in subsequent practice, the data base is set up according to the values of reviews of hair dryer handles image design of different brands, further to the design of the decision assisting system for future consumers. According to the preferences of each individual in handle image selection, and through the operation on the screen, the selection is being quickly completed. Within the works of the designer, it also conveys the design philosophy and the emotional feelings that consumers desire. The methodology of this study can be used in all types of product development.

Acknowledgements This research was financially supported by 2020 General Scientific Research Project of Zhejiang Provincial Department of Education, Y202045216, "Research on the Mechanism and Strategy of Hangzhou's Tourism Cultural and Creative Industry from the Perspective of "Digital Creative Industry", 2020.12-2023.12.

References

1. Berkowitz, M.: Product shape as a design innovation strategy. J. Prod. Innov. Manag. **4**(4), 274–283 (1987)
2. Shih, D., Lu, K., Wu, T., Shih, P., Yen, D.C.: What Features of a Product Can Catch Shoppers & Visual Attention Online? An Eye-Tracking Analysis. SSRN Electronic Journal (2022)
3. Norman, D.A.: The invisible computer: why good products can fail, the personal computer is so complex, and information appliances are the solution. MIT Press (1998)
4. Toffler, A.: The Third Wave. William Morrow, New York (1980)
5. Taghikhah, F., Voinov, A., Shukla, N.: Extending the supply chain to address sustainability. J. Clean. Prod. **229**, 652–666 (2019)
6. Shao, J., Ünal, E.: What do consumers value more in green purchasing? Assessing the sustainability practices from demand side of business. J. Clean. Prod. **209**, 1473–1483 (2019)
7. Herd, K.B., Mehta, R.: Head versus heart: The effect of objective versus feelings-based mental imagery on new product creativity. J. Con. Res. **46**(1), 36–52 (2019)
8. Elder, R.S., Krishna, A.: A review of sensory imagery for consumer psychology. J. Consum. Psychol. **32**(2), 293–315 (2022)
9. Park, H.J., Lin, L.M.: The effects of match-ups on the consumer attitudes toward internet celebrities and their live streaming contents in the context of product endorsement. J. Retail. Consum. Serv. **52**, 101934 (2020)
10. Shrestha, N.: Factor analysis as a tool for survey analysis. Am. J. Appl. Math. Stat. **9**(1), 4–11 (2021)
11. Fu, F., Luximon, Y.: Comfort and fit perception based on 3D anthropometry for ear-related product design. Appl. Ergonom. 103640 (2021)
12. Kuo, L., Chang, T., Lai, C.C.: Research on product design modeling image and color psychological test. Displays **71**, 102108 (2022)
13. Wang, D., Fu, J., Qian, Y.: Investigation on Research Model of Product Design Integrated with Big Data Technology. In International Conference on Human-Computer Interaction, pp. 312–323. Springer, Cham (2022)
14. Chen, C.T.: A fuzzy approach to select the location of the distribution center. Fuzzy Sets Syst **118**, 65–73 (2001)
15. Al-Jamimi, H.A., Saleh, T.A.: Transparent predictive modelling of catalytic hydrodesulfurization using an interval type-2 fuzzy logic. J. Clean. Prod. **231**, 1079–1088 (2019)

A Scheme for Determining Maintenance Task Priority

Minmin Qin, Lifang Liu, Qingfeng Zeng, and Xiaogang Qi

Abstract The determination of military equipment maintenance task priority is directly related to the smooth development of subsequent maintenance task allocation and scheduling. There are many factors affecting the priority of maintenance tasks, and comprehensive consideration of various factors is the key to reasonably determine the priority of maintenance tasks. Through the analysis of the characteristics of wartime military equipment maintenance support, a maintenance task evaluation index system is constructed. Using the Dematel method of current hot factor analysis and identification, the weight of each evaluation index is determined, and the value of each index with different dimensions is standardized, so as to determine the priority of maintenance tasks. In addition, the proposed maintenance task priority determination scheme is applied to an actual military case and the impact index is expanded. By comparison, it is found that the proposed scheme is simpler and clearer than previous schemes and can also obtain similar priority order, with strong robustness. The new method can provide a reasonable basis for equipment support commanders to determine the priority of equipment maintenance tasks in wartime.

1 Introduction

In wartime, a scientific and effective maintenance tasking program can effectively improve force combat effectiveness. Maintenance task refers to a certain level of maintenance squad to be able to repair equipment that cannot continue to participate

M. Qin · L. Liu (✉)
School of Computer Science and Technology, Xidian University, Xi'an 710071, Shaanxi, China
e-mail: lfliu@xidian.edu.cn

X. Qi
School of Mathematics and Statistics, Xidian University, Xi'an 710071, Shaanxi, China

Q. Zeng
31657 Troops, Huzhou 313000, Zhejiang, China

K. Nakamatsu et al. (eds.), *Advanced Intelligent Technologies for Information and Communication*, Smart Innovation, Systems and Technologies 365,
https://doi.org/10.1007/978-981-99-5203-8_17

in combat due to hit damage, technical failure, or natural damage within a specified period of time. Priority then refers to the wartime maintenance tasks combined with the maintenance tasks own impact indicator attributes by the maintenance squad to carry out repairs in the order of priority. Dynamic scheduling of maintenance operations refers to the activity of assigning operations of each dimension to maintenance squads in real time and rationally according to the actual situation on the wartime battlefield, which belongs to the category of workshop operations scheduling [1].

Based on the strong application value of maintenance scheduling tasks, it has attracted widespread attention in the industry. At this stage, the research on task priority has been quite in-depth. Li [2], Zhou [3], Wang [4] and others abstracted the priority determination scheme involving remote sensing measurement and control technology into a multi-objective and multi-task decision optimization problem, and studied the target problem using different optimization algorithms, and gave the corresponding optimization model. Lehtonen [5], Singh [6], Abaynew [7] and others studied the priority determination scheme from the perspective of resource optimization and scheduling. It is not difficult to find that the above-mentioned research objects are mainly in non-military fields such as aerospace measurement and control, enterprise projects, etc., but few are involved in the military field, and there are fewer related studies based on the priority determination of military equipment maintenance tasks.

According to the analysis of the maintenance support characteristics of military equipment in wartime, the maintenance task evaluation index system is constructed, and the Dematel method of factor analysis and identification, which is currently researched hotly, is used to determine the weight of each evaluation index, and normalize the value of each index with different dimensions. Prioritize maintenance tasks. Finally, the rationality of the proposed scheme is verified in a practical case.

2 Prioritization Model

There are m military equipment maintenance tasks A_1, A_2, A_3, ..., A_m, and n evaluation indicators B_1, B_2, B_3, ..., B_n can be used to evaluate each military equipment maintenance task. In addition, military equipment should be sturdy, portable, compact, easy to use, easy to maintain, and require minimal consumables [8].

Equipment importance A, that is, the importance of the equipment to be repaired to the mission, can reflect the contribution of the equipment to the combat mission. Military equipment is an important symbol of military modernization, an important basis for the preparation of military struggle, an important support for national security and national rejuvenation, and an important weight in the international strategic game [9]. However, the state of military core equipment is directly related to the victory of an operation or not, therefore, military core equipment is of great merit in the battlefield environment [10].

Repair time B, that is, the time for the equipment to be repaired to return to a combat-ready state from the start of repair. Combined with the demand resources, with a certain repair strategy in the repair center for the repair of damaged equipment, when the damaged equipment is repaired by the repair center and then restored to operational status time is called repair time [11].

Resource demand degree C, that is, the degree of resource demand of the equipment to be repaired. The equipment to be repaired is not only dependent on a single resource, but may also involve multiple resources, i.e., repairing a piece of equipment may require the participation of multiple resources at the same time. In addition, some of the equipment to be repaired may be limited by a resource, i.e. the equipment to be repaired can only be repaired for a fixed resource, and other resources cannot repair the equipment even if it is currently free [12].

The damage degree D, the damage degree of the equipment to be repaired, can be used to measure the size of the repair cost. Because of the complex military battlefield environment and the significant contribution of core equipment to combat, core equipment is also more vulnerable to attack. In addition, due to the cost, size, location, and material constraints, the degree of damage to equipment varies. Extremely severely damaged equipment is considered for replacement with new equipment, while equipment in a repairable state requires timely equipment repair [13].

Location E, the location of the equipment to be repaired, the closer to the location of the repair center, the lower the maintenance cost; otherwise, the farther the distance is, the higher the cost.

The two influencing factors connected by arrows in Fig. 1 have a certain correlation, and the numbers on them indicate the degree of influence of the factor pointed by the end of the arrow on the factor pointed by the arrow, and the numbers from 1 to 5 represent, in order, a little correlation, a small correlation, an average correlation, a large correlation and a great correlation, and the correlation is proportional to the degree of influence.

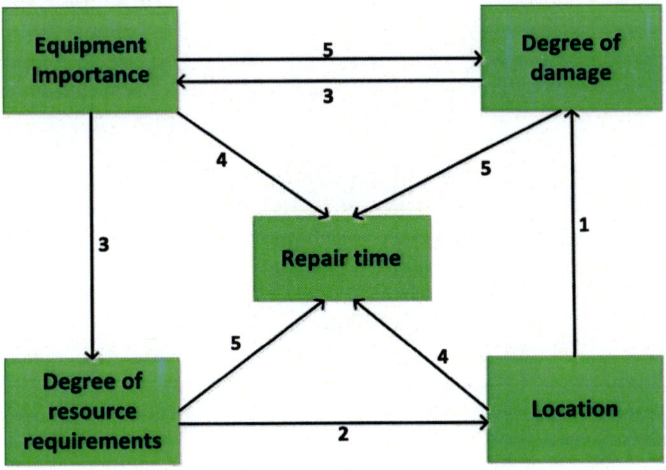

Fig. 1 The influencing factor relationship model of task priority

Fig. 2 Direct influence matrix M

Influencing Factors	A	B	C	D	E
A	0	4	3	5	0
B	0	0	0	0	0
C	0	5	0	0	2
D	3	5	0	0	0
E	0	4	0	1	0

Through the analysis of the influencing factors of the maintenance task priority and the correlation between the corresponding influencing factors, A, B, C, D, and E are used to replace the five influencing factors: equipment importance, repair time, resource demand degree, receiving damage, location. It is not difficult to obtain the direct influence matrix M shown in Fig. 2 through the relationship model of the influencing factors of task priority [14].

In order to eliminate the influence of the dimensions of different data indicators, the matrix is normalized. In the knowledge of probability theory, the standardized method is generally expressed as: $\frac{X-EX}{\sqrt{DX}}$, but here the method is optimized accordingly, as follows:

Assume that there are n influencing factors to form the following matrix [15]:

$$X = \begin{bmatrix} x_{11} & x_{12} & \cdots & x_{1n} \\ x_{21} & x_{22} & \cdots & x_{2n} \\ \vdots & \vdots & \ddots & \vdots \\ x_{n1} & x_{n2} & \cdots & x_{nn} \end{bmatrix} \tag{1}$$

Then, the matrix that normalizes it is denoted as Z, each element in Z:

$$z_{ij} = x_{ij} / \sqrt{\sum_{i=1}^{n} x_{ij}^2 (j = 1, 2, ..., n)} \tag{2}$$

The matrix M is standardized as above, and the standardized matrix shown in Fig. 3 is obtained M′.

The normalized matrix obtained above is not very good, and further data processing of the matrix M is needed to obtain the regularized influence matrix M″.

Fig. 3 Standardized matrix M′

Influencing Factors	A	B	C	D	E
A	0	0.4417	1.0000	0.9806	0
B	0	0	0	0	0
C	0	0.5522	0	0	1.0000
D	1.0000	0.5522	0	0	0
E	0	0.4417	0	0.1961	0

Regularized influence matrix M″ solution method:

(i) Find the sum of the elements of each row of the direct influence matrix M and take the maximum value.
(ii) Dividing each element of the direct influence matrix M by the maximum value, we obtain the regularized influence matrix M″ as in Fig. 4.

Based on the results obtained above, the integrated impact matrix is solved as shown in Fig. 5.

Through the comprehensive influence matrix it is easy to find that influence factor B, i.e. the repair time of the equipment to be repaired, is most influenced by the other elements, while influence factor A, i.e. the importance of the equipment to be repaired to the task, has the greatest influence on the other elements.

In addition, through the results of the above comprehensive influence matrix centrality calculation, it is easy to obtain the weighting ratio of each influence factor: from A-E in order: 26% for equipment importance, 28% for repair time, 13% for resource requirement, 23% for damage degree, and 10% for location.

Fig. 4 Formalization impact matrix M″

Influencing Factors	A	B	C	D	E
A	0	0.3333	0.2500	0.4167	0
B	0	0	0	0	0
C	0	0.4167	0	0	0.1667
D	0.2500	0.4167	0	0	0
E	0	0.3333	0	0.0833	0

Influencing Factors	A	B	C	D	E	Degree of influence	Centrality	Degree of cause
A	0.1174	0.7000	0.2793	0.4695	0.0466	1.6128	2.0367	1.1889
B	0	0	0	0	0	0	2.1548	-2.1548
C	0.0039	0.4805	0.0010	0.0155	0.1669	0.6678	1.0237	0.3119
D	0.2793	0.5917	0.0698	0.1174	0.0116	1.0698	1.7653	0.3743
E	0.0233	0.3826	0.0058	0.0931	0.0010	0.5058	0.7319	0.2797
Influenced Factors	0.4239	2.1548	0.3559	0.6955	0.2261			

Fig. 5 Integrated impact matrix

3 Task Priority Classification

The indicators in the set of priority indicators for equipment maintenance tasks are quantified. With n units of equipment to be repaired, there are n maintenance tasks correspondingly. Because of the different levels under each indicator, it is particularly important to convert the non-quantitative criteria of each indicator into quantitative criteria. Therefore, the indicators involved in this practical problem can be normalized as follows:

i. Gainful (the larger the value of the original indicator, the better)

$$y_i = \frac{x_i}{\max(x_i)} (i = 1, 2, ..., n) \tag{3}$$

For revenue-based metrics, the higher the value of the metric, the higher the priority.

ii. Costliness (the smaller the value of the original indicator, the better)

$$y_i = \frac{\min(x_i)}{x_i} (i = 1, 2, ..., n) \tag{4}$$

For cost-based metrics, the higher the value of the metric, the lower the priority.

4 Case Study

An armored battalion to carry out a certain combat mission, in the implementation of the mission, a total of 8 equipment damage, that is, corresponding to 8 maintenance tasks, the index value is shown in Table 1, the sample evaluation indicators for the normalization process, and then the data processing to obtain the normative value of each indicator is shown in Table 2.

It is not difficult to determine the following maintenance task priorities from the composite rating values obtained from the comprehensive maintenance task priority evaluation list in Table 3:

$$T7 > T1 > T4 > T8 > T3 > T6 > T2 > T5$$

Table 1 Indicator values for maintenance tasks of an armored battalion

Maintenance tasks	Equipment importance	Repair time	Degree of resource requirements	Degree of damage	Location
T1	0.56	0.70	0.08	0.58	800
T2	0.37	1.50	0.15	0.38	900
T3	0.35	0.80	0.24	0.36	900
T4	0.52	0.50	0.53	0.50	1200
T5	0.36	2.00	0.18	0.42	1400
T6	0.32	1.20	0.39	0.28	1400
T7	0.50	0.40	0.05	0.52	300
T8	0.38	1.20	0.36	0.44	200

Table 2 Normative values of each indicator for the maintenance tasks of an armored battalion

Maintenance tasks	Equipment importance	Repair time	Degree of resource requirements	Degree of damage	Location
T1	1.00	0.57	0.63	0.48	0.25
T2	0.66	0.27	0.33	0.74	0.22
T3	0.63	0.50	0.21	0.78	0.22
T4	0.93	0.80	0.09	0.56	0.17
T5	0.64	0.20	0.28	0.67	0.14
T6	0.57	0.33	0.13	1.00	0.14
T7	0.89	1.00	1.00	0.54	0.67
T8	0.68	0.33	0.14	0.64	1.00

Note Benefit-based metrics: equipment importance

Table 3 List of comprehensive evaluation of maintenance task priorities

Maintenance tasks	Equipment importance	Repair time	Degree of resource requirements	Degree of damage	Location	Comprehensive evaluation value
	0.26	0.28	0.13	0.23	0.10	
T1	1.00	0.57	0.63	0.48	0.25	0.64
T2	0.66	0.27	0.33	0.74	0.22	0.48
T3	0.63	0.50	0.21	0.78	0.22	0.53
T4	0.93	0.80	0.09	0.56	0.17	0.62
T5	0.64	0.20	0.28	0.67	0.14	0.43
T6	0.57	0.33	0.13	1.00	0.14	0.50
T7	0.89	1.00	1.00	0.54	0.67	0.83
T8	0.68	0.33	0.14	0.64	1.00	0.53

The priority ranking order obtained is basically consistent with that of the original example, which proves the rationality of the proposed method. In addition, the priority ranking order obtained by introducing the metric of damage degree on the basis of the original example is basically consistent with the original example, which can further demonstrate the robustness of the proposed method.

Cost-based indicators: repair time, resource requirement degree, damage degree, location.

5 Conclusion

By analyzing the characteristics of wartime military equipment maintenance and guarantee, constructing a maintenance task evaluation index system, applying the Dematel method of factor analysis and identification, which is hotly researched nowadays, determining the weights of each evaluation index, normalizing the values of each index with different magnitudes, and then proposing a scheme of military maintenance task priority, based on the actual analysis of military cases, finding the reliability The reliability and robustness of the proposed scheme are found to be strong based on the actual analysis of military cases.

In addition, since there are few studies related to the prioritization scheme based on military equipment maintenance tasks, Wang et al. [16] proposed the use of AHP to determine the weights of each evaluation index, and based on the characteristic of effective stripping the correlation between the indexes by using the improved TOPSIS method based on the Mahalanobis distance to calculate the closeness of each equipment maintenance task, so as to determine the priority of the equipment maintenance tasks, which has the disadvantage in that the complexity is high, the intermediate data processing process is error-prone, and the degree of influence by the evaluation index is large.

References

1. Xu, L.Z., Xie, Q.S.: Dynamic production scheduling of digital twin job-shop based on edge computing. J. Inf. Sci. Eng.: JISE **1**, 37 (2021)
2. Lin, L.I., Yuan, L., Wang, L., Zheng, R., Yanpeng, W.U., Wang, X.: Recent advances in precision measurement & pointing control of spacecraft. Chin. J. Aeronaut.: English (2021)
3. Peijie, Z., Jinfeng, L., Su, L., et al.: Performance evaluation of model predictive control based on priority strategy. J. Shanghai Jiaotong Univ. **49**(11), 1641–1654
4. Wang, Z., Wang, J., Han, J.: Spatial prediction of groundwater potential and driving factor analysis based on deep learning and geographical detector in an arid endorheic basin. Ecol. Indic. **142** (2022)
5. Lehtonen, J.M., Virtanen, K.: Choosing the most economically advantageous tender using a multi-criteria decision analysis approach. J. Public Procure. **22**(2), 164–179 (2022)
6. Singh, S., Dhote, P.R., Thakur, P.K., Chouksey, A., Aggarwal, S.P., Glade, T., et al.: Identification of flash-floods-prone river reaches in Beas river basin using GIS-based multi-criteria technique: validation using field and satellite observations (2021)
7. Alene, A., et al.: Identifying rainwater harvesting sites using integrated GIS and a multi-criteria evaluation approach in semi-arid areas of Ethiopia. Appl. Water Sci. **12**(10), 1–16 (2022)
8. Watson, E.L., Round, J.A.: Field anaesthesia and critical care equipment used by the British military. Anaesth. Intensive Care Med. (2022)
9. Shi, Y. et al.: Summary of the modernization of India-Russia weapons and equipment. J. School Electron. Eng. (2020)
10. Zhang, S., Liu, Y., Ling, P.: Multi-source military equipment knowledge association organization method. CN112328855A (2021)
11. He, Y.: Mean time to repair equipment accidental damage measurement method, evaluation method and application. CN112200324A (2021)
12. Guo, D., et al.: Assessment of satisfaction of natural resources survey demands by high score satellite. J. Remote Sens. **26**(3), 9 (2022)
13. Li, M., Gao, L., Xu, H.: Analysis and design of "damage unit" for equipment maintenance and assurance task measurement. J. Artillery Firing Contr. **43**(3), 6 (2022)
14. Lu, Y., Ding, W., Liu, M.: Research on the cause and mode of sustainable supply chain collaborative management of water diversion project. Sci. Technol. Manag. Res. **41**(3), 9 (2021)
15. Yu, J.: Data sensing based inventive system for multimedia and sensing of web applications. In: 2020 fourth international conference on inventive systems and control (ICISC) (2020)
16. Wang, X., Chen, C., Cao, Y., Zan, X., Mei, Y.: An equipment maintenance task prioritization method based on improved TOPSIS method. Comput. Measur. Contr. **26**(4), 5 (2018)

An Improved Channel Assignment and Topology Control Algorithm in Cognitive Radio Networks

Xu Yuelin, Ding Kai, and Xu Xingchun

Abstract Cognitive radio has emerged as a promising technology to exploit the unused portions of spectrum in an opportunistic manner. Spectrum assignment is a key mechanism that limits the interference between second users and primary users, enabling a more efficient usage of the wireless spectrum. The licensed channels are randomly occupied by primary users in the cognitive radio networks. The channel occupancy time of the primary user in different channels is different too. By assuming it is negative exponentially distributed and combining the channel selection criteria of the robust topology control algorithm (CRTCA), an improved channel assignment algorithm (ICAA) was proposed. The simulation results indicate that the channel assignment method of considering the channel occupancy time of primary user is more than CRTCA in throughput and less than CRTCA in collision rate.

1 Introduction

In the cognitive radio networks, Primary users (PUs) have the absolute priority of using licensed spectrum. Second users (SUs) detect spectrum holes periodically. SUs access the segment of spectrum in an opportunistic manner as long as they detect spectrum holes. To avoid interference with PUs, SUs must vacate the spectrum which is occupying when it is accessed by PUs. The problem how to share the spectrum resources and optimize the spectrum utilization in the premise of interference-constricted PUs and SUs is the key to improve the spectrum utilization efficiency. The performance of spectrum assignment algorithms can affect the performance of the entire network directly.

We divide the spectrum into several channels with different frequencies based on OFDM. Data transmission in different channels is non-interfering because of the orthogonality of different channels. However, the activity areas of PUs are larger

X. Yuelin · D. Kai (✉) · X. Xingchun
Science and Technology On Near-Surface Detection Laboratory, Wuxi 214035, China
e-mail: winfast113@sina.com

K. Nakamatsu et al. (eds.), *Advanced Intelligent Technologies for Information and Communication*, Smart Innovation, Systems and Technologies 365,
https://doi.org/10.1007/978-981-99-5203-8_18

than SUs', hence, when PUs access the spectrum that occupying by SUs, data transmissions in several links will be affected. Thus, SUs need to detect new spectrum resources and then carry out spectrum switching. They have to spend a considerable amount of time for spectrum sensing and neighbour discovery. These conditions will lead to packet delay and throughput decrease. Once a single node implements channel switching, its neighbours also need to carry out the corresponding channel switching strategies. It may cause a ripple effect. Moreover, it is difficult to predict when a PU will appear in a given spectrum, and hence it is hard to address these issues.

Channel assignment models are generally divided into centralized and distributed approaches. A fusion center coordinates and manages SUs' spectrum using in the centralized model. The fusion center can collect spectrum sensing information from all SUs, build an available spectrum library and then realize the global optimization of channel assignment. In the distributed model, SU only can detect surrounding PUs and communicate with its neighbors. This model is more flexible than the first one. However, its performance is not so good due to the influence of the shadow effect, the multi-path effect and the hidden terminal problem. Thus, the model we select is the centralized model.

SUs have several half-duplex radios. A radio has the ability to receive and send data. However, it cannot send and receive data simultaneously and it only can utilize one channel in one certain time-slot. Most papers [5] assume that each node has only one radio which can switch among multiple channels. If PUs access the channel being used by SUs currently, SUs' data transmission will be interrupted. Some papers [6] assume that the activity of PU is predictable or SU has the ability to quit the channel in time when PU is active. But in the actual network environment, these assumptions are difficult to obtain. Authors in paper [7] have studied the average performance of the network with local and global information and proposed a distributed channel assignment algorithm when nodes are removed or added to the network. Experimental studies have shown that channel switching can result in up to 3% dropout rate. Thus, in order to avoid or reduce channel switching as far as possible, SU must take credible strategy choice to find other available channels. We assume that one SU has multiple radios. If one channel is occupied by PU, the source node can also use other unoccupied channels for data transmission to the destination node. One node represents one SU in the next descriptions. However, due to the condition of the hardware limitations, the assembly of radios consumes a lot, so it is necessary to make a trade off between the cognitive network performance and the number of radios. Authors in paper [3] have proposed a centralized robust topology control algorithm (CRTCA) based on creating a robust and stable topology and minimizing channel interference. In CRTCA, without assuming a predictable PU activity, data transmission can be re-routed along an unaffected path during the spectrum switching time, and thus packet dropping or packet delay can be avoided. However, this approach does not consider the channel occupancy time of PU. The channel selection rule only selects the least used channel from the set of current available channels, which will likely select the channel frequently used by PU, and then the collision-rate between PU and SU will increase. It also can cause frequent channel switching of SUs and

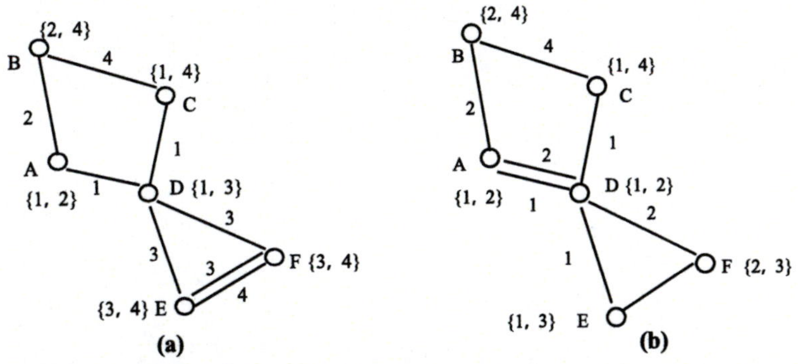

Fig. 1 Topology channel. **a** Non-robust topology channel assignment. **b** Robust topology channel assignment

result in the increase of delay and the decrease of system throughput. We will make a tradeoff between the interference among SUs and the interference between SUs and PU, and give an improved channel assignment algorithm (ICAA) following the principle of "The spectrum of higher quality has the more priority to be selected".

Figure 1 is a topology diagram with six nodes and four available channels. Capital letters represent the number of nodes; numbers in parentheses represent the channels currently assigned to that node; marks beside edges represent the channel currently assigned to that link. In Fig. 1a, if PU occupies channel 1 or channel 3, the graph will be disconnected, that is, the channel assignment approach is of non-robustness. In Fig. 1b, no matter which one channel is occupied by PU, the graph is connected, that is, the channel assignment approach is of robustness.

2 System Model

The model we select is the centralized model. There is only one PU that randomly accesses licensed channels. Only one channel is permitted to access in one certain time slot. The fusion center receives sensing information from SUs periodically and makes judgment whether those channels are idle or not, then SUs access channels in an opportunistic manner.

Channel model is a Markov chain of two states, they are idle (off) and busy (on) states respectively. In the idle state, channels are not occupied by PU, SUs can utilize these channels to implement data transmission. In the busy state, channels are being occupied by PUs, SUs are not entitled to use these channel. SU will conflict with PU if SU does not quit the channel promptly when channel state turns idle into busy (Figs. 2 and 3).

We assume the two channel states are negative exponentially distributed with parameter λ_i and μ_i respectively [8], thus the idle state distribution function of channel i is:

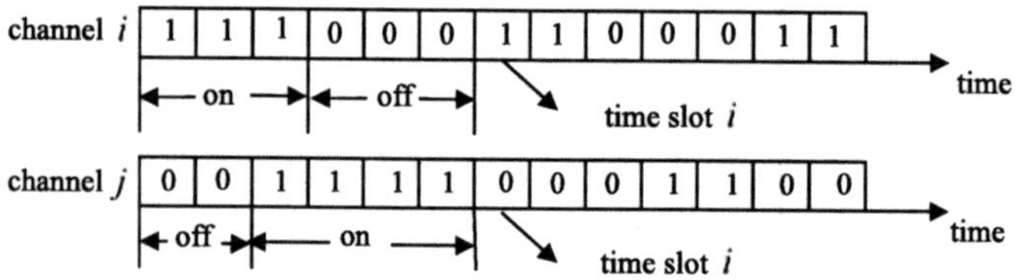

Fig. 2 The PU channel model

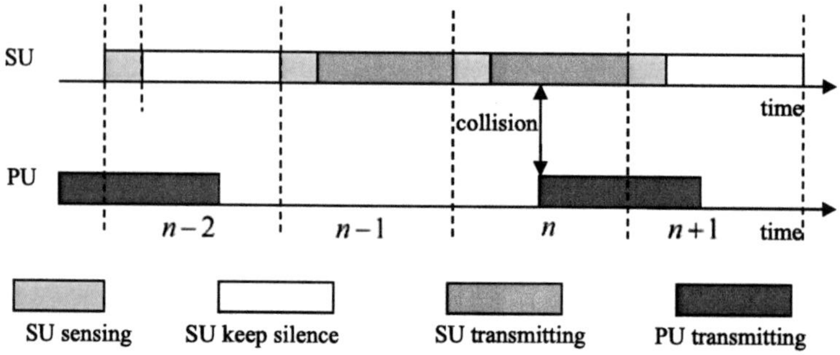

Fig. 3 The collision model

$$f_{i,1}(t_i) = \begin{cases} \lambda_i \exp(-\lambda_i t_i) & t_i \geq 0 \\ 0, & t_i < 0 \end{cases} \quad (\lambda_i > 0) \tag{1}$$

the busy state distribution function of channel i is:

$$f_{i,2}(t_i) = \begin{cases} \mu_i \exp(-\mu_i t_i) & t_i \geq 0 \\ 0, & t_i < 0 \end{cases} \quad (\mu_i > 0) \tag{2}$$

Thus, the average idle state duration time of channel i is:

$$T_{i,1} = \int_{-\infty}^{+\infty} t_i f_{i,1}(t_i) dt_i = \frac{1}{\lambda_i} \tag{3}$$

And the average busy state duration time of channel i is:

$$T_{i,2} = \int_{-\infty}^{+\infty} t_i f_{i,2}(t_i) dt_i = \frac{1}{\mu_i} \tag{4}$$

The availability of channel i is defined as the ratio of the average idle state duration time of channel i to the sum of the average idle state duration time and the busy state duration time of channel i. Then we have

$$U_i = \frac{T_{i,1}}{T_{i,1} + T_{i,2}} = \frac{1}{\lambda_i} / \left(\frac{1}{\lambda_i} + \frac{1}{\mu_i} \right) = \frac{\mu_i}{\lambda_i + \mu_i} \tag{5}$$

CRTCA selects the least used channel from the set of current channels to reduce the interference among SUs. But the algorithm does not consider the channel occupancy time of PU. The selected channels maybe frequently used by PU, i.e. the availability of channel U_i is relatively small. It will cause frequent channel switching of SUs, packet loss and the decrease of network throughput. However, channels with larger U_i will be selected frequently if we only consider the availability of channels and then not only the probability of the co-channel but also the interference among SUs will increase. So we have to make a tradeoff in these two aspects. Therefore, we propose an improved channel assignment algorithm (ICAA).

Let set C be the set of current available channels and E be the set of edges. $I_i(e, e')$ represents whether edges e, e' $(e, e' \in E)$ occupy channel i at the same time. $I_i(e, e') = 1$ only if the two edges e, e' are assigned the same channel i when one edge is in the interference scope of the other; under the other circumstance $I_i(e, e') = 0$.

The interference coefficient of channel i is defined as:

$$R_i = \frac{\sum\limits_{e,e' \in E} I_i(e, e')}{\sum\limits_{i \in C} \sum\limits_{e,e' \in E} I_i(e, e')} \tag{6}$$

The coefficient is updated constantly with the channel assignment process. For every channel $i \in C, R_i = 0$ in the beginning. Through formula (6), we can easily calculate $R_1 = 3/5$, $R_2 = 2/5$, $R_3 = 0$, $R_4 = 0$ respectively in the Fig. 1b.

Generally, the bigger the U_i, the more the channel is suitable for access and the smaller the R_i, the more the channel is suitable for access too. Integrating these two aspects, the channel access probability is defined as:

$$Access_i = \alpha U_i + (1 - \alpha)(1 - R_i) \tag{7}$$

where α is the channel access parameter. The selection of it is related to the specific data business. Some businesses require little interference among SUs, α is smaller by this time. The others require little interference between SU and PU, α is larger at the moment. The larger the $Access_i$, the more the channel is suitable for access.

Robust Topology Control Algorithm:

1: Randomly generate a topology $G = (V, E)$ within a limited area, and V is the set of nodes, E is the set of edges.

2: Compute $p(e)$(the total number of edges in the interference area of $e = (u, v)$) for each edge $e \in E$).

3: Sort E in the descending order of $p(e)$.

4: For each edge e in the sorted order of E do

5: Call the channel selection rule for e and add e to queue LQ.

6: while $LQ \neq \Phi$ do

7: $e' \leftarrow LQ.pop(e)$.

8: if connectivity test fails (whether there exists a different path between the two nodes of e)

9: then call the channel selection rule to assign a new different backup channel c.

10: end if

11: end while

12: end for

13: Assign nodes having extra radios with channels that are least used by their neighbors

ICAA Channel Selection Rule for $e = (u, v)$.

1: if $|A(u)| < Q(u)\&\&|A(v)| < Q(v)(A(u)(A(v))$ denotes the set of channels assigned to $u(v), Q(u)(Q(v))$ denotes the radio numbers of $u(v)$)then

2: $c \leftarrow$ the channel with the maximum $Access_i$ in C.

3: $A(u) \leftarrow A(u) \cup c$; $A(v) \leftarrow A(v) \cup c$.

4: end

5: if $|A(u)| < Q(u) \,|||A(v)| < Q(v)$ then

6: $c \leftarrow$ the channel with the maximum $Access_i$ in $A(u)$ { assume $|A(u)| = Q(u)$}

7: end.

8: if $A(u) \cap A(v) \neq \Phi$.

9: $c \leftarrow$ the channel with the maximum $Access_i$ in $A(u) \cap A(v)$.

10: end.

11: if $(|A(u)| = Q(u)\&\&|A(v)| = Q(v))\&\&(A(u) \cap A(v) \neq \Phi)$.

12: $c \leftarrow$ the channel with the maximum $Access_i$ in $A(u) \cup A(v)$.

13: $c' \leftarrow$ the channel with the minimum $Access_i$ in $A(v)$\{assume $c \in A(u)$\}

14: Adjust c' to c in $A(v)$.

15: for each previously assigned link $e' = (v, w)$ of which $c' \in A(e')$.

16: if $A(v) \cap A(w) = \Phi$.

17: push e' into LQ.

18: adjust c' to c in $A(w)$ {This may cause channel adjustment for links connecting to w, and the adjustment is a recursive process}

19: end if

20: end for

21: end if

3 Simulation Results

The basic assumptions:

1. PU occupies only one channel in a certain timeslot.
2. The radio of one SU occupies only one channel in a certain timeslot, and data reception and transmission cannot be performed simultaneously.
3. The source nodes and the destination nodes are selected randomly, and the route is selected from all the available links which has the minimum R_i.

Simulation parameters: 25 nodes are randomly placed in the $900 * 900$ m^2 simulation area. The transmission radius is $250m$ and the interference radius is 500 m,λ_i and μ_i are randomly selected between [3 and 20]s^{-1}. The channel updating round is $T = 100$ ms and the channel sensing time is $t = 10$ ms.The maximum transmission rate of each channel is set to $BMbps$. The number of radios of each node is set to Q. The bit stream is $F(F = \{f_1, f_2, \ldots f_k\})$ and the flow rate of each stream is 1 Mbps. If a node has multiple radios connected to the next node in the path, it dynamically forwards the packet using the channel that is least interfered by other transmissions. The measurement of the network performance is the average number of 1000 times randomly topology.

The measurement of the network performance:

1. The network throughput is defined as the total data sizes of all available links per unit time.
2. Interference index is defined as the sum of all co-channel interference coefficients in one stable network state, i.e. $\sum_{i \in C} \sum_{e,e' \in E} I_i(e, e')$. The index is normalized with the ratio of its maximum value.

As we can see from Fig. 4, within its transmission range, a node may has many neighbors, but the number of radios of each node are limited. This will cause the same channel assignment on many different links. The robustness of the system and the throughput of the network will increase when it is allocated more radios, but this will cost a lot. As it is proved in theorem 1 paper [3], there exists a channel assignment strategy satisfying the robustness constraint if each node has at least two radios. In this paper we assume that each SU has two radios.

Fig. 4 Random topology

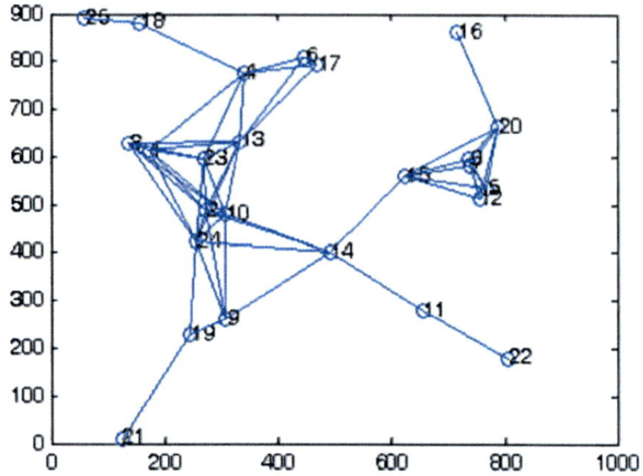

We make tradeoff between channel availability U_i and channel interference coefficient R_i. Here, we have four simulation diagrams with different values of α and we analyze the trend of curves in the diagrams.

As we can see from Fig. 5, the less number of channels will lead to the higher probability of the co-channel on different links. Then SUs' data transmission will be affected when PU accesses channels, and the probability of co-channel interference among SUs will also increase. This will result in the decrease of network throughput. With the increase of the number of channels, both of these two probabilities decrease. Hence, the network throughput is increased correspondingly. ICAA considers the channel occupancy time of PU. This method reduces the interference between PU and SU in the selection of channels. Thus it increases the network throughput to some extent. Besides, ICAA can reach the maximum throughput faster than CRTCA in different values of α.

As we can see from Figs. 6 and 7, with the increase of value of α, the collision rate is decreasing, but the interference index is increasing. The reason is that the proportion

Fig. 5 Throughput versus number of channels ($k = 2,\ Q = 2, B = 2$)

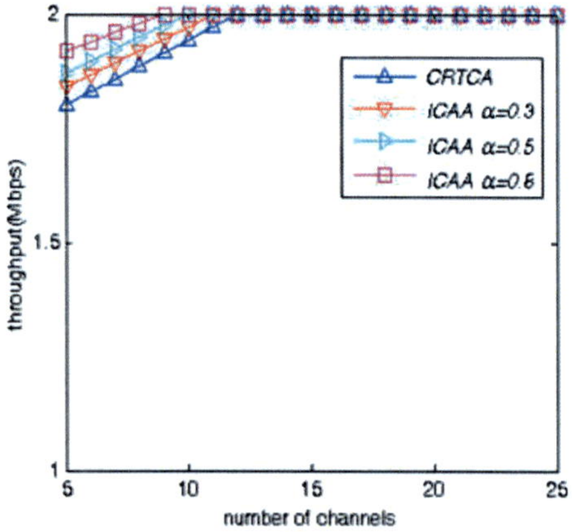

of the channel occupancy time in the channel access probability is increasing and the proportion of the interference index is decreasing. This will lead to the decrease in the collision rate between PU and SU and the increase of co-channel interference among SUs. With the increase in the number of channels, the collision rate and the interference index are decreasing and tend to be stable. The reason is that the increase of the available channel resource gives more choices for SUs to select channels. Therefore, the co-channel probabilities in the same interference zone will decrease, and then both the collision rate and the interference index will decrease too. The robust network topology can be achieved with several channels; too many channels will not lead to the change of the collision rate and the interference index. In the actual network environment, we can select the appropriate value of α

adaptively based on the specific network business.

In Fig. 8, with the increase of the number of bit streams, the throughput is almost linearly increasing. However, the interference among SUs is also increasing. Thus,

Fig. 6 Collision rate vs. number of channels $(k = 2,\ Q = 2, B = 2)$

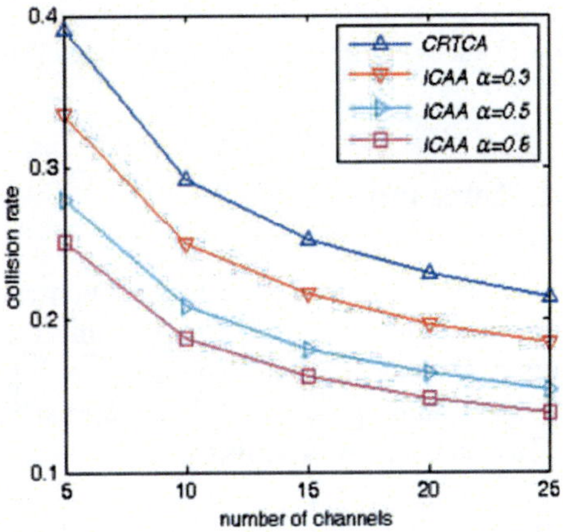

Fig. 7 Interference index versus number of channels $(k = 2,\ Q = 2, B = 2)$

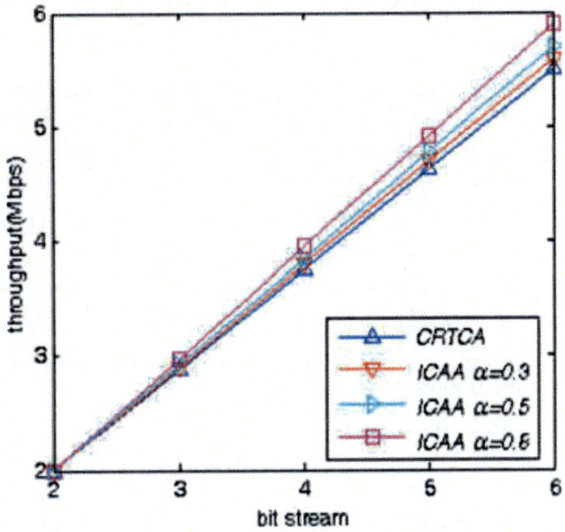

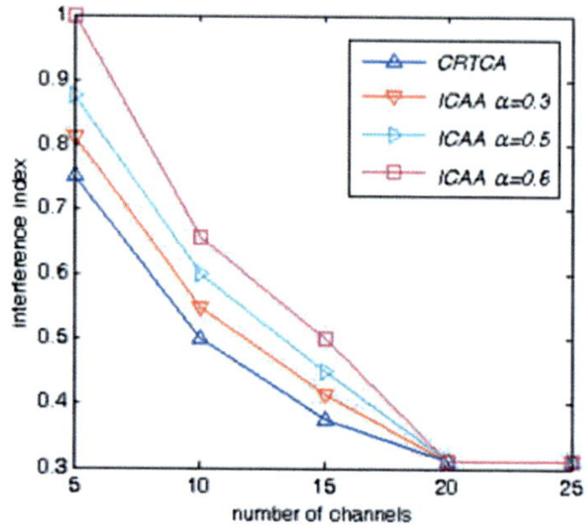

Fig. 8 Throughput versus bit streams ($|C| = 20$, $Q = 2$, $B = 2$)

the network throughput will be affected to some extent. R_i will be smaller when the channel occupancy time of PU increases in the same value of $Access_i$, and then the co-channel interference can be reduced. The network throughput will increase.

4 Summary

This paper studies the channel assignment issues in the multi-interface multi-hop cognitive radio network. By assuming that the channel occupancy time of PU is negative exponentially distributed, we make tradeoff between the two indicators U_i and R_i and propose ICAA. This approach is more than CRTCA in throughput and less than CRTCA in collision rate.

References

1. Song, M., Xin, C., Zhao, Y., Cheng, X.: Dynimic spectrum access: from cognitive radio to network radio. In: IEEE Wireless Communications (2012)
2. Liang, Y., Chen, K., Li, G.Y.: Petri Mähönen. Cognitive radio networking and communications: an overview. IEEE Trans. Veh. Technol. **60**(7), 3386–3402 (2011)
3. Zhao, J., Cao, G.: Robust topology control in muti-hop cognitive radio network. In: Proceedings of IEEE Infocom, Orlando (2012)
4. Ma, Z., Cao, Z.: The effective throughput optimization in the distributed cognitive radio networks[J]. J. Tsinghua Univ. **48**(4), 506–509 (2008)
5. Salameh, H.B.: Rate-maximization channel assignment scheme for cognitive radio network. In: IEEE Globecom (2010)
6. Feng, W., Cao, J., Zhang, C., liu, C.: Joint optimization of spectrum handoff scheduling and routing in mutihop muti-radio cognitive networks. In: IEEE ICDCS (2009)
7. Komali, R.S., Thomas, R.W., DaSilva, L.A., MacKenzie, A.B.: The price of ignorance: distributed topology control in cognitive networks[J]. IEEE Trans. Wireless Commun. **9**(4), 1434–1445 (2010)

8. Guo, C., Zeng, Z., Feng, C., Liu, F.F.: A new active spectrum selection algorithm in the cognitive radio networks. J. Xi'dian Univ. **35**(6), 1–6 (2008)
9. Tang, J., Xue, G., Zhang, W.: Interference-aware topology control and qos routing in muti-channel wireless mesh network. In: ACM MobiHoc (2005)

LFPS-HSDN: Link Failure Protection Scheme in Hybrid SDNs

Jiahui Li, Xiaogang Qi, Haoran Zhang, and Lifang Liu

Abstract Software-Defined Networks (SDNs) decouple the control plane from the data plane of forwarding devices, which can significantly improve network functions, including failure protection. In the current Internet architecture, legacy networks are gradually migrated to pure SDNs to provide flexible services, which form the hybrid SDNs. Then how to select the fewest legacy devices to upgrade to SDN switches for protecting all failed links and how to select appropriate protection paths have become the urgent problems to be tackled. In this paper, we consider the failure protection difficulty of links and the protection capability of nodes jointly, and propose the FP-SCS approach for SDN Candidate Selection (SCS) to deploy the fewest SDN switches. Based on these SDN switches, we further propose the FP-PPS algorithm for Protection Path Selection (PPS), which can flexibly select the most appropriate protection paths according to the respective importance of the interactive path parameters. The simulation results show that our proposed approaches have excellent performance compared with the state-of-the-art approaches.

1 Introduction

Software-Defined Networks (SDNs) have developed as a new networking paradigm to improve network functions, including flexible routing, traffic engineering and access control, etc. [1, 2]. Due to significant time and considerable financial requirements, organizations are unwilling to install SDN infrastructure from scratch. Thus, hybrid SDNs are derived that coexist with SDN switches and legacy routers [3, 4].

J. Li · X. Qi (✉) · H. Zhang
School of Mathematics and Statistics, Xidian University, Xi'an 710071, China
e-mail: xgqi@xidian.edu.cn

L. Liu
School of Computer Science and Technology, Xidian University, Xi'an 710071, China

K. Nakamatsu et al. (eds.), *Advanced Intelligent Technologies for Information and Communication*, Smart Innovation, Systems and Technologies 365,
https://doi.org/10.1007/978-981-99-5203-8_19

Rapid failure protection for hybrid SDNs can effectively improve the resilience of networks [5]. The following two goals are required to achieve during failure protection by deploying SDN switches in legacy networks: (1) deploying the fewest SDN switches that can protect all failed links, i.e. SDN Candidate Selection (SCS); (2) selecting appropriate protection paths to guarantee the network performance, i.e. Protection Path Selection (PPS).

The existing researches of failure protection for hybrid SDNs have conducted extensive investigations on SCS and PPS [6–8]. However, some deficiencies still exist and need to be urgently improved. Specifically, the existing researches only focus on one of the failure protection capability of nodes and the protection difficulty of links for SCS, and fail to consider these factors comprehensively. Moreover, these researches also have limitations on PPS, which are mainly manifested in the fact that the Protection Path Length (PPL) and the Maximum Link Utilization (MLU) of paths are not flexibly considered as mutually influencing path parameters. As a result, the selected protection paths cannot fully satisfy the actual requirements of users.

Observing the above deficiencies, we propose an improved protection scheme of link failure on SCS and PPS in hybrid SDNs. Our contributions are summarized as follows.

(1) Considering the failure protection difficulty of links and the protection capability of nodes jointly, we propose the FP-SCS approach for SCS that can select the fewest SDN switches to protect all links.
(2) Based on the SDN switches selected by FP-SCS, we further propose the FP-PPS algorithm with the destination-based tunneling mechanism for PPS. FP-PPS fully considers the interactive path parameters such as PPL and MLU, and can flexibly select the most appropriate protection paths according to the respective importance of these parameters.
(3) The performance of our proposed FP-SCS and FP-PPS approaches is evaluated by simulation experiments. Moreover, the conducted comparison experiments demonstrate that our proposed approaches outperform the state-of-the-art.

2 Related Work

Failure protection is an important component of network management, and many researches focus on SCS and PPS of failure protection in hybrid SDNs.

Chu et al. [6] leveraged Greedy strategies to sequentially select the nodes with the maximum protection capability to deploy SDN switches for protecting all links. Moreover, they further proposed Greedy-LB to select the protection paths for achieving load balancing, which focuses on selecting the paths with minimum MLU by the link-based tunneling mechanism. However, Greedy does not involve the failure protection difficulty of links, and Greedy-LB's path selection lacks flexibility and consideration of path length. In order to improve these defects, Yang et al. [7] suggested that when multiple nodes with the maximum protection capability exist, the nodes with shorter average protection path length are preferentially selected to upgrade to SDN switches. For PPS, they proposed the RPR-DT algorithm with the destination-based tunneling mechanism to minimize PPL, and RPR-LB algorithm that relaxes the requirements for PPL to minimize MLU. According to the failure protection difficulty of links, Li et al. [8] leveraged the search tree to find all feasible SDN candidate schemes and chose the one with the highest reliability degree. Considering the path length and load balancing, the SDN switch with largest utility will be chosen for each link. However, when the link fails, it can only be protected with the chosen SDN switch.

Inspired by the above concerns, we jointly consider the failure protection difficulty of links and the protection capability of nodes to propose the FP-SCS approach for SCS and FP-PPS algorithms for PPS.

3 Model Formulation

The network topology can be denoted by a directed graph $G = (V, E)$, where V is a node set and E is a link set. The bidirectional link between nodes i and j includes two directed links, $e_1 =< i, j >$ and $e_2 =< j, i >$. In addition, we assume that the network topology is a two-edge connected graph to ensure the existence of protection paths. When link e fails, the paths that cannot transmit packets are affected paths, which require to be replaced by alternative protection paths without passing through e (Table 1).

We establish the optimization model (1)–(4) to determine the minimum required SDN switches for SCS. The symbols used in the model are presented in Table 1. Moreover, the requirements of our model are as follows:

(1) At least one designated SDN switch can be found for the end router of the failed link e to transmit packets, as shown in Eq. (2).
(2) Only one designated SDN switch is used when e fails, as shown in Eq. (3).
(3) The nodes selected as designated SDN switches must have been upgraded from legacy routers, as shown in Eq. (4).

Table 1 Symbol descriptions

Symbols	Descriptions
ρ_x^e	Binary, $\rho_x^e = 1$ if node x is the end router of link e; $\rho_x^e = 0$, otherwise
$\gamma_{s,d}^e$	Binary, $\gamma_{s,d}^e = 1$ if the path from node s to d includes link e; $\gamma_{s,d}^e = 0$, otherwise
$\psi_{x,k}^e$	Binary, $\psi_{x,k}^e = 1$ if node k is selected as the designated SDN switch of node x when link e fails; $\psi_{x,k}^e = 0$, otherwise
$N_{k,m}^e$	Binary, $N_{k,m}^e = 1$ if node m is a one-hop neighbor of node k; $N_{k,m}^e = 0$, otherwise
u_k	Binary, $u_k = 1$ if node k is selected and upgraded to an SDN switch; $u_k = 0$, otherwise

Minimize:

$$\sum_k u_k \tag{1}$$

Subject to:

$$\sum_k \psi^e_{x,k}(1-\gamma^e_{x,k})\left[\sum_m N_{k,m}(1-\gamma^e_{k,m})(1-\gamma^e_{m,d})\right] \geq \rho^e_x \gamma^e_{x,d} \tag{2}$$

$$\sum_k \psi^e_{x,k} \leq \rho^e_x \tag{3}$$

$$\psi^e_{x,k} \leq u_k \tag{4}$$

Similarly, an optimization model can also be constructed for the PPS problem under the objective and constraints of optimizing path parameters [6].

4 Failure Protection Scheme for Hybrid SDNs

In this section, we describe our proposed failure protection scheme for hybrid SDNs in detail, including FP-SCS for SCS and FP-PPS for PPS.

4.1 SDN Candidate Selection Approach

Considering the deployment overhead and device downtime during the migration to pure SDNs, we focus on minimizing the SDN switches required to protect all links. The framework of our proposed FP-SCS approach for SCS is shown in Fig. 1.

For the construction of SDN candidate table **tSDN**, the purpose is to identify the locations of eligible candidate SDN switches for each link *e*. If node δ_τ can protect *e*, **tSDN**(e, τ) is marked as '1'; otherwise, it is marked as '0'. Based on **tSDN**, the nodes are directly selected and added into ***sSDN***, which are responsible for protecting the links with the highest protection difficulty that can only be protected by one node. Then, the links that have been protected by ***sSDN*** are removed from the SDN candidate table to update the table. Further, the candidate set of nodes upgraded to SDN switches will be narrowed to ***csSDN*** according to the unprotected links with the highest failure protection difficulty. Borrowing from the greedy strategy, FP-SCS then selects the node in ***csSDN*** with the maximum protection capability for unprotected links and adds it into ***sSDN***. Until the selected nodes ***sSDN*** can protect all links, FP-SCS approach terminates.

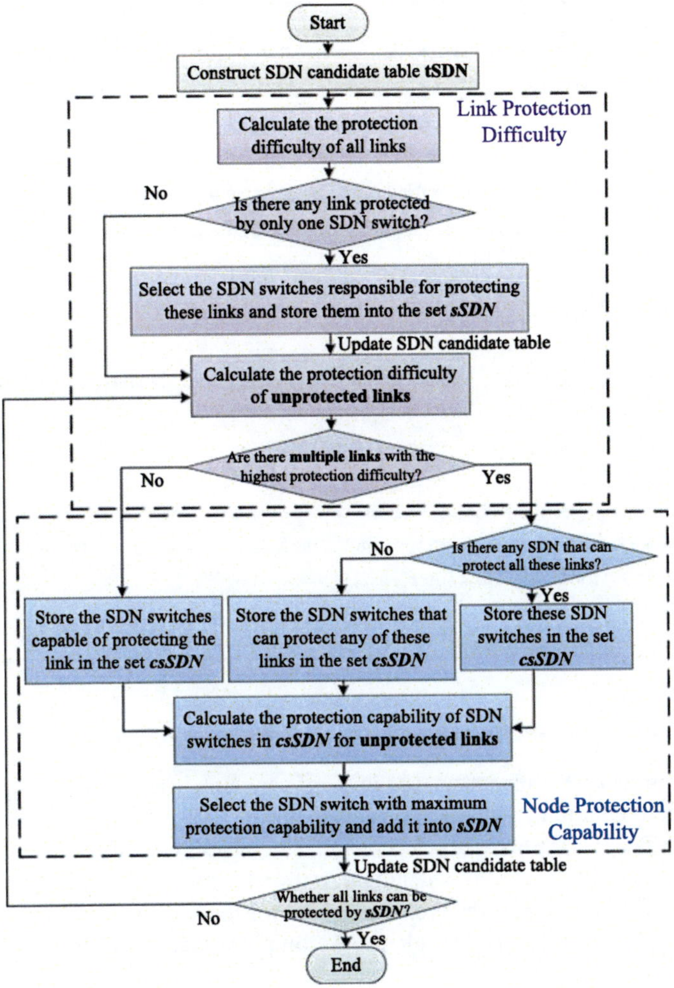

Fig. 1 FP-SCS approach framework

4.2 Protection Path Selection Algorithm

After the fewest required SDN switches are determined by FP-SCS, we focus on leveraging the routing flexibility of these switches to further select the protection paths. Specifically, we propose the FP-PPS algorithm with the destination-based tunneling mechanism for PPS. FP-PPS fully considers the interactive path parameters such as PPL and MLU, and aims to flexibly select the most appropriate protection paths according to the respective importance of these parameters.

FP-PPS Algorithm.

Input: $G=(V,E)$, SDN switch set $\Delta=\{\delta_1,\cdots,\delta_{|\Delta|}\}$, traffic matrix T_N

Output: Protection paths $P=\{p_{sd}^{e'}\}$ for all affected paths

1: **for each** failed link $e\in E$ **do**

2: Find affected path set $P_{aff}^e=\{p_{sd}^e\}$, $\forall s\in V, d\in V\setminus\{s\}$ and p_{sd}^e is the affected path from s to d

3: Based on T_N, monitor all link traffic T_L after transmitting the traffic on unaffected paths

4: Sort the paths of P_{aff}^e in descending order of the traffic that requires to be tunneled to form $P_{aff,sort}^e$

5: **for each** path $p_{sd}^e\in P_{aff,sort}^e$ **do**

6: **for each** SDN switch $\delta_i\in\Delta$ **do**

7: Construct candidate protection path $pc_{sd}^{e'}=p_{s\delta_i}\cup p_{\delta_i m_r}\cup p_{m_r d}$, where m_r is the neighbor of δ_i

8: Calculate the path score by Eq. (5)

9: **end for**

10: Select the protection paths $p_{sd}^{e'}$ with the lowest score. If there is a tie, the winner is the paths with the smaller parameter corresponding to the larger of the weights α and β.

11: Update the traffic of all links on $p_{sd}^{e'}$ in T_L

12: **end for**

13: **end for**

The details of the FP-PPS Algorithm are described as follows.

- On the basis of T_N, the traffic data in T_L of each link is updated after the normal traffic transmission on unaffected paths in line 3.
- Considering that the order of selecting protection paths for different affected paths affects the overall traffic distribution of networks, line 4 sorts the affected paths and successively selects the protection paths for these paths.
- Lines 7–8 construct all available protection paths and calculate the path score according to Eq. (5).

$$ps_{sd}^{e'}=\alpha\times l_{sd}^{e'*}+\beta\times U_{sd}^{e'*} \tag{5}$$

where α and β are variable weights set according to the respective importance of PPL and MLU by the variable-weight Multi-Attribute Decision-Making (vm-MAMD) [9, 10], and $\alpha+\beta=1$. Moreover, $l_{sd}^{e'*}$ and $U_{sd}^{e'*}$ are the normalized results of the protection path length $l_{sd}^{e'}$ and the maximum link utilization $U_{sd}^{e'}$ of paths from s to d [11], which are in the same order of magnitude.

- Line 9 selects the protection paths with the lowest path score. If there is a tie and α is larger, it means that PPL is dominant to user requirements, and the winner is the protection path with smaller PPL. Similarly, if β is larger, the winner is the protection path with lower MLU.

5 Performance Evaluation

To evaluate the performance of our proposed FP-SCS and FP-PPS, we conduct extensive experiments under various network topologies and parameters in this section.

5.1 Simulation Setup

For the number of required SDN switches for SCS, we compare FP-SCS with Greedy [6], SCS-ST [7] and Search-tree [8]. Moreover, FP-PPS is compared with Greedy-LB [6], RPR-DT [7], RPR-LB [7] and MASSS [8] in terms of path parameters for PPS.

We adopt Python at Windows 7 operating system with 3.60GHz Intel Core i7 CPU to verify the effectiveness of our scheme. The used real network topologies are Internet2 (I2), Abilene (Ab), US, Germany(GM), 30-node scale-free (SF(30)) networks [12], and ER random networks (ER(0.1)) with 50 nodes and node connection probability of 0.1 [13].

Moreover, the traffic in the adopted random traffic matrices from node i to node j is uniformly distributed in [1,100] Mbytes/second when node $i \neq j$; otherwise, the traffic is 0.

5.2 Number of SDN Switches

To be specific, we study the performance of our proposed FP-SCS in the networks with various network topologies and sizes (Tables 2 and 3).

Adaptability of Network Topologies. The number of required SDN switches and the running time of SCS for various network topologies are shown in Tables 2 and 3. Among the four algorithms, FP-SCS is able to select the fewest SDN switches for all network topologies, and only needs 0.0028–0.1045s more than Greedy. However, it takes 0.0047–0.3842s and 0.0323–0.476s less than SCS-ST and Search-tree, respectively.

Table 2 Number of SDN switches for SDN candidate selection

Scheme topology	FP-SCS	Greedy	SCS-ST	Search-tree
I2	5	5	5	5
Ab	4	6	4	4
US	3	3	3	3
GM	7	7	7	7
SF (30)	3.2	3.4	3.4	3.2
ER (0.1)	3	3.3	3.3	3

Table 3 Running time for SDN candidate selection (s)

Scheme topology	FP-SCS	Greedy	SCS-ST	Search-tree
I2	0.0660	0.0632	0.0707	0.0983
Ab	0.0720	0.0671	0.0762	0.1102
US	0.1736	0.1688	0.1809	0.2535
GM	0.2604	0.2499	0.3448	0.3932
SF (30)	0.7219	0.7047	0.7556	0.9962
ER (0.1)	1.3999	1.2954	1.7841	1.8759

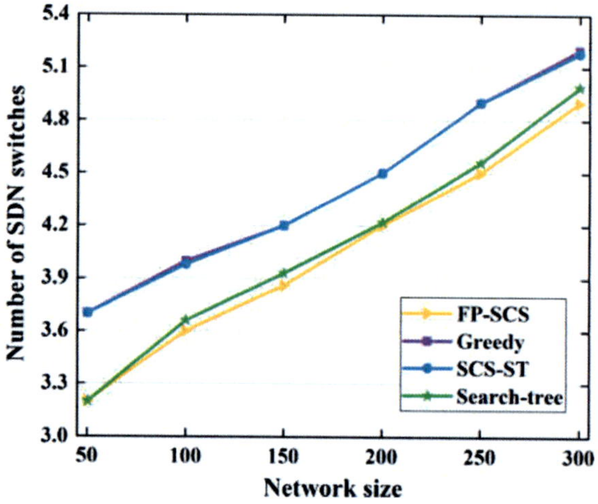

Fig. 2 Impact of network size on SDN candidate selection

Impact of Network Size. The impact of network size on the required SDN switches is illustrated in Fig. 2. As the network size expands, the required SDN switches gradually increase. Moreover, FP-SCS takes 9.62, 9.46 and 1.19% fewer SDN switches on average than Greedy, SCS-ST and Search-tree, respectively.

5.3 *Average Protection Path Length and Maximum Link Utilization*

We explore the suitability of protection paths obtained by FP-PPS in this section. Two evaluation metrics are adopted, Average Protection Path Length (*APPL*) and Average Maximum Link Utilization (*AMLU*), which are evaluated following Eqs. (6) and (7).

$$APPL = \frac{\sum_{e \in E} \frac{\sum_{s,d \in V} l_{sd}^{e'}}{\varepsilon_e}}{2|E|} \tag{6}$$

$$AMLU = \frac{\sum_{e \in E} MLU_e}{2|E|} \tag{7}$$

where ε_e is the number of paths affected by the failed link e, $l_{sd}^{e'}$ is the length of the protection path from s to d and MLU_e is the MLU of networks after protecting e.

The comparison results are shown in Fig. 3. When PPL is used as the only evaluation basis, the *APPL* of FP-PPS ($\alpha = 1,\ \beta = 0$) and RPR-DT is similar, but the *AMLU* of FP-PPS is 0.34–6.38% lower than that of RPR-DT. According to the MLU of protection paths, the *AMLU* of FP-PPS ($\alpha = 0,\ \beta = 1$) is slightly lower than that of Greedy-LB and RPR-LB by 0.13–4.83% and 0.11–7.74%. Moreover, compared with these two algorithms, the *APPL* of FP-PPS has significant reduction of 1.33–24.4% and 0.14–6.13%, respectively.

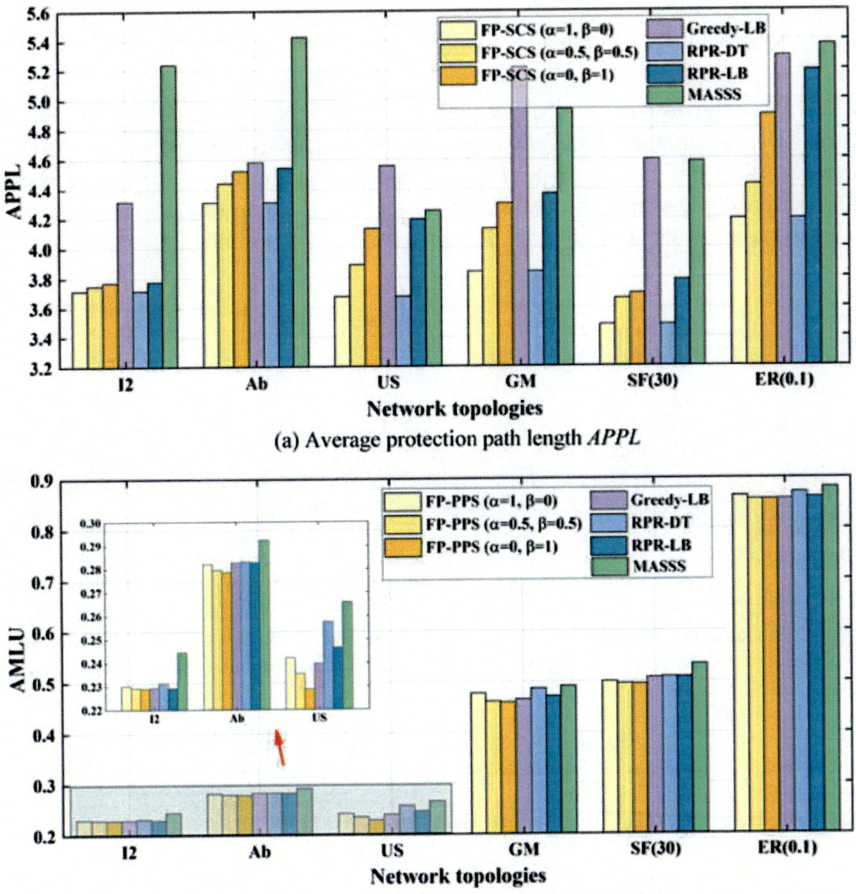

(a) Average protection path length *APPL*

(b) Average maximum link utilization *AMLU*

Fig. 3 Comparisons of the suitability of protection paths

Considering PPL and MLU jointly, we take the same importance of these two parameters ($\alpha = 0.5$, $\beta = 0.5$) as a specific case, and compare it with MASSS that also considers these two parameters concurrently. MASSS adopts the link-based tunneling mechanism to construct candidate paths, which is independent from the affected destination and causes the limitation of candidate protection paths. Different from MASSS, our FP-PPS adopts the destination-based mechanism and can dynamically adjust the weights to promote the selection of more appropriate protection paths. Hence, compared with MASSS, FP-PPS reduces $APPL$ by 9.34–39.78% and $AMLU$ by 2.72–13.03%.

6 Conclusion

Aiming at the SCS and PPS issues encountered in failure protection of hybrid SDNs, we propose the FP-SCS approach for SCS and FP-PPS algorithm for PPS. Considering the protection difficulty of links and the protection capability of nodes jointly, FP-SCS can rapidly select the fewest SDN switches for protecting all links. Based on these SDN switches, FP-PPS fully considers the interactive path parameters and can flexibly select the most appropriate protection paths. Extensive experiments show that our proposed scheme outperforms the comparison schemes. In future research, we will promote our scheme for wider applications, such as node failure protection.

References

1. Seddiqi, H., Babaie, S.: A new protection-based approach for link failure management of software-defined networks. IEEE Trans. Netw. Sci. Eng. **8**(4), 3303–3312 (2021)
2. Rasool, Z. I., Abd Ali, R. S., Abdulzahra, M.M.: Network management in software-defined network: A survey. In: Materials Science and Engineering in IOP Conference Series, pp. 012055. IOP Publishing (2021)
3. Huang, X., Cheng, S., Cao, K., Cong, P., Wei, T., Hu, S.: A survey of deployment solutions and optimization strategies for hybrid SDN networks. IEEE Commun. Surv. & Tutor. **21**(2), 1483–1507 (2019)
4. Khorsandroo, S., Sánchez, A.G., Tosun, A.S., Arco, J.M., Doriguzzi-Corin, R.: Hybrid SDN evolution: A comprehensive survey of the state-of-the-art. Comput. Netw. **192**, 107981 (2021)
5. khan, N.: Failure recovery of data plane in SDN. TechRxiv. Preprint. https://doi.org/10.36227/techrxiv.19336481.v1. Last accessed 2022/06
6. Chu, C. Y., Xi, K., Luo, M., Chao, H. J.: Congestion-aware single link failure recovery in hybrid SDN networks. In: 34th IEEE Conference on computer communications (INFOCOM), pp. 1086–1094. IEEE (2015)
7. Yang, Z., Yeung, K.L.: SDN candidate selection in hybrid IP/SDN networks for single link failure protection. IEEE/ACM Trans. Networking **28**(1), 312–321 (2020)
8. Li, N., Shi, Y., Zhang, Z., Martinez, J. F., Yuan, X.: Search-tree based SDN candidate selection in hybrid IP/SDN network. In: 28th IEEE International conference on network protocols (ICNP), pp. 1–6. IEEE (2020).

9. Li, N., Yuan, X., Martinez, J. F., Zhang, Z.: The variable-weight MADM algorithm for wireless network. In: 5th International workshop on systems and network telemetry and analytics (SNTA), pp. 51–55 (2022).
10. Liu, S., Li, L.: Learning general temporal point processes based on dynamic weight generation. Appl. Intell. **52**(4), 3678–3690 (2022)
11. Prasetyowati, S.A.D., Ismail, M., Budisusila, E.N., Purnomo, M.H.: Dataset feasibility analysis method based on enhanced adaptive LMS method with Min-max normalization and fuzzy intuitive sets. Int J Elect Eng Infor. **14**(1), 55–75 (2022)
12. Knight, S., Nguyen, H.X., Falkner, N., Bowden, R., Roughan, M.: The internet topology zoo. IEEE J. Sel. Areas Commun. **29**(9), 1765–1775 (2011)
13. Zacharis, A., Margariti, S. V., Stergiou, E., Angelis, C.: Performance evaluation of topology discovery protocols in software defined networks. In: 7th IEEE conference on network function virtualization and software defined networks (NFV-SDN), pp. 135–140. IEEE (2021)

Fractional Gradient Descent Algorithm for Nonlinear Additive Systems Using Weierstrass Approximation Method

Yingjiao Rong, Fei Peng, Rongqi Lv, and Shanshan Li

Abstract In this study, a fractional gradient descent (FGD) algorithm for nonlinear additive systems with unknown structure is proposed. The nonlinear model is turned into a linear-parameter model using the Weierstrass approximation approach. Then, a FGD algorithm is created to estimate all of the system's unknown parameters. A preconditioned matrix approach is also presented to improve the FGD algorithm's convergence rate. To demonstrate the effectiveness of the proposed algorithms, simulation examples are provided.

1 Introduction

Additive system, as a special kind of nonlinear systems, has both nonlinear input and output. The additive system are widely existed in engineering practices [1, 2]. For the nonlinear system identification, there are plenty of identification algorithms [3–8]. However, it has quite heavy computational efforts. It is the last and unwilling choice when the GD and LS algorithms are inefficient for the nonlinear systems with complex structures.

The LS technique offers quicker convergence rates, but it requires an iterative function to be solved [9, 10]. The iterative function may not have analytical answers if the investigated model has a complicated structure. The GD algorithm is an excellent alternative for resolving this quandary. Its fundamental concept is to update the parameters depending on a direction and a step size [5, 11]. A method of this type does not need solving an iterative function and requires less computational work. Despite the fact that the GD algorithm has significant benefits over the LS and intelligent algorithms. Because of its zigzagging structure, it has a somewhat sluggish convergence rate [12, 13]. In this research, we want to create a modified GD method known as the fractional gradient descent (FGD) algorithm. This nonlinear system, shown in Fig. 1, is written by

Y. Rong · F. Peng (✉) · R. Lv · S. Li
Science and Technology on Near-Surface Detection Laboratory, Wuxi 214028, China
e-mail: 38349735@qq.com

K. Nakamatsu et al. (eds.), *Advanced Intelligent Technologies for Information and Communication*, Smart Innovation, Systems and Technologies 365,
https://doi.org/10.1007/978-981-99-5203-8_20

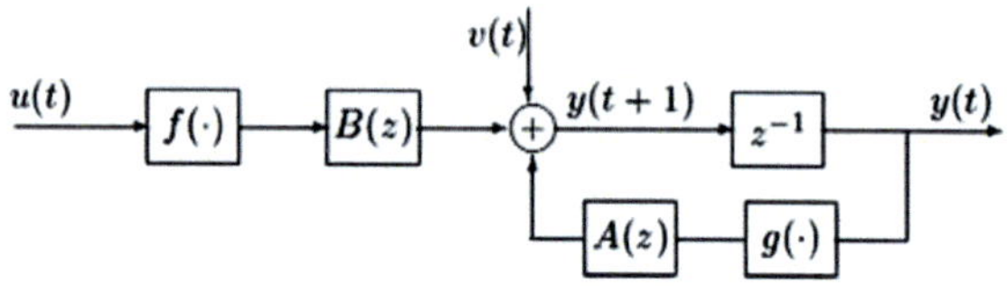

Fig. 1 The additive nonlinear system

$$y(t+1) = B(z)f(u(t)) + A(z)g(y(t)) + v(t) \tag{1}$$

We might refer to a nonlinear model as a black-box model when its structure is unclear [14, 15]. A linear-parameter model may be used to estimate this sort of model using the Weierstrass approximation approach [16, 17]. The Weierstrass technique, like the Taylor series, is a linear-parameter model with nonlinear behavior. It has a simpler structure than the Taylor series and may thus be widely employed in modern enterprises. The Weierstrass approximation approach, on the other hand, results in the curse of dimensionality. It is widely recognized that identifying a large-scale system is more challenging, for example, due to substantial computing costs and sluggish convergence rates [18, 19].

The FGD method is a modified GD algorithm that has a higher rate of convergence than the GD algorithm [20, 21]. It features an additional fractional gradient direction when compared to the GD algorithm. The GD algorithm's convergence rates are boosted with the aid of the extra direction. The FGD algorithm has become a prominent topic in the recent decade due to its benefits. For example, a two-stage FGD technique for a linear system is described, in which the linear system is split into two pieces and these two portions are identified repeatedly [22]. Naveed et al. proposed two FGD methods for signal modeling that require less computing effort [23]. Xu et al. proposed momentum-based FGD and adaptive-based FGD methods for time-delayed models, which have fast convergence rates and require low computing effort [24]. However, the above work assumed that the models are linear and have known structures.

An FGD approach for a nonlinear additive model with uncertain structure is proposed. The model is turned into a linear-parameter model with known structure using the Weierstrass approximation approach. The FGD approach is then used to update the linear-parameter model's parameters. A preconditioned matrix-based FGD (PM-FGD) method is also created to improve convergence rates. In comparison to typical algorithms, the approach described in this study makes the following contributions: (1) the approach provides faster convergence rates; (2) the model under consideration has an uncertain structure.

2 The Nonlinear Additive Model with Unknown Structure

Rewrite the system as

$$y(t+1) = B(z)f(u(t)) + A(z)g(y(t)) + v(t) \tag{2}$$

The FGD approach cannot be used to directly identify this system. The Weierstrass approximation theorem is introduced to help with this challenge:

Theorem of Weierstrass approximation: There is an algebraic polynomial p such that,

$$p(x) = a1x + a2x2 + a3x3 + \cdots + anxn. \tag{3}$$

Weierstrass approximation theorem demonstrates that the nonlinear input can be roughly represented by:

$$f(u(t)) = c1u(t) + c2u2(t) + c3u3(t) + \cdots + cpup(t).$$

When p is sufficiently large, $c1u(t) + c2u2(t) + c3u3(t) + \cdots + cpup(t)$ is close to $f(u(t))$, and the nonlinear output $g(y(t))$ is written by

$$g(y(t)) = g1y(t) + g2y2(t) + g3y3(t) + \cdots + gqyq(t).$$

Thus,

$$\begin{aligned} y(t+1) = {} & B(z)(c1u(t) + c2u2(t) + c3u3(t) + \cdots + cpup(t)) \\ & + A(z)(g1y(t) + g2y2(t) + g3y3(t) + \cdots + gqyq(t)) \\ & + v(t). \end{aligned} \tag{4}$$

Equation (4) shows that the additive model is turned to a linear-parameter model. Define

$$\begin{aligned} \theta := {} & [algl, \ldots, a1gp, a2g1, \ldots, a2gp, \ldots, ang1, \ldots, angp, \\ & b1c1, \ldots, b1cp, b2c1, b2c2, \ldots, b2cp, \ldots, bnc1, bnc2, \ldots, bncp] \\ & \mathrm{T} \in \mathrm{Rnp + nq}, \end{aligned} \tag{5}$$

$$\begin{aligned} \varphi(t) := {} & [y(t), \ldots, yq(t), y(t-1), \ldots, y(t-n+1) \\ & , \ldots, yq(t-n+1), u(t-1), u2(t-n), \ldots, u(t-2), u2(t-2) \\ & , \ldots, up(t-2), \ldots, u(t-n), u2(t-n), \ldots, up(t-n)] \\ & \mathrm{T} \in \mathrm{Rnp + nq} \end{aligned} \tag{6}$$

Then, we have

$$y(t) = \varphi \mathrm{T}(t)\theta + v(t). \tag{7}$$

Without losing generality, assume that the first coefficients b_1 and a_1 equal 1, i.e., $b_1 = 1$ and $a_1 = 1$.

Define

$$\begin{aligned}
&\theta := [g1, \ldots, gq, a2g1, \ldots, a2gq, \ldots, ang1, \ldots, angq, c1, c2 \\
&, \ldots, cp, b2c1, b2c2, \ldots, b2cp, \ldots, bnc1, bnc2, \ldots, bncp]T, \\
&\varphi(t) := [y(t), \ldots, yq(t), y(t-1), \ldots, yq(t-1) \\
&, \ldots, y(t-n+1), \ldots, yq(t-n+1), u(t-1), u2(t-1) \\
&, \ldots, up(t-1), \ldots, u(t-2), u2(t-2), \ldots, up(t-2) \\
&, \ldots, u(t-n), u2(t-n), \ldots, up(t-n)]T.
\end{aligned} \tag{8}$$

Therefore, the linear-parameter model is listed as follows:

$$Y(L) = \Phi(L)\theta + V(L). \tag{9}$$

where

$$\begin{aligned}
&Y(L) := [y(L), \ldots, y(1)]\mathrm{T} \in RL, \\
&\Phi(L) := [\varphi(L), \ldots, \phi(1)]\mathrm{T} \in RL \times (nq + np), \\
&V(L) := [v(L), \ldots, V(1)]\mathrm{T} \in RL.
\end{aligned}$$

3 The FGD Algorithm

3.1 GD Algorithm—Review

Define

$$J(\theta) = \frac{1}{2}[Y(L) - \Phi(L)\theta]^{\mathrm{T}}[Y(L) - \Phi(L)\theta] \tag{10}$$

The direction is set as

$$d_k = \Phi^{\mathrm{T}}(L)[Y(L) - \Phi(L)\theta_k - 1],$$

while the step size is computed by

$$0 < \gamma < \frac{2}{\lambda T_{\max}}$$

These two elements are crucial to the GD algorithm. One should try to identify a better direction, such as the conjugate GD method [25], or try to compute an appropriate step size, such as the forgetting factor GD algorithm [26].

Remark 1

Because it only employs the first-order Taylor expansion to get an approximation of the cost function, the GD method has a somewhat sluggish convergence rate. The Newton technique approximates the cost function using the second-order Taylor expansion, in contrast to the GD algorithm, and as a result, the Newton method converges more quickly than the GD algorithm. However, each iteration of the Newton approach requires computing the matrix inverse, which is time-consuming.

Can we create an nth-order Taylor expansion to get an approximation of the cost function, where $1 < n < 2$? This is a natural question. Compared to the GD algorithm, this technique converges more quickly and requires less computer work than the Newton method.

3.2 *Fractional Derivative*

The fractional derivative method can obtain an *nth*-order Taylor expansion. To help the readers to follow the paper, three traditional fractional derivative methods are introduced [27–29]:

1: RL fractional derivative method

$$\begin{aligned} {}^{RL}_{\alpha}D^{\alpha}_{x} f(x) &\triangleq D^{n}\left({}_{\alpha}D^{\alpha-n}_{x} f(x)\right) \\ &= \frac{1}{\Gamma(n-\alpha)} \frac{d^{n}}{dx^{n}} \int_{\alpha}^{x} \frac{f(t)}{(x-t)^{\alpha-n+1}} dt, \end{aligned} \tag{11}$$

where $\Gamma(\alpha) = \alpha \times \Gamma(\alpha - 1)$.

2: CAP fractional derivative method

$${}^{C}_{a}D^{\alpha}_{x} f(x) = \frac{1}{\Gamma(n-\alpha)} \int_{a}^{x} \frac{f^{(n)}(t)}{(x-t)^{\alpha+1-n}} dt. \tag{12}$$

3: GL fractional derivative method

$$
\begin{aligned}
& {}_{a}^{GL}D_{x}^{\alpha} f(x) \\
& \lim_{h \to 0} \frac{1}{n^{\alpha}} \sum_{k=0}^{\left[\frac{x-\alpha}{h}\right]} \frac{(-1)^{k}\Gamma(\alpha+1)}{\Gamma(k+1)\Gamma(\alpha-k+1)} f(x-kh) \\
& = \lim_{h \to 0} \frac{1}{n^{\alpha}} \sum_{k=0}^{\left[\frac{x-\alpha}{h}\right]} \frac{\Gamma(k-\alpha)}{\Gamma(k+1)\Gamma(-\alpha)} f(x-kh)
\end{aligned} \tag{13}
$$

Take $f(x) = x$ for example, the α-order fractional derivative of $f(x)$ is written by

$$
f^{\alpha}(x) = \frac{\Gamma(2)}{\Gamma(2-\alpha)} x^{1-\alpha}. \tag{14}
$$

3.3 FGD Algorithm

For the cost function J(θ), J(θ) $= \frac{1}{2}[Y(L) - \Phi(L)\theta]^{\mathrm{T}}[Y(L) - \Phi(L)\theta]$ the GD algorithm uses the first-order Taylor expansion, that is

$$
J(\theta) = J(\theta_k - 1) + \nabla J(\theta_k - 1)[\theta - \theta_k - 1] \tag{15}
$$

while the Newton method utilizes second-order Taylor expansion, which means

$$
\begin{aligned}
J(\theta) = {} & J(\theta_{k-1}) + \nabla J(\theta_{k-1})\left[\theta - \theta_{k-1}\right] \\
& + \frac{1}{2} J''(\theta_{k-1})\left[\theta - \theta_{k-1}\right]^{2}
\end{aligned} \tag{16}
$$

Without loss of generality, the FGD algorithm applies the n*th*-order Taylor expansion, it follows that

$$
\begin{aligned}
J(\theta) = {} & J(\theta_{k-1}) + \nabla J(\theta_{k-1})\left[\theta - \theta_{k-1}\right] \\
& + \frac{1}{\alpha!} J^{\alpha}(\theta_{k-1})\left[\theta - \theta_{k-1}\right]^{\alpha}.
\end{aligned} \tag{17}
$$

Letting the derivative function equal to zero yields

$$
\theta_k = \theta_{k-1} - r1 \frac{\partial J(\theta)}{\partial \theta} - r2 \frac{\partial^{\alpha} J(\theta)}{\partial \theta^{\alpha}}, \tag{18}
$$

where $\frac{\partial J(\theta)}{\partial \theta} = -\Phi^{\mathrm{T}}(L)\left[Y(L) - \Phi(L)\theta_{k-1}\right]$, is the gradient descent direction, and $\frac{\partial^{\alpha} J(\theta)}{\partial \theta^{\alpha}} = -\Phi^{\mathrm{T}}(L)\left[Y(L) - \Phi(L)\theta_{k-1}\right]\frac{\partial^{\alpha}\theta}{\partial \theta^{\alpha}}$ is the additional direction.

Based on Equation (14), it follows that

$$\frac{\partial^{\alpha}\theta}{\partial\theta^{\alpha}} = \frac{1}{\Gamma(2-\alpha)}\theta^{1-\alpha}.$$

Then, the FGD algorithm is transformed into

$$\begin{aligned}\theta_k &= \theta_{k-1} + r_1\Phi^{\mathrm{T}}(L)\big[Y(L) - \Phi(L)\theta_{k-1}\big]\\ &+ r_2\Phi^{\mathrm{T}}(L)\big[Y(L) - \Phi(L)\theta_{k-1}\big]\frac{\partial^{\alpha}\theta}{\partial\theta^{\alpha}}\\ &= \theta_{k-1} + \left[r_1 + r_2\frac{1}{\Gamma(2-\alpha)}\theta_{t-1}^{1-\alpha}\right]\\ &\Phi^{\mathrm{T}}(L)\big[Y(L) - \Phi(L)\theta_{k-1}\big].\end{aligned} \tag{19}$$

The FGD algorithm is listed as follows:

$$\begin{aligned}\theta_k &= \theta_{k-1} + \left[r1 + r2\frac{1}{\Gamma(2-\alpha)}\theta_{t-1}^{1-\alpha}\right]\\ &\Phi^{\mathrm{T}}(L)\big[Y(L) - \Phi(L)\theta_{k-1}\big]\end{aligned} \tag{20}$$

$$Y(L) = [y(L), \ldots, y(1)]^{\mathrm{T}} \in \mathbb{R}^{L}, \tag{21}$$

$$\Phi(L) = [\varphi(L), \ldots, \varphi(1)]^{\mathrm{T}} \in \mathbb{R}^{L\times(nq+np)}. \tag{22}$$

Remark 2

Based on Eq. (20), we can conclude that the FGD algorithm is a special kind of GD algorithm when let $\left[r_1 + r_2\frac{1}{\Gamma(2-\alpha)}\theta_{t-1}^{1-\alpha}\right] = \gamma$. Clearly, the step size γ should be smaller than $\frac{2}{\lambda_{\max}[\Phi^{\mathrm{T}}(L)\Phi^{\mathrm{T}}(L)]}$.

Remark 3

Equation (20) also shows that the FGD algorithm has an additional direction, and this additional direction brings more information.

3.4 Properties of the FGD Algorithm

Rewrite the FGD algorithm as

$$\begin{aligned}\theta_k &= \theta_{k-1} + \left[r_1 + r_2\frac{1}{\Gamma(2-\alpha)}\theta_{t-1}^{1-\alpha}\right]\\ &\Phi^{\mathrm{T}}(L)\big[Y(L) - \Phi(L)\theta_{k-1}\big].\end{aligned} \tag{23}$$

Then

$$e_k = \left[I - \left[r_1 + r_2 \frac{1}{\Gamma(2-\alpha)} \theta_{t-1}^{1-\alpha} \right] \Phi^{\mathrm{T}}(L)\Phi(L) \right] e_{k-1}, \tag{24}$$

where

$$e_k = \theta_k - \theta.$$

To keep the FGD algorithm convergent, one should guarantee that

$$0 < \left[r_1 + r_2 \frac{1}{\Gamma(2-\alpha)} \theta_{t-1}^{1-\alpha} \right] < \frac{2}{\lambda_{\max}\left[\Phi^T(L)\Phi(L)\right]}.$$

In addition

$$\left[r_1 + r_2 \frac{1}{\Gamma(2-\alpha)} \theta_{t-1}^{1-\alpha} \right] = \frac{2}{\lambda_{\max}\left[\Phi^{\mathrm{T}}(L)\Phi(L)\right] + \lambda_{\min}\left[\Phi^{\mathrm{T}}(L)\Phi(L)\right]}$$

Thus

$$\mu = \frac{\tau - 1}{\tau + 1},$$

where

$$\tau = \frac{\lambda_{\max}\left[\Phi^{\mathrm{T}}(L)\Phi(L)\right]}{\lambda_{\min}\left[\Phi^{\mathrm{T}}(L)\Phi(L)\right]} \geq 1$$

Remark 4

For the convergence factor μ, the function $\frac{\tau-1}{\tau+1}$ is a monotonically increasing function. If $\tau = 1$, the convergence factor $\mu = 0$; on the other hand, if $\Phi^{\mathrm{T}}(L)\Phi(L)$ is singular, $\mu = 1$.

4 Preconditioned Matrix-Based FGD Algorithm

When the condition number $\tau = 1$, the convergence factor is $\mu = 0$. In this case, we have

$$e_k = 0 \times e_k - 1.$$

$\tau = 1$ is the best condition number.

Then

$$Y(L) = \Phi(L)\theta + V(L). \tag{25}$$

Introducing a preconditioned matrix **D** on the FGD algorithm, we have

$$\theta k = \theta k - 1 + \gamma \mathbf{D}\Phi \mathrm{T}(L)[Y(L) - \Phi(L)\theta k - 1]. \tag{26}$$

The best preconditioned matrix **D** is

$$\mathbf{D} = [\Phi \mathrm{T}(L)\Phi(L)] - 1.$$

However, when the order $np + nq$ is larger, to compute the inverse is difficult. We can use iterative algorithm to obtain its inverse.

We can get

$$\mathbf{D}_k = \mathbf{D}_{k-1} - r_{k-1}\{[\Phi(L)\Phi^{\mathrm{T}}(L)]\mathbf{D}_k - \mathbf{I}\} \tag{27}$$

$$0 < r_{k-1} < \frac{2}{\lambda \max[\Phi(L)\Phi^{\mathrm{T}}(L)]}. \tag{28}$$

Then

$$E_k = \{[\Phi(L)\Phi^{\mathrm{T}}(L)]\mathbf{D}_k - \mathbf{I}\},$$

where $E_k = \mathbf{D}_k - \mathbf{D}$. Substituting Eq. (27) into

$$\begin{aligned} Ek &= Ek - 1 - rk - 1[\Phi(L)\Phi \mathrm{T}(L)]Ek - 1 \\ &= \{\mathbf{I} - rk - 1[\Phi(L)\Phi \mathrm{T}(L)]\}Ek - 1. \end{aligned} \tag{29}$$

If $0 < r_{k-1} < \frac{2}{\lambda_{max}[\Phi(L)\Phi^T(L)]}$, we can obtain

$$0 < \frac{\|E_k\|}{\|E_{k-1}\|} = \|\mathbf{I} - r_{k-1}[\Phi(L)\Phi^{\mathrm{T}}(L)]\| < 1 \tag{30}$$

Once we obtain the estimates $\mathbf{D}_k$ of $[\Phi(L)\Phi^T(L)]^{-1}$.

Remark 5

By using the preconditioned matrix-based FGD (PM-FGD) algorithm, the condition number τ can be reduced. Then, the convergence factor can be reduced.

5 Example

5.1 A Simulation Example

Consider the following system,

$$y(t+1) = [q^{-1} + 0.9q^{-2}]f(u(t)) + g(y(t)) + v(t),\ f(u(t) = u(t) - 1 + eu(t),$$

$$g(y(t)) = sin(y(t)),$$

$$u(t) \sim N(0, 1), v(t) \sim N(0, 0.12).$$

By using the Weierstrass approximation method, the nonlinear input can be expressed as

$$f(u(t)) = c_1u(t) + c_2u^2(t) + c_3u^3(t) = 2u(t) + 0.5u^2(t) + 0.26u^3(t),$$

and the nonlinear output

$$g(y(t)) = g_1y(t) + g_2y^2(t) + g_3y^3(t) = -0.32y(t) - 0.1y^2(t).$$

Then, we have.

$$\begin{aligned}\theta &= [g_1, g_2, c_1, c_2, c_3, b_2c_1, b_2c_2, b_2c_3]^{\mathrm{T}}\\ &= [-0.32, -0.1, 0.5, 0.26, 1.8, 0.45, 0.234]^{\mathrm{T}}\\ \varphi(t) &= [y(t), y^2(t), u(t-1), u^2(t-1), u^3(t-1), u(t-2), u^2(t-2), u^3(t-2)]^{\mathrm{T}}.\end{aligned}$$

Figure 2 displays the results of the linear-parameter and nonlinear models.

Use the GD, FGD, and PM-FGD methods to calculate the parameters of this system. Figure 3 shows the parameter estimation errors.

From this illustration, we may infer the following:

(1) Table 1 and Fig. 3 demonstrate that the PM-FGD algorithm converges at the fastest rates, and the GD algorithm converges at the slowest rates;
(2) Figure 2 demonstrates that the linear-parameter model captures the dynamics of the nonlinear model well, demonstrating the utility of the Weierstrass approximation method for nonlinear models with unknown structures.

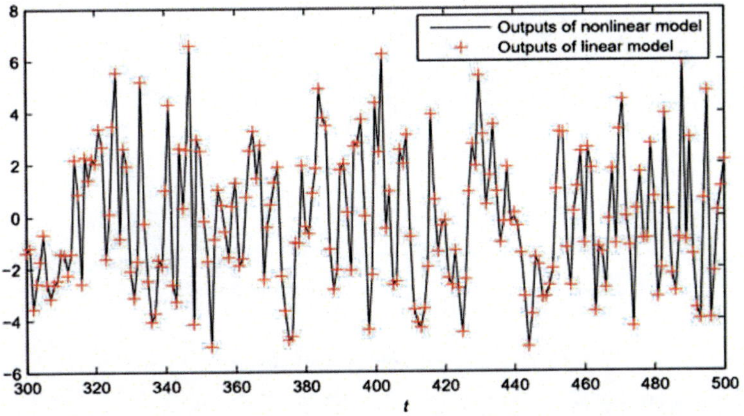

Fig. 2 The outputs of the two models

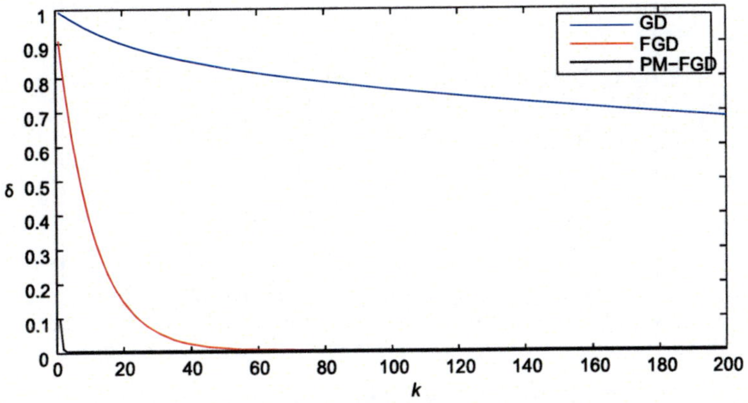

Fig. 3 The parameter estimation errors δ versus k

5.2 A Tank Heater Example

We examine a continuous stirred tank heater (CSTH) [30] in this portion. Figure 4 depicts the construction of the CSTH, where input u(t) represents the position of the steam valve, output x(t) represents genuine temperature, and y represents polluted temperature (t). The following nonlinear model can be used to approximate such a CSTH model.

$$\begin{aligned} y(t) &= g(y(t-1)) + f(u(t)) + v(t) \\ &= 0.6y(t-1) + 0.2y2(t-1) + 0.32u(t) + 0.145u2(t) + v(t). \end{aligned}$$

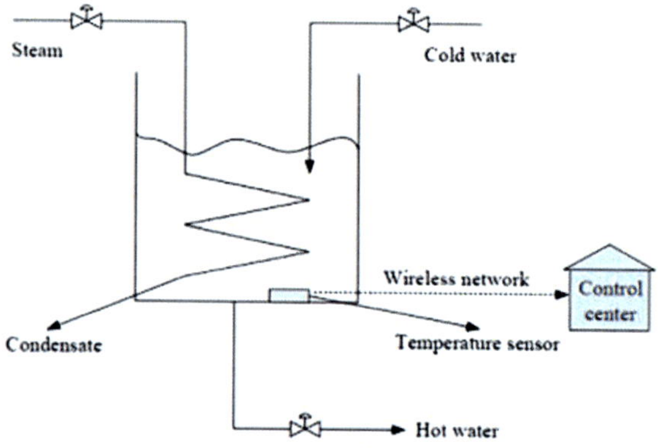

Fig. 4 The CSTH system

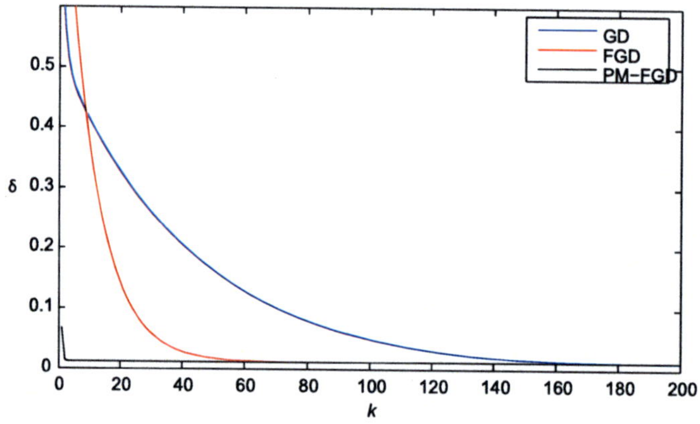

Fig. 5 The parameter estimation errors δ versus k

For this CSTH model, use the GD, FGD, and PM-FGD algorithms. Figure 5 depicts the parameter estimate mistakes. The results also suggest that the FGD and PM-FGD algorithms work well for the nonlinear model.

6 Conclusions

An FGD identification technique for a nonlinear additive system is presented in this research. The structure of such an additive nonlinear system is unknown. It is possible to convert the nonlinear system model into a linear-parameter model by utilizing the Weierstrass approximation theorem. The suggested FGD method can then estimate all of the system's parameters.

Acknowledgements This work was supported by the Funds of the Science and Technology on Near-Surface Detection Laboratory (Numbers 61424140207 61424140202).

References

1. Bai, E.W.: Identification of nonlinear additive FIR systems. Automatica **41**(7), 1247–1253 (2005)
2. Bai, E.W., Chan, K.S.: Identification of an additive nonlinear system and its applications in generalized Hammerstein models. Automatica **44**(2), 430–436 (2008)
3. Jiao, M., Wang, D.Q., Yang, Y., Liu, F.: More intelligent and robust estimation of battery state of-charge with an improved regularized extreme learning machine. Eng. Appl. Artif. Intell. **104**(2), 104407 (2021)
4. Xu, L., Ding, F., Yang, E.F.: Auxiliary model multiinnovation stochastic gradient parameter estimation methods for nonlinear sandwich systems. Int. J. Robust Nonlinear Control **31**(1), 148–165 (2021)
5. Ding. F., Ma. H., Pan. J., Yang. E.F.: Hierarchical gradient and least squares-based iterative algorithms for input nonlinear output-error systems using the key term separation. J. Frankl. Inst. **358**(9), 5113–5135 (2021)
6. Li, J.M., Ding, F.: Identification methods of nonlinear systems based on the kernel functions. Nonlinear Dyn. **104**(3), 2537–2552 (2021)
7. Li, J.H., Zong, T.C., Gu, J.P., Hua, L.: Parameter estimation of Wiener systems based on the particle swarm iteration and gradient search principle. Circuits Syst. Signal Process. **39**(7), 3470–3495 (2020)
8. Zhao, C.Y., Liu, X.G., Ding, F.: Melt index prediction based on adaptive particle swarm optimization algorithm-optimized radial basis function neural networks. Chem. Eng. Technol. **33**(11), 1909–1916 (2010)
9. Söderström, T., Stoica, P.: Systen Identification. Prentice-Hall, Englewood Cliffs, NJ (1989)
10. Bianchi, F., Breschi, V., Piga, D., Piroddi, L.: Model structure selection for switched NARX system identification: a randomized approach. Automatica **125**, 109415 (2021)
11. Chen, J., Gan, M., Zhu, Q.M., Mao, Y.W.: Varying infimum gradient descent algorithm for agentCserver systems with uncertain communication network. IEEE Trans. Instrum. Meas. **70**, 9510511 (2021)
12. Ji, Y., Kang, Z.: Three-stage forgetting factor stochastic gradient parameter methods for a class of nonlinear systems. Int. J. Robust Nonlinear Control. **31**(3), 971–987 (2021)
13. Sinquin, B., Verhaegen, M.: K4SID: large-scale subspace identification with kronecker modelling. IEEE Trans. Autom. Control **64**(3), 960–975 (2019)
14. Chakrabarti, K., Gupta, N., Chopra, N.: Iterative pre-conditioning for expediting the distributed gradient-descent method: the case of linear least-squares problem. Automatica **137**, 110095 (2022)
15. Goodwin, G.C., Sin, K.S.: Adaptive Filtering, Prediction and Control. Prentice-Hall, Englewood CliPs, NJ (1984)
16. Chen, J., Wang, X.P.: Identification of hammerstein systems with continuous nonlinearity. Inf. Process. Lett. **115**(11), 822–827 (2015)
17. Wang, X., Ding, F., Alsaedi, A., Hayat, T.: Auxiliary model-based iterative parameter estimation for a nonlinear output-error system with saturation and dead-zone nonlinearity. Int. J. Robust Nonlinear Control **31**(9), 4262–4286 (2021)
18. Ding, F., Ding, T.: Hierarchical least squares identification and convergence for large-scale multivariable systems. Tsinghua Sci. Technol. **7**(3), 299–304 (2002)
19. Ding, F., Yang, J.B.: Hierarchical identification of large scale systems. Acta AutomaticaSinica **25**(5), 647–654 (1999)

20. Naveed, I.C., Muhammad, A.Z.R. et al.: Design of fractional hierarchical gradient descent algorithm for parameter estimation of nonlinear control autoregressive systems. Chaos, Solitons Fractals **157**, 111913 (2022)
21. Naveed, I.C., Muhammad, A.Z.R., et al.: Design of multi innovation fractional LMS algorithm for parameter estimation of input nonlinear control autoregressive systems. Appl. Math. Model. **93**, 412–425 (2021)
22. Muhammad, A.Z.R., Naveed, I.C.: Two-stage fractional least mean square identification algorithm for parameter estimation of CARMA systems. Signal Process. **107**, 327–339 (2015)
23. Naveed, I.C., Syed, Z., Muhammad, A.Z.R.: A new computing approach for power signal modeling using fractional adaptive algorithms. ISA Trans. **68**, 189–202 (2017)
24. Xu, T.Y., Chen, J., Pu, Y., Guo, L.X.: Fractional-based stochastic gradient algorithms for timedelayed ARX models. Circuits Syst. Signal Process. **41**(4), 1895–1912 (2022)
25. Abbasbandy, S., Jafarian, A., Ezzati, R.: Conjugate gradient method for fuzzy symmetric positive definite system of linear equations. Appl. Math. Comput. **171**(2), 11841191 (2005)
26. Fan, Y.M., Liu, X.M.: Data filtering-based multi-innovation forgetting gradient algorithms for input nonlinear FIR-MA systems with piecewise-linear characteristics. J. Frankl. Inst. **358**(18), 9818–9840 (2021)
27. Almeida, R.: A caputo fractional derivative of a function with respect to another function. Commun. Nonlinear Sci. Numer. Simul. **44**, 460–481 (2017)
28. Cheng, S.S.: Study on Fractional Order LMS Adaptive Filtering Algorithm. University of Science and Technology of China, Anhui (2018)
29. Hartley, T.T., Lorenzo, C.F.:Fractional-order system identification based on continuous orderdistributions. Signal Process. **83**(11), 2287–2300 (2003)
30. Xie, L., Yang, H.Z., Huang, B.: FIR model identification of multirate processes with random delays using EM algorithm. AIChE J. **59**(11), 4124–4132 (2013)

Design and Implementation of a Low-Code and No-Code Platform Applied in Meteorological Monitoring Service

Da Xu, Chao Sun, Guaiguai Liu, Jinting Bai, Yingjie Wang, and Mingjing Duan

Abstract With the development of technology, workers' demand for software is increasing sharply, and the problem of software development platform has become the focus in the IT department. In order to deal with this problem, National Meteorological Information Center (NMIC) designed a visual integrated development environment where non-developers who are in charge of meteorological business can develop and design webs for their work. The purpose of this paper is to provide an environment that is based on low-code and no-code [1] execution platform. It supports various templates and components to shield different software structure and program skills while developing. The most important feature of future meteorological monitoring service development is that it is helped people who are non-developer worker to spend more time on business logic and algorithm. In the future, we will promote the platform in nationwide and help the local meteorological departments to develop their own meteorological monitoring system.

1 Introduction

The Mirror was established by the National Meteorological Information Center (NMIC) to meet the requests of meteorological observed data and weather predict product monitoring. The main duty of The Mirror is monitoring the timeliness and integrity of the data and products in various links, such as collection, user distribution, decoding and warehousing, data synchronization and interface services. In 2018, it has been deployed in 31 provinces to help local meteorological department to monitor their business systems. However, in recent years, with the accelerated construction of information digitization and modernization, more and more local meteorological departments begin on upgrading their business systems to provide monitoring services for weather forecast and important government activities. Because of the

D. Xu · C. Sun (✉) · G. Liu · J. Bai · Y. Wang · M. Duan
National Meteorological Information Center, Zhonguancun South Road, Beijing 46, China
e-mail: sunc@cma.gov.cn

K. Nakamatsu et al. (eds.), *Advanced Intelligent Technologies for Information and Communication*, Smart Innovation, Systems and Technologies 365,
https://doi.org/10.1007/978-981-99-5203-8_21

development history and climate characteristics, local meteorological departments will have their own special requirements for meteorological business system, especially in their own unique personalized needs. Among them, the software development platform and software architecture is not only the focus of the computer industry but also is the focus in the meteorological field. With the growing demands for meteorological business monitoring, the development speed of programmers can't be satisfied, and the people who are work on weather phenomenon analysis are not familiar with various technical frameworks, which are used in software development, therefore a large number of application development tasks overstock. In order to help local meteorological operators can easily build their own business monitoring systems, NMIC has designed a visual integrated development environment, which is called Simulation and Co-develop Platform (SCP) to solve the traditional software development problems, which are low reusability, code redundancy and difficult to maintenance.

It efficiently scales the work of related departments without writing code, users can drag and drop application components through visual editing and configuration to design and build powerful their own systems. The advantage is that it can make the development style consistency and shield the underlying technical framework and dependencies, and it will help the operators focus on the implementation of business logic and strategy.

This paper is about environment-based low-code and no-code execution platform and execution method that combines back-end and front-end in web service. In detail, this paper describes the Low-Code/No-Code execution structure in SCP. Simultaneously, SCP supports various templates and components, which can help people to absorb and learn from different ideas. For example, user who is working in Shanxi Meteorological Bureau designed an UI for Weather Modification System. Other local meteorological departments that have same business requirements can use the UI templates from Shanxi. Low-Code/No-Code is developed as a front-end technology of the web and can be run in a typical browser, which has the effect that what you see is what you get. The browser can run a Low-Code/No-Code app instead of program IDE tools.

2 Simulation and Co-Develop Platform

2.1 Concept of Low-Code/No-Code

Low-Code was mentioned by Forrester Research in 2014. It is a kind of technology mainly used in the field of software development. According to Gartner, a global research firm headquartered in Stamford, CT, USA, low-code platforms will be used in 65% of application development work by 2024 [2]. With a low code platform, developers do not need to code but still can generate the common functions of applications and design more extended features. No-Code focused on metadata and business management requirements, which designed the data structure and user interaction through the visualization Low-Code and No-Code are both fast development technologies, which is a secondary development software framework that users can build the information system by writing less or no coding. It will let users build management applications like building blocks.

2.2 Structure of Low-Code/No-Code

Simulation and Co-develop Platform generate the scaffold, which is a maven project. Developer can use visual interface to set the attribute of project. The application framework code will submit on GitLab automatically, which is a version management tool. Developers drag and drop application components from Components Libs to design their business logic algorithm and UI. When developers saved their scheme, SCP build the back-end by Springboot [3] and build the front-end with React [4] at same time. The render engine will register data service as API and integrate the UI. Finally, developers deployed the application on the server and users can visit the website by all kinds of clients. Low-Code/No-Code technology [5] make relational database and spreadsheet have a corresponding relationship. Generally, the three-level architecture model is adopted to apply the whole business of meteorological monitoring service. It is divided into visual layer, logical layer and data layer (Fig. 1), and adopts the concept, which is "high cohesion and low coupling". The idea of low coupling makes the interface, business logic and data access mutually independent, and data can be encrypted and transmitted between layers. The deployment of business logic layer and database access layer can work in different server clusters.

2.3 Development Mode Comparison

The traditional development mode is technology-oriented development. The people who are familiar with business have to describe the business requirements to the developer and the developer builds the system according to their own understanding. Because of the inadequate information communication and the deviation of developers' understanding of the business, the system often does not match the actual business and needs. Usually, it has to be revised and adjusted repeatedly, and cost large labor consumption. The development mode of Low-Cod/No-code is business-oriented [6], which is managed by the people who have more experience lead the development and design, and the developer only needs to focus on the business logic functions [7] that used to require a lot of coding now only need to be implemented through the platform configuration can be completed, and let more researchers and designers can act as the developer. They do not need to rely on technicians to build a management application. Compared with the traditional applications (Table 1),

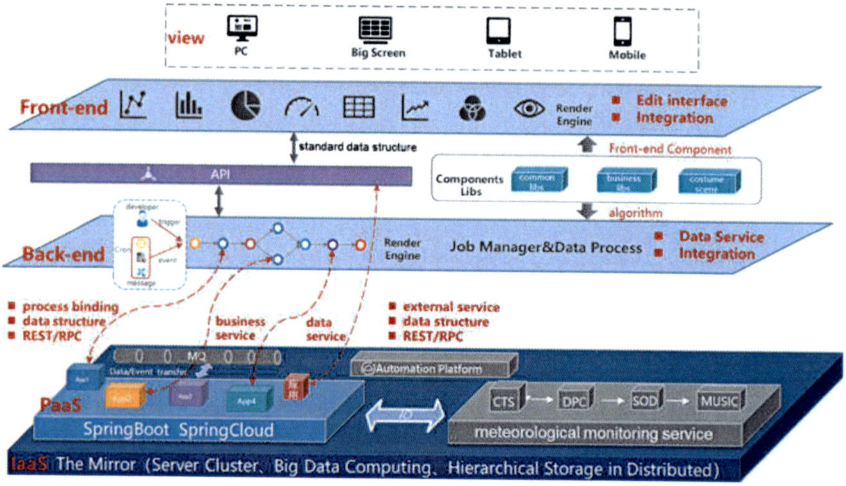

Fig. 1 Structure of Low-Code/No-Code platform

the development cycle of realizing functional requirements would be less than one-tenth that of traditional development, and greatly shorten the development cycle and improve the rate of online published. For example, recently New York City officials build a coronavirus (COVID-19) Engagement Portal, which was up and running in 3 days—all without writing any code [2].

Table 1 Comparison of development mode characteristics

Number	Features	Traditional development	Low Code/No Code development
1	Development framework	Front end and back end separation, code	Front-end and back-end integrated in visual design framework, no code
2	Development environment	High performance for CPU and memory Software develop kit	Normal computer hardware, web browser only
3	Development compiling	Building, package, upload to server	Refresh the website
4	Development deployment	Hot-deployment, reboot service	Hot-deployment, real-time update
5	Development cycle	Single function mode 3–5 days	Single function mode 1 day
6	Development	Java, Maven, Springboot	No requirement
7	Skill	React, Vue	Experts with business
8	User	Programmer	Easy
	Ease of use	Difficulty	

3 Key Technology

3.1 Process Engine and Service Orchestration Technology

On the one hand, the process engine should ensure that the corresponding calculation rules are accurately obtained, on the other hand, it should also ensure that the variables referenced by the expression can be obtained correctly. At the same time, the scheduling ability of the process engine is encapsulated to implement the process scheduling of the data processing layout, and the calculation results can be directly returned to the front-end page [8, 8]. SCP supports cron expression processing mechanism to realize intelligent matching of artificially triggered, event-triggered and timing triggered, and match the corresponding service component configuration through automatic intelligent job scheduling. At the same time, it provides a completely service-oriented process management architecture, in which the process engine totally supports that the SCA (Service Component Architecture) service components. What's more, it supports the preparation of components and services into business processes, and it can drive work collaboration and coordinate the execution between manual tasks and automatic tasks. Data processing uses node connection to express business execution logic. The process level not only provides basic process models such as order, branch, concurrency, loop, nested sub-process, multi-path selection, and multi-path merging but also provides conditional routing, free flow, rollback, activation strategy, completion strategy, parallel countersigning, serial countersigning, assignment, multi-real example process, process monitoring and other special process models. Therefore, rational service orchestration can solve the dependency problem between service calls in business systems.

The data required for a business function may come from different business systems or different hardware devices. There is a sequential dependence in the processing of data. An atomic service provides an atomic data. Therefore, the arrangement of services can ensure the reasonable input and output of data and form a reasonable data processing chain. Service orchestration can be realized by using the service assembly framework designed by SCP, which include the following functions:

- Based on service-oriented architecture, component implementation and assembly details are separated to achieve true loose coupling. SCP-SCA (SCP-Service-Component-Architecture) is a set of service system construction framework protocol for SOA (Service Oriented Architecture). It integrates the idea of IOC (Inversion of Control), and transforms the object-oriented reuse from code reuse to business module component reuse. At the same time, it completely separates the service interface, implementation, deployment and call, and flexibly assembles and binds them in the form of configuration.
- Businesses can be arbitrarily disassembled and assembled using the building block principle. Traditional applications are like ordinary toys, which cannot be disassembled and spliced. The building block toys can be assembled according to the

user's own ideas because it is composed of standard tiny components. Applications based on SCP service assembly are composed of smaller service components, such as building block toy modules. With building blocks toys can build a variety of different shapes, just like the SCP service assembly architecture can achieve different applications; block toy modules are based on standardization, so they can be reused.
- Creating flexible business solutions for greater organizational agility and flexibility. SCP service assembly supports top-down development to create business solutions with a flexible service assembly model. Each component can be replaced by another component with the same interface. This component group can be adjusted according to IT architecture requirements, such as service connection, transport protocol, transaction, security and message reliability. SCP provides optional transport bindings that enable developed solutions to accommodate a wider range of deployment needs.

3.2 Component Management Technology

Realizing the unified management and automatic verification of various components that need the specification requirements will enter the audit and authorization link. The components are encrypted, which can only run in the production environment after being authorized. SCP supports that third-party developers can develop the required function, which the business components and page components have been already provided, and SCP manager can use the upload module to update the developed components to the corresponding component library of the unified access framework. During the upload process, the components are checked for standard and normative. After the administrator approved component, the uploaded component can be used through the service visualization or page visualization module.

Through the version management of the component, the original configured data service process and page can be not affected after the component upgrade. That is, the original referenced component will not be automatically upgraded when the component has been upgraded.

Audited components can be shared components, page sharing, data service sharing and sharing of the entire application, but also to the designated user or provincial units to share (Table 2).

3.3 Electron Frame

Electron is an open source library with HTML, CSS and JavaScript to build desktop applications, which can be packaged for Mac, Windows, Linux system. This runtime environment contains Node and Chromium that can work on web applications with a Node.js environment. In addition, it supports developers to build GUI using Web

Table 2 Audited components list

Number	Group	Components	Description
1	Server status	Server status details	Query server CPU usage, memory usage, disk usage, total disk capacity, disk free status and other data, return the latest data
2		Disk idle situation	Query the disk-free status of the server, returning the free status of each disk
3		Disk usage	Query server disk usage, return the latest data
4		Process situation	Query the process on the server
5	Data access	ElasticSql	Query data from ElasticSearch
6 7 8		File reading and writing HTTP component	Read local file data in a variable To get HTTP request return data
9		SQL query component	Querying relational databases through SQL, such as MySQL
10	Data process	Collection batch assignment component	Batch assignment for list arrary map set
11		Data filtering component	Filtering for the value of the specified field on the data
12		Data extraction component	Extraction operations for the specified fields of data
13		Data deduplication artifacts	Deduplication for specified fields of data
14		Data merge	Combine the two data
15		Data calculation	The calculation of data by expression
16		Data aggregation component	Used to read data in the cache
17	Data storage	Elastic storage	Save the data to ElasticSearch
18		File writing	Write the data to the file in the form of JSON
19		Mysql data storage component	Storage for data from the specified data source
20	Front-end	Iframe	Integrate content outside the site into the page, support configuration styles and interactive actions
21		Button	Implement interactive functions such as opening links, changing component visibility, invoking services, and assigning values to variables
22		Line chart	The multi-series data configuration, showing the same kind of different data changes, to a combination of broken lines and regions, intelligent display of multi-dimensional data trends

(continued)

Table 2 (continued)

Number	Group	Components	Description
23		Histogram	Basic bar chart component supports custom y-axis intervals and multiple series of data configurations to more intelligently display multidimensional data differences but takes up more space in visualization applications
24		Bar chart	Bar chart is a kind of histogram, which supports custom x-axis interval, multi-series data configuration and stacked data display. It can intelligently display multi-dimensional data differences in a small visual application space
25		Area graph	Supports custom y-axis intervals, multi-series data configurations and a series of digital flip-flop displays to intelligently display trends in multi-dimensional real-time data in a combination of polylines and zones
26		Table	It is used to display intuitive data of some table types. Manual configuration display, icon display, numerical display

pages and get the operating system function by APIs. While the provincial client starting, two services will be started locally, one of the services is front page, and the other is for the front request processing. In order to solve the problem that the provincial resources cannot be accessed under the national deployment (Fig. 2), the data storage environment for the service configuration is based on the national node, and the client will deploy a service environment at the provincial node to provide provincial users to use provincial resource [10].

4 Application

Smart-gridded forecast is the important content and support for the China Meteorological Administration modern weather operation [11]. Sometimes the observation elements need to be revised, and the weather forecasters want to know which province revised the elements in their smart-gridded forecast production. In order to detect the elements' value modification in time, our team designed the monitor system by SCP, and we just spend 4 h to finish the project. Here is how we applied the project through this magic tool. First of all, the smart-gridded forecast system needs to provide the data which include all the important information, such as province name, product name, element's name, element's code, data time, update frequency, etc. This monitoring data store in ElasticSearch [12], which is an open distributed database. Secondly, we use the mature components to obtain data from the database

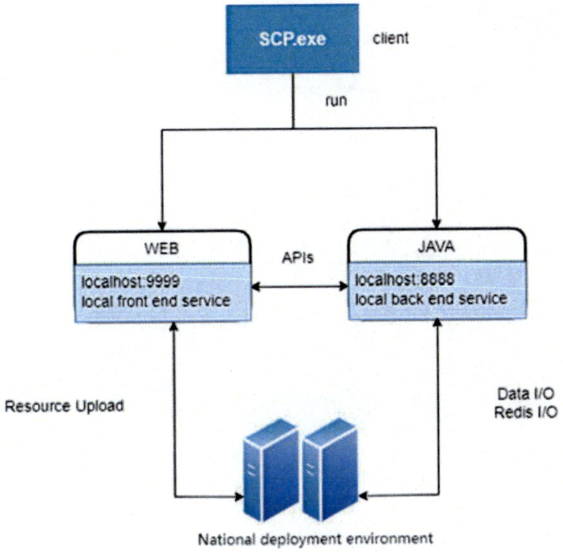

Fig. 2 Structure of SCP client

and process it with the meteorological business logic, and then, encapsulate the result into the interface services (Fig. 3). Next, we choice the table component for the front-end interface display, and make the back-end interface services in the table corresponding display (Fig. 4). What's more, the SCP also supports interactive features, the front-end to pass a time parameter to the back-end service to query the history records. Finally, the forecaster has to check it whether the monitoring information is correct before package the project.

5 Conclusions

Smart-gridded forecast data revised monitoring system has been running and deployed in the application of provincial meteorological bureau. Compared with the traditional development mode, SCP has the advantage of agile development, which can quickly design the prototype and help the forecasters to realize the functions that they focus on. It can also reduce the development cost and improve the reusability of the program. In the future, we will design more common components for meteorological business development. And we believe that the application scenarios of rapid development platforms such as low-code and no-code, more enterprises or enterprise information service providers will adopt a zero-code platform because of lower technical barriers and higher development efficiency. Without code development, they can build a good business application with a good user experience, and make "development is configuration, what you see is what you get" become reality. What's

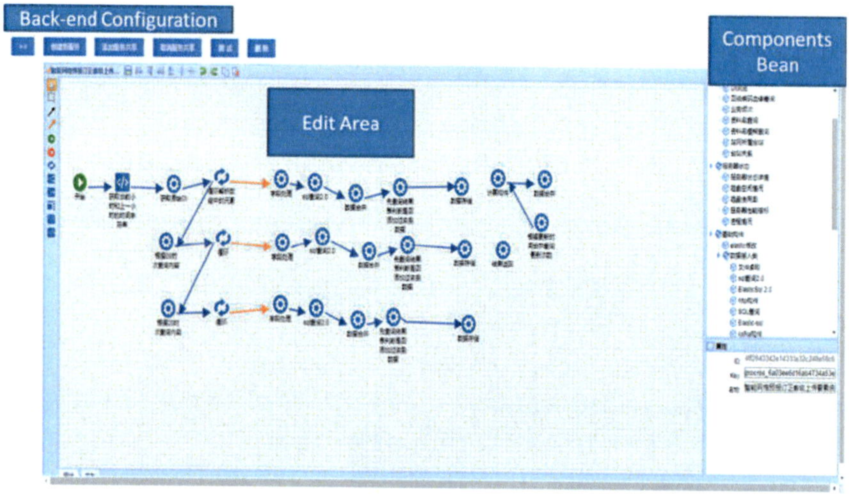

Fig. 3 SCP back-end workstation

Fig. 4 SCP front-end workstation

more, using the "smart and simple development" model to promote the rapid launch of enterprise information systems and promote the process of enterprise informatization effectively.

Acknowledgements The project is supported by "Meteorological Integrated Business Intelligent Monitoring" Innovation Team of National Meteorological Information Center (NMIC-202011-05).

References

1. Design and implementation of a Low-Code/No-Code system. Int. J. Adv. Smart Converg. **8**(4) (2019)
2. Woo, M.: The rise of No/Low code software development—no experience needed? Engineering **6**(9) (2020)
3. Spring Project [Internet]. Available from: https://spring.io/projects
4. React Frame [Internet] Available from: https://react.docschina.org
5. A study on the design of low-code and no code platform for mobile application development. Int. J. Adv. Smart Converg. **6**(4) (2017)
6. da Cruz Mauro, A.A., de Paula Heitor, T.L., Caputo Bruno, P.G., Mafra Samuel, B., Pascal, L., Rodrigues Joel J.P.C.: OLP—A RESTful open low-code platform. Futur. Internet **13**(10) (2021)
7. Waszkowski, R.: Low-code platform for automating business processes in manufacturing. IFAC PapersOnLine **52**(10) (2019)
8. Bock Alexander C., Ulrich, F.: Low-code platform. Bus. Inf. Syst. Eng. **63**(6) (2021)
9. Waszkowski, R.: Low-code platform for automating business processes in manufacturing. IFAC PapersOnLine (10) (2019)
10. Kinney, S.: Electron in Action. Manning Publications (2018)
11. H, et al.: Design and implementation of gridded forecast application analysis platform. Meteor. Mon. **46**(10), 1340–1350 (2020). (in Chinese)
12. ElasticSearch [Internet]. https://www.elastic.co/cn/

Model-Data Driven Fusion Method Considering Charging Rate and Temperature to Predict RUL of Lithium-Ion Battery

Hailin Feng and Anke Xu

Abstract Predicting the remaining useful life (RUL) of lithium-ion batteries is critical to the safe and reliable use of batteries. Due to the highly non-linear degradation properties, the conventional approach requires sophisticated models and large amounts of operational data to accurately predict the RUL. A fused model data-driven approach ISE-PF-pSVR is presented in this paper. The semi-empirical degradation model is improved by considering the charging rate and temperature characteristics. And the use of particle filter (PF) algorithm updates degradation model parameters. Since the filtering algorithm obtains the observations by recursion, there is error accumulation. P-step support vector regression (SVR) is introduced to estimate the observation iteratively, so as to boost the precision of dynamic parameter update. In addition, the performance of ISE-PF-pSVR model is verified using three datasets with inconsistent degradation forms. It shows that the established RUL prediction model has remarkable universality, applicability and accuracy.

1 Introduction

Due to extensive working temperature scope, low self-discharge rate and environmental performance, lithium-ion batteries are used in many applications [1]. Yet, the internal reactions of li-ion batteries during use are accompanied by performance degradation. Studying the li-ion batteries remaining useful life (RUL) can prevent the capacity degradation mutation in the use of batteries and exclude potential risks [2, 3]. Thus, precise prediction of li-ion battery RUL provides a strong guarantee for the application.

RUL prediction is classified into three general approaches, model-based method, data-driven method and fusion method [4]. The model-based approach uses models to reflect the relationship between the state of health (SOH) and internal or external characteristics. This method has lower computational cost and clearer description

H. Feng · A. Xu (✉)
Xidian University, Xi'an 710126, China
e-mail: 1035400582@qq.com

K. Nakamatsu et al. (eds.), *Advanced Intelligent Technologies for Information and Communication*, Smart Innovation, Systems and Technologies 365,
https://doi.org/10.1007/978-981-99-5203-8_22

of degradation process. Among them, the commonly used models include empirical degradation model (EDM) and semi-empirical degradation model (SEDM). EDM collects a great deal of test data to accommodate the degradation trend. Qin et al. [5] used double exponential EDM to better fit li-ion battery decay through neural network (NN). However, the model is susceptible to changes in current and temperature. Under different conditions of operation, it is challenging to obtain exact mechanism models, and it is necessary to resort to machine learning algorithms. SEDM is a balance between electrochemical model and data-driven approach. More important physical parameters are usually considered, and updated through data-driven methods. Data-driven methods predict RUL by training a large amount of data with machine learning algorithms. The algorithms used include Gaussian process regression (GPR) [6], neural network (NN) [7], and time series prediction method [8]. Severson et al. [9] established a linear model based on capacity cycle difference and used elastic network to predict li-ion batteries RUL. Guo et al. [10] extracted eight effective features based on li-ion battery charging data and established a prediction model based on relative vector machine. Tian et al. [11] extracted health features based on the temperature difference data in the constant current charging phase of li-ion batteries and quantified the indeterminacy of capacity decline using vector machine. The data-driven method has great prediction precision, but it needs to include all possible operating conditions, requires a vast range of data, and lacks a reasonable explanation of the degradation process. Therefore, the accuracy of this method is poor in different environments or when the training sample size is insufficient.

The fusion model for RUL prediction is the combination of model and data-driven, and requires neither a precise physical model nor a lot of data. This method makes full use of the best properties of a single model. Chen et al. [12] fused grey model and NN to forecast the RUL of battery through particle filter (PF) framework. Zheng et al. [13] built a capacity prediction framework for li-ion batteries driven by empirical model and data and used extended Kalman filter (EKF) twice to accurately predict the online capacity. Han et al. [14] integrated SEDM and PF and considered the effect of resistance on degradation to enable RUL forecasting online under different working conditions. However, most of the current methods use two filters or multiple machine learning algorithms for prediction. Moreover, the internal resistance data that are noisy and difficult to measure are used to enhance the precision of the model. These problems make RUL prediction model with high complexity and data cost. Therefore, this paper proposes a dynamic and applicable model data-driven fusion model for RUL prediction of li-ion batteries. The main contributions include the following three aspects: (1) Firstly, the semi-empirical degradation model is improved, and the ISE model considering the charging rate and temperature is proposed. The model can track the degradation of li-ion batteries under various charging conditions and environmental circumstances. (2) The p-step support vector machine (SVR) is proposed to predict capacity values. A non-linear regression is established on the capacity data of the li-ion battery degradation process to predict the observed values of the p-step iterations. This improves the accuracy of the degradation model parameter updates. (3) Finally, the PF algorithm updates the ISE model parameters and the observed data simultaneously to complete the dynamic model of the capacity degeneration of li-ion batteries. The ISE-PF-pSVR model predicts the RUL, making full use of the mechanistic model and partial data, and the prediction model is less complex and more practical. Results of the experiment show that the proposed ISE-PF-pSVR model effectively improves the prediction precision of li-ion batteries RUL.

The remainder is shown below. Section 2 presents the improved SEDM and Sect. 3 builds the degradation model based on the PF algorithm. In Sect. 4, SVR is incorporated to update the observations. The results of the trials show that the accuracy of parameter updating is improved. Finally, Sect. 5 gives the conclusions.

2 Improved SEDM—ISE

This section discusses SEDM for li-ion batteries that consider charge rate and temperature. Based on three li-ion battery datasets, the relationship between the cycles and the capacity degradation of li-ion batteries with different operating conditions is investigated.

2.1 Relevant Indicators for Li-Ion Batteries

The li-ion battery SOH based on capacity variation is defined as,

$$SOH_k = \frac{C_k}{C_0}, k = 1, 2, \ldots, N \tag{1}$$

where SOH_k is the SOH at the *kth* cycle, C_0 is the li-ion battery initial capacity, C_k is the li-ion battery actual capacity at the *kth* cycle, and N is the total charge/discharge cycles.

The li-ion batteries' capacity degradation ratio can be defined as:

$$Q_k = 1 - SOH_k = \frac{C_0 - C_k}{C_0}, k = 1, 2, \ldots, N \tag{2}$$

where Q_k is the capacity degradation ratio at the *kth* cycle.

The RUL of li-ion battery indicates the number of available charge/discharge cycles from the current cycle to the threshold of end of life (EOL). In this paper, the capacity degradation ratio is used as a metric, and li-ion battery EOL is usually determined 0.2.

2.2 Description of Li-Ion Battery Datasets

Dataset 1: 18,650 li-ion battery tested at NASA Ames Research Center [15]. In this paper, B0005 and B0006 batteries with nominal capacity of 2.0 Ah are selected, which were tested at 24 °C. The charging process is completed by charging to 4.2 V at 1.5 A (i.e. 0.75 C) and then charging at 4.2 V until the current drops to 20 mA. Discharging at 2A (1 C), B0005 and B0006 drop to 2.7 and 2.5 V, respectively. The li-ion batteries' capacity degeneration profile in this dataset is shown in Fig. 1a.

Dataset 2: Kokam li-ion batteries with a nominal capacity of 740mAh tested at Oxford University [16]. The test was performed at 40 °C for characteristic testing, driving cycle and life judgment. This paper studies Cell 1 and Cell 7. The charging process is charged at 740 mA (1 C) to 4.2 V. Discharge to 2.7 V at 740 mA (1 C). The capacity decay is shown in Fig. 1b.

Dataset 3: Commercial li-ion batteries tested under rapid charging circumstances by Severson et al. [9] at MIT. The test was conducted at 30 °C with a nominal battery capacity of 1.1Ah. The batteries were charged to 3.6 V using either one-step or

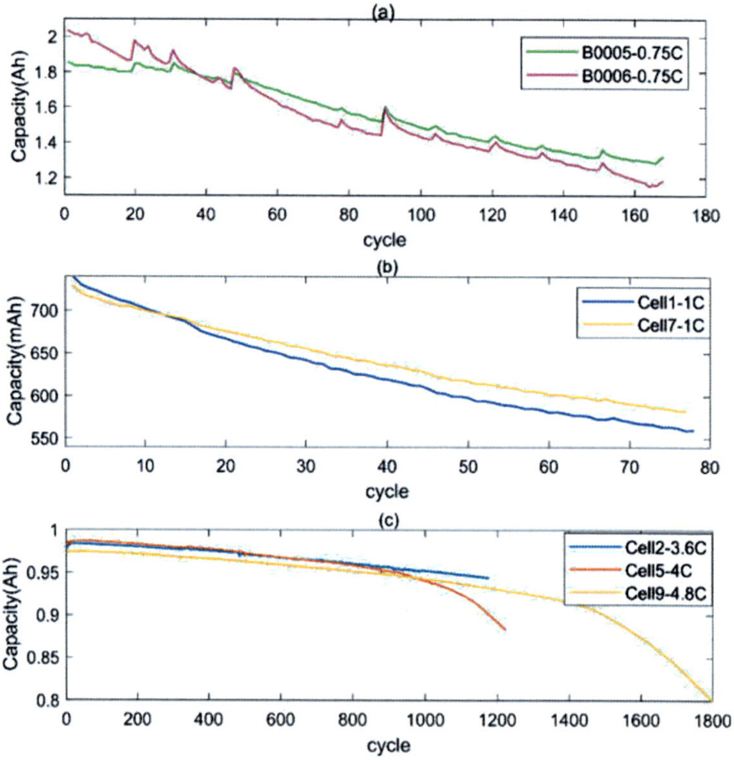

Fig. 1 Li-ion battery capacity degradation **a** NASA **b** Oxford **c** MIT

two-step rapid charging strategy, discharged to 2.0 V at 4 C. The one-step charging strategy is considered for this study, selecting cell 2 and cell 5 from the '2017–5–12' batch, and '2018–4–12' batch of cell 9, with charging rate of 3.6, 4 and 4.8 C, respectively. The capacity degeneration is in Fig. 1c.

2.3 *Semi-empirical Degradation Model Considering Charging Rate and Temperature*

Degradation of li-ion batteries is usually caused by chemical side reactions during cycling to create a solid electrolyte interface [17]. Arrhenius law effectively reflects the effect of the temperature factor on the rate of chemical reactions and is used to represent the chemical side reactions that occur during the cycling of li-ion batteries. The conventional SEDM was obtained according to Arrhenius law [18].

$$Q = A\exp\left(-\frac{E_a}{RT}\right)t^z \tag{3}$$

where Q represents the capacity degradation ratio, A is the exponential prefactor of the Arrhenius equation, Ea is the reaction activation energy, R means the standard gas constant, T represents the absolute temperature, t is time and z is the power factor.

Unlike conventional SEM, this paper uses the product of the charge rate and the cycles instead of time. At a fixed charging rate, the capacity is proportional to the time. It [18] was found that the reaction activation energy is inversely related to the rate, i.e. a higher rate accelerates the chemical process of irreversible active lithium depletion. Therefore, the enhanced SEDM (denoted ISE model) created in the paper introduces the charging rate of li-ion batteries as follows,

$$Q = A\exp\left(-\frac{E_a}{RT}\right)(C_{\text{rate}}k)^z \tag{4}$$

where C_{rate} represents the li-ion battery charge rate.

The ISE model can show the power rate relationship at different charge rates to simulate the li-ion batteries' capacity degradation.

3 Degradation Modeling of Li-Ion Batteries Based on PF

Analyzing the capacity degradation process of li-ion battery in Sect. 2, the fixed parameter ISE model is unable to accurately track the sudden changes in degradation. To better evaluate the degradation, the degradation model parameters are dynamically renewed using the particle filter algorithm.

3.1 Model Derivation

In the ISE model, the parameters A, $\frac{E_a}{R}$, z impact the capacity degradation ratio. Further the variation of these parameters reflects the li-ion battery degeneration.

Let $X = [A, \frac{E_a}{R}, z]^T$, then the state space equation is,

$$X_{k+1} = X_k + w_k \tag{5}$$

The observed equation for capacity degradation is shown in (6).

$$y_k = A_k \exp\left(-\frac{E_a}{R} \cdot \frac{1}{T}\right)(C_{\text{rate}}k)^{zk} + v_k \tag{6}$$

where w_k means the process noise of X_k and v_k means the measurement noise, satisfying the Gaussian distribution with zero mean and variance of Q and R respectively.

3.2 ISE Parameter Update Based on PF

PF algorithm can estimate dynamic parameters and time-varying systems with great precision. The parameters of the degradation model are observed online using PF. The algorithm steps are as follows.

Step 1: Particle initialization. The number of particles is M. At the moment $k = 0$, a particle population $\{X_0\}^M$ is generated, with all particles having the $\frac{1}{M}$ weight.

Step 2: Prediction step. Prediction of particle states based on state space equations.

Step 3: Update step. Updating particle weights based on observations.

$$w_k^i = \exp\left(-\frac{1}{2R}(y_k - y_k^i)\right)\frac{1}{\sqrt{2\pi R}} \tag{7}$$

Step 4: Normalized weights.

$$w_k^i = \frac{w_k^i}{\sum_{i=1}^M w_k^i} \tag{8}$$

Step 5: Resample. Duplicate particles with larger weights and delete particles with smaller weights to obtain a new set of particles with all particle weights of $\frac{1}{M}$.

Step 6: $k = k + 1$, go to step 2.

Based on the observation equation (6), the capacity decay state of the future cycle is predicted from the known capacity decay ratio. The recurrence equation is:

$$y_{k+n} \approx y_k + n \cdot [A \exp\left(-\frac{E_a}{RT}\right)]^{\frac{1}{z}} \cdot z \cdot C_{\text{rate}} \cdot y_k^{\frac{z-1}{z}} \tag{9}$$

4 RUL Prediction with the Introduction of pSVR

The precision RUL prediction is critical for the wider application of li-ion batteries. According to the recursive observation update parameters in Sect. 3, there are recursive equation errors accumulated leading to inaccurate capacity tracking. Therefore, p-step SVR prediction observations are added to the proposed ISE-PF model to make the PF parameter updates match the actual degradation and boost the precision of RUL prediction.

The general framework of the ISE-PF-pSVR model developed in this paper is shown in Fig. 2.

The ISE-PF-pSVR model is divided into two main parts.

(1) pSVR estimation observations. First, depending on the li-ion battery capacity data obtained by data preprocessing, the dataset is reconstructed to obtain the training set with input x and output Y. Secondly, the model is trained, in which the optimal parameters are obtained by particle swarm optimization. Finally, the established SVR model is used to estimate the capacity observations.
(2) PF updates the parameters of ISE model. Firstly, the parameters of the ISE model are initialized to generate the particle set. After the prediction step to update the particle state, the update step updates the particle weights using the pSVR

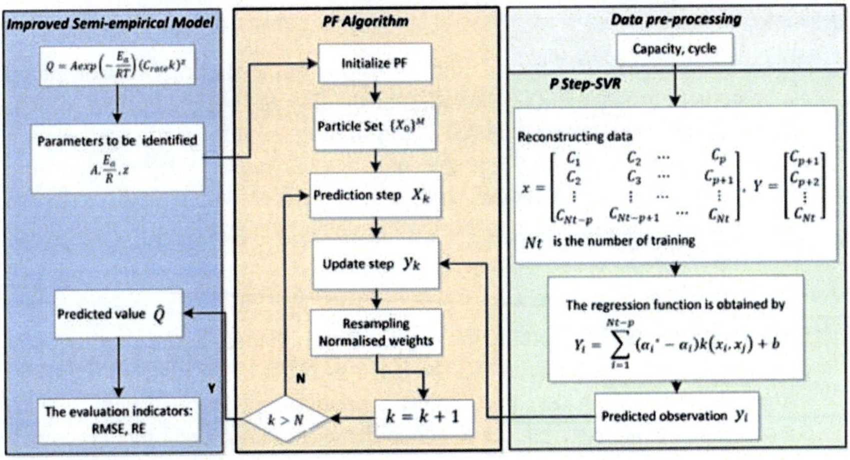

Fig. 2 ISE-PF-pSVR model framework diagram

observations. Then resampling to get new particles, return to the prediction step to continue the recursion. Until the last cycle N, so as to derive the capacity degradation ratio of the entire life cycle of li-ion battery.

Finally, Q and RUL of the proposed method predictions are evaluated using root mean square error (RMSE) and relative error (RE), respectively.

4.1 Model Evaluation Indicators

$$RMSE = \sqrt{\frac{1}{n}\sum_{i=1}^{n}(Q_i - PQ_i)^2} \tag{10}$$

where n is the samples number in the test set, Q_i is the capacity degradation ratio for the *ith* cycle and PQ_i is the capacity degradation ratio predicted value.

$$RE = \frac{|RUL - PRUL|}{RUL} \times 100\% \tag{11}$$

where RUL is the actual value of RUL and $PRUL$ represents the predicted RUL.

4.2 Experimental Verification of ISE-PF-PSVR

To validate the effectiveness of the presented approach, this section compares the proposed model ISE-PF-pSVR with the ISE-PF model. The experiments are validated using li-ion batteries from NASA, Oxford and MIT datasets.

(1) NASA dataset

Figure 3a shows the RUL prediction for B0005 at the starting point of the 60 cycle prediction. Since B0006 degrades too quickly and is already close to EOL near 60 cycles, 40th cycle is chosen as the starting point for prediction, as seen in Fig. 3b. The red line in the graph indicates the capacity degradation ratio predicted by the proposed model, compared to the green curve predicted by the ISE-PF model, the prediction is more accurate and the predicted RUL is closer to the true value.

(2) Oxford dataset

Figure 4 shows the RUL prediction for the Oxford battery at the starting point of the 20-cycle. It can be seen that the ISE-PF-pSVR better matches the li-ion battery degradation process and is able to predict the RUL well with only 20 training data, demonstrating the applicability of the proposed model on small datasets.

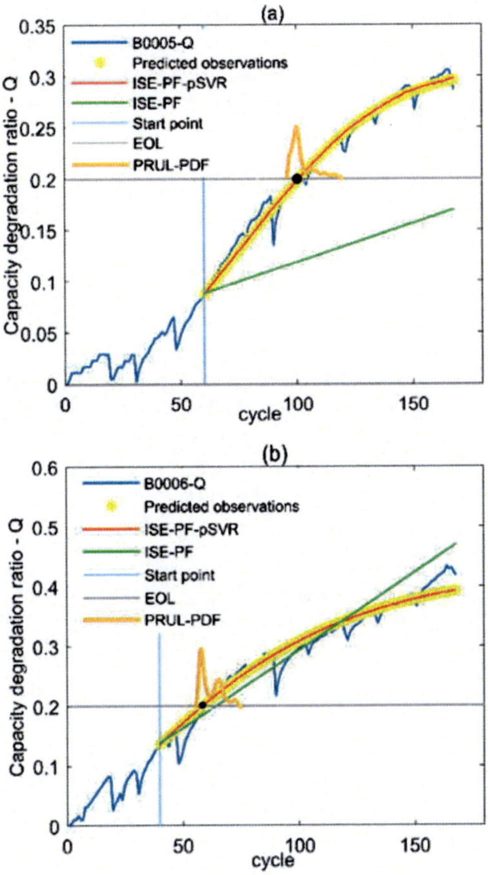

Fig. 3 RUL prediction for the NASA dataset, **a** B0005, **b** B0006

(3) MIT dataset

The validation results for the MIT battery data are shown in Fig. 5. The Cell 2 and 5 with approximately 1200 cycles in the test data use 300 cycles as the starting point for prediction. Cell 9 with 1800 cycles is using 500 cycles. The batteries in this dataset show a long cycle life. As the amount of data increases, the proposed model predicts a capacity degradation ratio that is closest to the actual capacity degradation. In contrast, the ISE-PF model, which lacks estimates of observations, has the cumulative error and the results deviate from the actual degradation. It is demonstrated that the ISE-PF-pSVR model is also applicable to long-cycle li-ion batteries with high-rate charging.

The numerical results of the ISE-PF-pSVR model proposed in this paper are given in Table 1 for the three datasets. When contrasted with the ISE-PF model results, the presented model predicts the smaller RMSE and RE of RUL for the

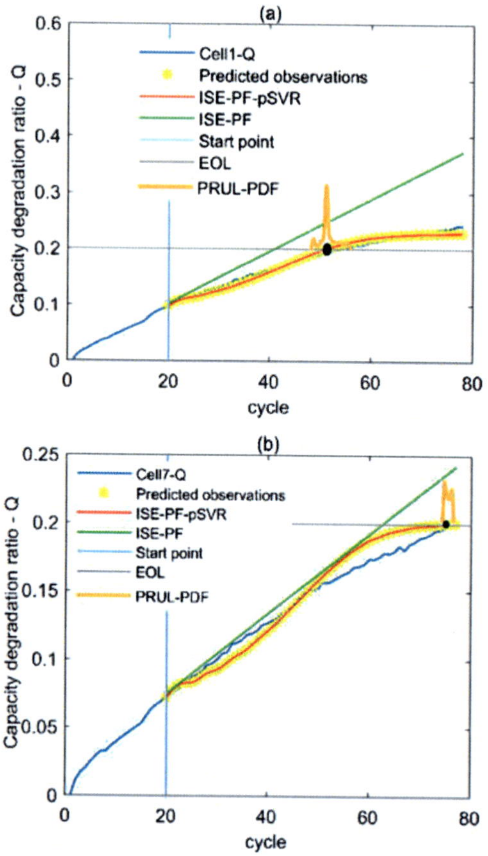

Fig. 4 The Oxford dataset RUL prediction, **a** Cell 1, **b** Cell 7

capacity degradation ratio. On three datasets with different charging and discharging conditions, the proposed model yields RMSE of less than 0.02 and RE of less than 5%. The accurate prediction of the ISE-PF-pSVR model indicates that the fusion approach compensates for the drawbacks of the single model. The ISE model is applicable to the li-ion batteries' degradation under different conditions, while the PF algorithm with the introduction of pSVR is more accurate in tracking the li-ion batteries capacity deterioration.

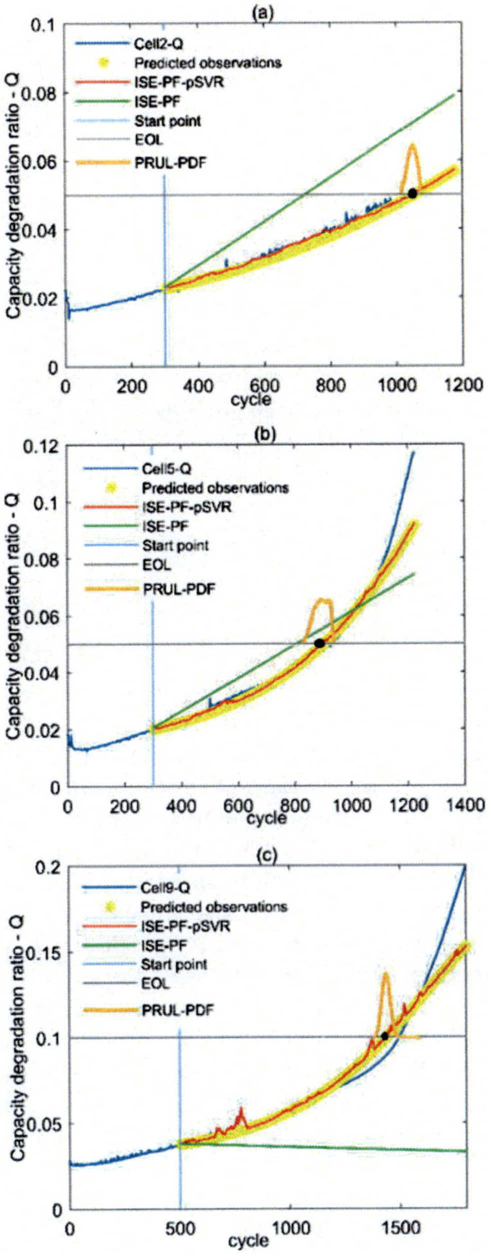

Fig. 5 RUL prediction for the MIT dataset, **a** Cell 2, **b** Cell 5, **c** Cell 9

Table 1 Comparison of RUL prediction results for the three datasets

Li-ion battery		EOL	Model	RMSE	RE (%)
NASA	B0005	0.2	ISE-PF	0.0974	/
			ISE-PF-pSVR	0.0084	0
	B0006	0.2	ISE-PF	0.0265	6.55
			ISE-PF-pSVR	0.0193	3.27
Oxford	Cell1	0.2	ISE-PF	0.0687	24.1
			ISE-PF-pSVR	0.0061	3.70
	Cell7	0.2	ISE-PF	0.0191	17.1
			ISE-PF-pSVR	0.0089	1.32
MIT	Cell2	0.05	ISE-PF	0.0147	30.8
			ISE-PF-pSVR	0.0008	0.29
	Cell5	0.05	ISE-PF	0.0111	13.3
			ISE-PF-pSVR	0.0054	4.83
	Cell9	0.1	ISE-PF	0.0615	/
			ISE-PF-pSVR	0.0113	4.40

5 Conclusion

In this paper, a fusion model ISE-PF-pSVR is created for the li-ion batteries RUL predication, which combines the advantages of semi-empirical degradation and data-driven model with high forecast accuracy. The semi-empirical degradation model is first improved to take into account the charge rate and temperature, and the ISE model is proposed to be applicable to the li-ion batteries' degradation at different working situations. Then, the p-SVR model is built using capacity data to provide observations for the filter algorithm. Finally, simultaneous updating of capacity observations and ISE model parameters is achieved in the PF algorithm. The problem that fixed parameter SEDM is not applicable to dynamic non-linear systems is addressed. The accuracy of PF in predicting RUL is improved by updating the observation through pSVR.

A comparison with the ISE-PF model under the three datasets shows that the proposed ISE-PF-pSVR model is superior, with more accurate prediction results and higher robustness for estimating the li-ion batteries RUL under various operating circumstances.

Acknowledgements This work was supported by the Natural Science Basic Research Program of Shaanxi Province (Grants Number 2021JZ-19).

References

1. Zhang, Y., Qin, C., Lai, X., et al.: A noval capacity estimation method for lithium-ion batteries using fusion eatimation of charging curve sections and discrete Arrhenius aging model. Appl. Energy **251**, 113327 (2019)
2. Liu, C., Wang, Y., Chen, Z.: Degradation model and cycle life prediction for lithium-ion battery used in hybrid energy storage system. Energy **166**, 796–806 (2019)
3. Lyu, Z., Gao, R., Chen, L.: Li-ion battery state of health estimation and remaining useful life prediction through a model-data-fusion method. IEEE Trans. Power Electron. **36**(6), 6228–6240 (2021)
4. Chen, D., Meng, J., Huang, H., et al.: An empirical-data hybrid driven approach for remaining useful life prediction of lithium-ion batteries considering capacity diving. Energy **245**, 123222 (2022)
5. Qin, W., Lv, H., Liu, C., et al.: Remaining useful life prediction for lithium-ion batteries using particle filter and artificial neural network. Ind. Manag. Data Syst. **120**(2), 312–328 (2020)
6. Son, S., Jeong, S., Kwak, E., et al.: Integrated framework for SOH estimation of lithium-ion batteries using multiphysics features. Energy **238**(A), 121712 (2022)
7. Feng, F., Teng, S., Liu, K., et al.: Co-estimation of lithium-ion battery state of charge and state of temperature based on a hybrid electrochemical-thermal-neural-network model. J. Power Sour. **455**, 227935 (2020)
8. Aggab, T., Avila, M., Vrignat, P., et al.: Unifying model-based prognosis with learning-based time-series prediction methods: application to li-ion battery. IEEE Syst. J. **15**(4), 5245–5254 (2021)
9. Severson, K., Attia, P., Jin, N., et al.: Data-driven prediction of battery cycle life before capacity degradation. Nat. Energy **4**, 389–391 (2019)
10. Guo, P., Cheng, Z., Yang, L.: A data-driven remaining capacity estimation approach for lithium-ion batteries based on charging health feature extraction. J. Power Sour. **412**, 442–450 (2019)
11. Tian, J., Xiong, R., Shen, W.: State-of-health estimation based on differential temperature for lithiumion batteries. IEEE Trans. Power Electron. **35**(10), 10363–10373 (2020)
12. Chen, L., Ding, Y., Liu, B., et al.: Remaining useful life prediction of lithium-ion battery using a noval particle filter framework with grey neural network. Energy **244**, 122581 (2022)
13. Zheng, Y., Cui, Y., Han, X., et al.: A capacity prediction framework for lithium-ion batteries using fusion prediction of empirical model and data-driven method. Energy **237**, 121556 (2021)
14. Han, X., Wang, Z., Wei, Z.: A novel approach for health management online-monitoring of lithium-ion batteries based on model-data fusion. Appl. Energy **302**, 117511 (2021)
15. Saha, B., Goebel, K.: Battery data set. In: NASA Ames Prognostics Data Repository, NASA Ames Research Center, Moffett Field, CA (2007). http://ti.arc.nasa.gov/project/prognostic-data-repository
16. Birkl, C., Roberts, M., McTurk, E., et al.: Degradation diagnostics foe lithium-ion cells. J. Power Sour. **341**, 373–386 (2017)
17. Chen, Z., Xu, K., Wei, J., et al.: Voltage fault detection for lithium-ion battery pack using local outlier factor. Measurement **146**, 544–556 (2019)
18. Wang, J., Liu, P., Hicks, J., et al.: Cycle-life model for graphite-$LiFePO_4$ cells. J. Power Sour. **196**, 3942–3948 (2011)

Design of Indoor Air Purification Control System Based on STM32

Xiao Yan Li, Kan Hong Cao, and Zhen Yu Lai

Abstract This paper is about an air purifier control system that can monitor the indoor air environment in real time and can adjust its working mode according to the indoor air condition, which can keep the indoor air fresh. The system is based on the STM32 microcontroller control to develop an indoor air purifier to solve indoor pollution problems. The system is based on the STM32 microcontroller control and is used to solve the indoor pollution problem. The system adds intelligent functions such as air quality detection and remote control to the previous air purifier to achieve smart home control and remote control. It has been tested and proven to be effective.

1 Introduction

As people's expectations of the air quality in their working and living environments continue to rise. And the particulate matter, bacteria and viruses in indoor air are also rising, affecting people's health. Intelligent air purifier systems have also become a subject of research. With the development of science and technology, the country's economy is growing and the standard of living of people is gradually improving. People are beginning to pay more and more attention to their health. According to statistics, modern people spend more than 85% of their time indoors, so indoor air quality has become one of the environmental issues of concern in many countries [1]. Indoor air quality has become a top concern. The enclosed nature of indoor areas is also increasing nowadays, with an increase in pollution sources such as various decoration materials and car emissions. Air is something that people cannot live without, and there are a large number of air pollutants in the rooms where we usually live or work. Examples include dust, germs, formaldehyde, etc. These pollutants may cause a wide variety of respiratory diseases and may lead to serious lung diseases or death.

X. Y. Li (✉) · K. H. Cao · Z. Y. Lai
Guangdong University of Science & Technology, Dongguan 523000, China
e-mail: 365820634@qq.com

K. Nakamatsu et al. (eds.), *Advanced Intelligent Technologies for Information and Communication*, Smart Innovation, Systems and Technologies 365,
https://doi.org/10.1007/978-981-99-5203-8_23

With the rise of the Internet of Things and Big Data, smart home appliances are closer to people's lives and are increasingly preferred by consumers for their simple and intelligent operation. This design is to design an intelligent air purifier control system, which can automatically adjust the working state of the air purifier according to the indoor air condition in real time, and also through the manual mode to adjust the working state of the air purifier [2]. It is convenient for people to use and can also effectively save energy waste.

2 Solution Design

The design of this thesis consists of a microcontroller, a gas detection sensor, an A/D analogue-to-digital conversion section, a human–machine interface display section, a motor drive section, a Bluetooth transmission module and an alarm circuit drive section. The system structure diagram is shown in Fig. 1.

The gas detection module of the air purifier is mainly composed of the SGP30 gas detection module and the GPY1014AU dust detection module. It is used to detect the air quality in the room and transmit the data to the STM32F407 chip for data processing. The chip then adjusts the working state of the motor according to the data detected by the air sensor to control the flow of air into the air purifier filter to purify the indoor air. The system also detects the humidity in the room air and the room temperature by means of a temperature and humidity sensor, which is displayed on a TFTLCD screen. The time and date can also be displayed in real time on the TFTLCD screen. The time and date displayed on the TFTLCD screen can be changed via the keypad module or by connecting your mobile phone to Bluetooth. The data displayed by the air purifier can be seen in the mobile app and the working mode of the air purifier can be modified via the mobile app [3].

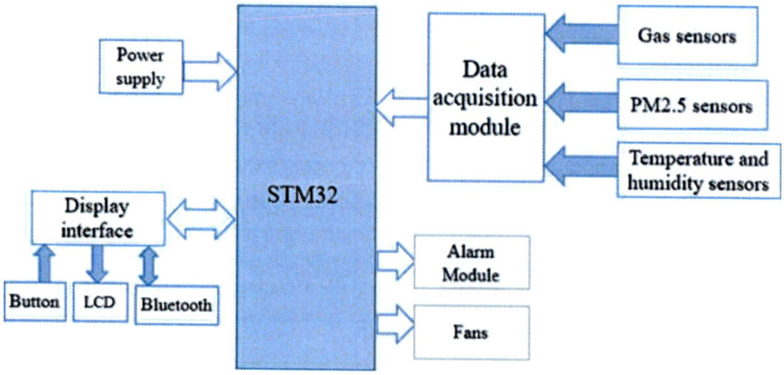

Fig. 1 System framework diagram

3 Hardware Selection

The STM32F407ZET6 is used as the central control unit for this project. This micro-controller meets the functional requirements of this design. The STM32F407 series has the ARM 32-bit CortexTM-M4 as its core and has many functions. For example, the ADC is used for analogue-to-digital conversion and up to four channels of each timer are used for input capture/output comparison of PWM or counting pulses. A wide range of communication is also supported, e.g. I2C, SPI and CAN. Very low power consumption with a maximum operating frequency of 168 MHz.

The DHT11 temperature and humidity sensor is selected as the temperature and humidity detection module in this design. The digital module acquisition technology and temperature and humidity sensing technology of the DHT11 temperature and humidity sensor is its unique technology, very high reliability and terrific stability are its advantages [4]. The DHT11 is a temperature and humidity sensor that has been calibrated with the digital signal before output, and its measurement accuracy humidity range is ±5%RH, ±%2 °C for temperature, a range of 20–90%RH for humidity and a range of 0–50 °C for temperature.

The GP2Y1010AU0F dust sensor was chosen. The GP2Y1010AU0F sensor mainly uses particles entering the light-sensing area so that they affect the output voltage through reflection [5]. The GP2Y1010AU0F dust sensor utilises an optical sensor. The measuring part of the sensor is made up of an infrared light emitting diode (IRED) and a photocell arranged diagonally, which mainly determines the amount of dust in the air by detecting the light reflected back from the dust in the air [6].

The driver chip used to design the L298N driver board is the L298N dual H-bridge DC motor driver chip [7]. The module contains two H-bridge high voltage high current bridge drivers that can be used to drive inductive loads such as DC motors, stepper motors and relay coils. The motor chosen is a brushless motor with dual micro ultra-fine bearings, which provides low noise, high airflow and easy control, making it easy to install on air purifiers and better able to draw polluted air from outside into the air purifier filter.

4 Software Design

The system software program design is composed of TFTLCD display program, PM2.5 sensor driver, data processing program, key processing program, Bluetooth data transmission processing program, temperature and humidity sensor processing program, interrupt processing program, fan motor driver and mode control program, etc. to form the air purifier control system program. To realise the control function of the air purifier, the flowchart of the software design is shown as follows (Fig. 2).

After the system is powered up, the microcontroller is initialised first followed by the initialisation of each sensor. The sensors upload the detected data to the STM32 microcontroller and display it on the TFTLCD screen. The default mode at the start

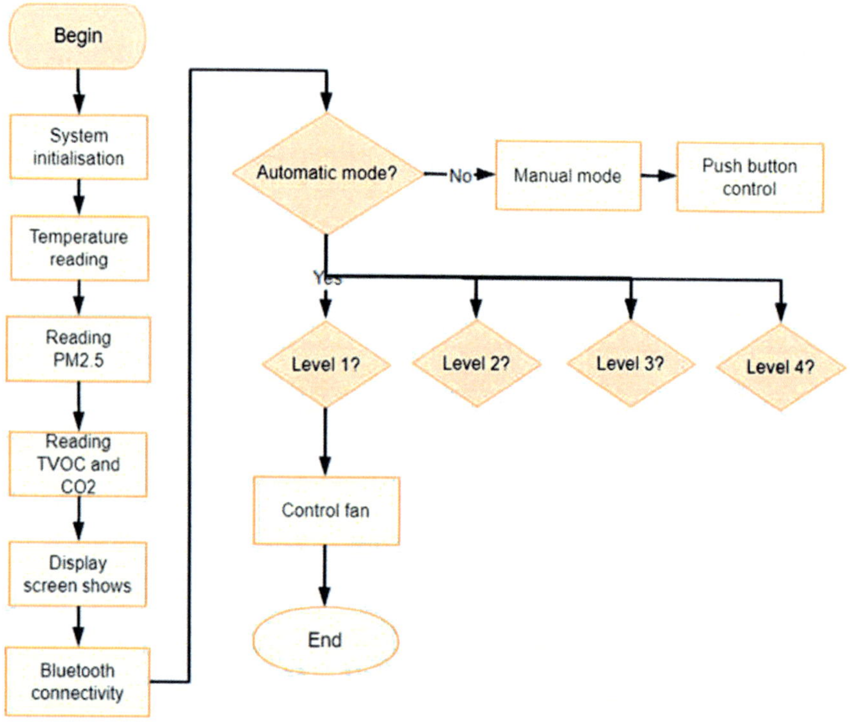

Fig. 2 Software flowchart

of the power-up is automatic mode. The microcontroller adjusts the working state of the air purification according to the concentration of particles or volatile gases detected by the sensors to improve the air purification efficiency, and the working state of the air purifier can also be controlled by real-time monitoring of the indoor air dye value and manual mode via the mobile phone.

5 Simulation and Demonstration

The design is based on the STM32 intelligent air purifier control system, which can display the data acquired by each sensor, can be adjusted to manual mode or automatic mode according to the customer's requirements, and the data collected by the sensor can be uploaded to the mobile phone and can control the air purifier in the mobile phone [8].

The system initialisation diagram is shown in Fig. 1. When the system is first powered on, the microcontroller and peripherals are initialised and the data collected by the sensors are uploaded to the microcontroller. Then, it is displayed through the TFTLCD screen, while the TFTLCD screen also displays the time, date and the working status of the air purifier in real time. The default operating state after system initialisation is automatic working mode (Fig. 3).

The system diagram for manual mode is shown in Fig. 4. The operating mode of the system can be adjusted by pressing KEY2. When the system is adjusted to manual mode, the operating level of the fan can be adjusted by pressing KEY1 and KEY0 to add or subtract, respectively.

The physical diagram of setting the time and date is shown in Fig. 5. The time or date of the system can be changed by pressing a button or by connecting the phone to Bluetooth.

The App page diagram is shown in Fig. 6. The data collected by the microcontroller can be transferred to the mobile phone App via the Bluetooth module, and the time on the screen can be modified via the keys on the mobile phone App, and the working mode and working status of the air purifier can also be switched via the mobile phone terminal.

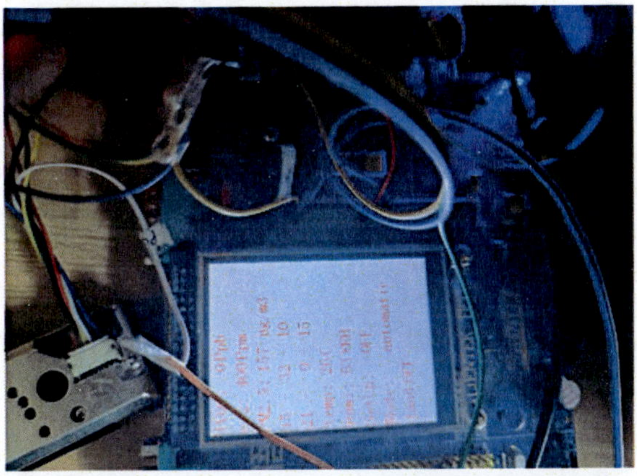

Fig. 3 System initialisation diagram

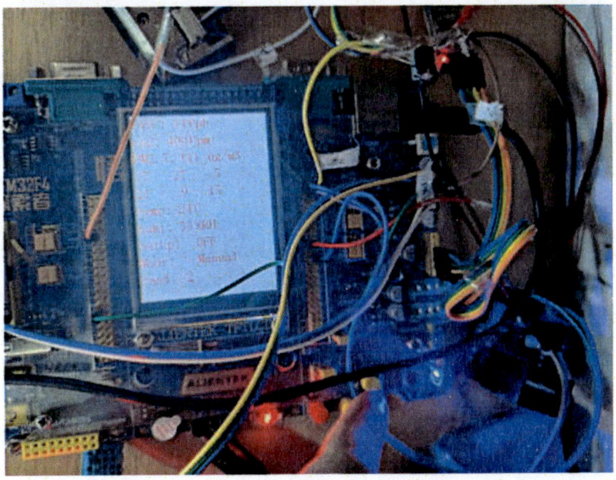

Fig. 4 Manual mode

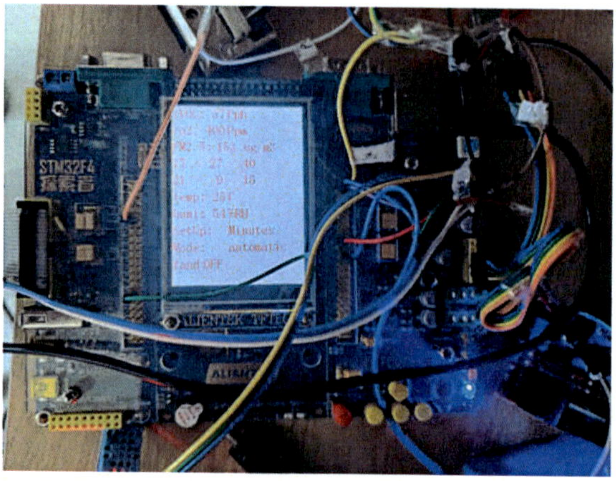

Fig. 5 Setting the time and date

Fig. 6 App page display

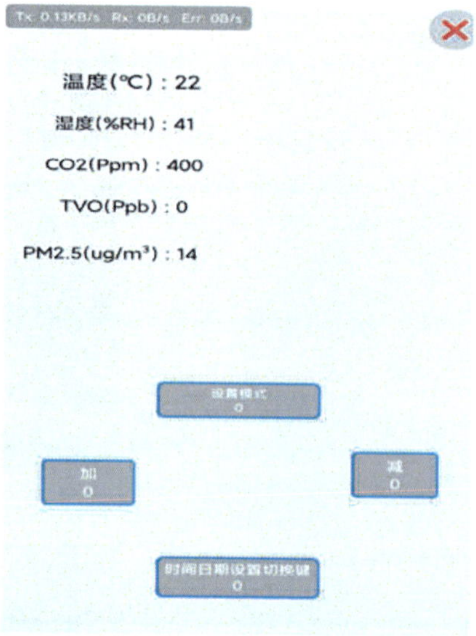

The system diagram when simulating serious pollution is shown in Fig. 7. When the air purifier control system is in automatic mode, the air purifier will adjust the working status of the fan according to the quality of the air, when the air is in serious pollution, the system will adjust the speed of the fan to the maximum and there is an alarm sound to prompt people that the air pollution is too serious at this time, so pay attention to protect yourself.

Through the previous work, the control part of the purifier can work normally, the functional part of which includes the detection, transmission and display part, what

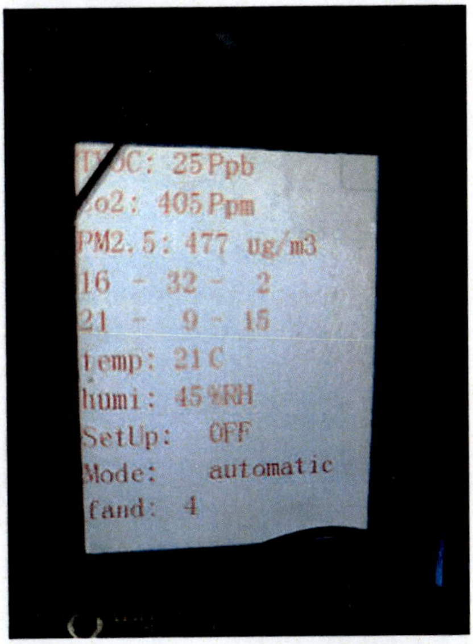

Fig. 7 Simulating the display when pollution is high

is the detection data and reliability of the sensor? The data from Foshan on March 6 and March 8 were compared and analysed, and data were obtained as shown in Tables 1, 2, 3 and 4.

Table 1 Foshan meteorological bureau data for March 6th

Item	08:00	09:00	10:00	14:00	15:00
Temperature (°C)	20	21	21	24	25
Humidity (%)	59	56	55	56	53
PM2.5(ug/m^3)	24	9	14	14	12

Table 2 Indoor testing data on March 6th

Item	08:00	09:00	10:00	14:00	15:00
Temperature (°C)	24	24	24	25	26
Humidity (%)	59	56	55	56	53
PM2.5(ug/m3)	30	15	23	25	20

Table 3 Foshan meteorological bureau data for March 8th

Item	19:00	20:00	21:00	22:00	23:00
Temperature (°C)	22	20	18	17	16
Humidity (%)	43	45	45	48	53
PM2.5 (ug/m^3)	14	11	14	14	14

Table 4 Indoor testing data on March 8th

Item	19:00	20:00	21:00	22:00	23:00
Temperature (°C)	22	22	21	21	22
Humidity (%)	44	41	41	45	40
PM2.5 (ug/m^3)	19	13	15	17	34

As this design uses special sensors for learning, the data difference with the industrial grade sensors is a little big, in the design of PM2.5 value 0–80 ug/m^3 air state for excellent, PM2.5 value 80–200 ug/m^3 for light pollution, PM2.5 value 200–300 ug/m^3 for moderate pollution, PM2.5 value 300–400 ug/m^3 for severe pollution [9]. By comparing the data above, it can be seen that there is a difference between the data tested and the data tested by the weather station, but the air pollution level tested is the same as that of the weather station.

6 Conclusion

This design makes full use of the STM32 to drive various peripherals through the I/O ports of the STM32. The TFTLCD screen is used as the display and the Bluetooth module is used as the bridge between the system and the mobile phone. The air quality of the room is displayed on the screen via gas sensors and temperature and humidity sensors and transmitted to the mobile phone app page via Bluetooth sensors. The time and date are displayed on the screen in real time. In automatic mode, the air purifier adjusts its operating status according to the level of air pollution. It is also possible to adjust the air purifier to manual mode according to people's needs. The air purifier can also be controlled via the mobile phone, which can be used to adjust the time and date of the air purifier in real time, and to adjust the working mode of the air purifier. It is also possible to check the air condition in the room through the mobile phone. The drawbacks and shortcomings of this design are due to the fact that the air quality sensor chosen is a learning version of the sensor and its accuracy is not up to the standard of industrial sensor design, resulting in a certain gap between the readings and the actual data.

7 Funding Project

1. Dongguan SCI-Tech Commissoner Program (20,211,800,500,462).
2. Guangdong Institute of Science and Technology Quality Project: Guangdong University of Science and Technology—Shenzhen Balun Technology Co.Company Electricity Professional Practice Teaching Base.

3. Guangdong University of Science and Technology Student Innovation and Entrepreneurship Training Program Project (S202113719013).
4. Guangdong Institute of Science and Technology Quality Project: Electronic Information Engineering Curriculum Civics Demonstration Major (GKZLGC2021249).

References

1. Yue, C., Tengfei, C., Mengmeng, Y., Lei, L.: Research on the regulation machine of indoor air purifier. Heilongjiang Sci. **12**(08), 40–41 (2021)
2. Di, L.: Research and Implementation of Intelligent Home Indoor Air Quality Monitoring System. Anhui University of Technology (2019)
3. Yang, F.: Design and implementation of indoor air quality detection and alarm system. Electron. Prod. (Z1), 29–30+28 (2020). https://doi.org/10.16589/j.cnki.cn11-3571/tn.2020.z1.010
4. Zhang, P., Hsien-Ying, F., Huo, R.: STM32-based multifunctional air purifier control system development. Electron. Technol. Appl. **43**(03), 80–83 (2017). https://doi.org/10.16157/j.issn.0258-7998.2017.03.020
5. Zhou, W.-P.: STM32-Based Indoor PM2.5 Monitor Design. Anhui University (2016)
6. Jian, R., Chen Kang, L., Min, D.: Intelligent monitoring and purification system design of PM2.5 and harmful gases based on embedded. Jiangsu Sci. Technol. Inf. (12), 67–69 (2016)
7. Xucai, S.: Application of L298N in DC motor PWM speed control system. J. Weifang College **9**(04), 19–21 (2009)
8. Liu, Z.: Design and Implementation of Wireless Gateway Based on STM32 Smart Home. Zhejiang University of Technology (2017)
9. Fumo, Y., Yongliang, M., Kebin, H.: Fine atmospheric particulate matter PM_(2.5) and its research overview. World Environ. (04), 32–34 (2000)

Decentralized Oracle with Mechanism for Reputation and Incentive

Liao Chen, Haomin Zhuang, and Yuxiang Feng

Abstract Blockchain was originally designed as a peer-to-peer network to handle cryptocurrency transactions. And with the advent of smart contracts, more and more applications are using Blockchain as an infrastructure. The smart contract is a mechanism for autonomously executing code based on predetermined criteria without the need for any other third-party interaction. However, Blockchain has a non-negligible drawback in that it cannot use the off-chain data. This is why we need trusted entities called oracle, which is in charge of writing external data into Blockchain. To avoid security issues such as malicious oracles writing incorrect data on Blockchain, this paper introduced a decentralized oracle based on reputation and incentive mechanism. And we also ran a simulation-based experiment to verify the feasibility and trustworthiness of our schema. The findings revealed that our system is capable of effectively defending against attacks and penalizing malicious oracles.

1 Introduction

Blockchain, the underlying support technology for Bitcoin, first appeared in Satoshi Nakamoto's "Bitcoin: A Peer-to-Peer Electronic Cash System" [1], a text that describes in detail how to create a new, decentralized, peer-to-peer transaction system that does not require a foundation of trust. A Blockchain network is a peer-to-peer (P2P) network in which each user has access to the same ledger data. The data from transactions are organized into a list of blocks based on their timestamp. The decentralized nature of Blockchain technology is its distinctive feature. Nodes can make reliable transactions without having to trust each other using cryptographic

These authors contributed to the work equally and should be regarded as co-first authors.

L. Chen · H. Zhuang · Y. Feng (✉)
School of Software Engineering, South China University of Technology, Guangzhou 510000, Guangdong, China
e-mail: yxfeng@scut.edu.cn

K. Nakamatsu et al. (eds.), *Advanced Intelligent Technologies for Information and Communication*, Smart Innovation, Systems and Technologies 365,
https://doi.org/10.1007/978-981-99-5203-8_24

algorithms, timestamps, tree structures, consensus mechanisms, and reward mechanisms, which solves the problems of the current centralized model's poor reliability, low security, and low efficiency.

The Smart contract is a program that can be executed based on defined conditions. It enables, for example, selling when the price of Bitcoin reaches a certain value. Smart contracts are extremely important in the Blockchain world. They not only give programmability to Blockchain's underlying data but they also encapsulate the complex behavior of each node in the network, providing a user-friendly interface for building upper-layer Blockchain applications.

A primary limitation of Blockchain is its inability to interact with the "outside world". This is because Blockchain is deterministic and consistent while the outside world is uncertain which means applications on Blockchain do not have direct access to data from the outside world.

To overcome this limitation of smart contracts, a mechanism for writing external data to the Blockchain is needed. And this mechanism is known as Oracle [2–4]. These oracles are stored on the Blockchain as smart contracts, and other contracts can invoke the interface provided by the oracles to request data. The main function of oracles is to write off-chain data into Blockchain. External data can be a stock price at a certain time, game results, weather conditions, and so on. Oracle collects information and sends it into the Blockchain in the form of transactions to enable the iteration of the Blockchain with off-chain data. In summary, oracle is the middleware between the real world and Blockchain.

2 Related Methods

2.1 Centralized Oracle

Data from a single source are used in the centralized trust model. The centralized trust model is efficient, but it has a single point of failure because the data's availability, accessibility, and level of certainty regarding its authenticity are all dependent on one node. The mechanism of the centralized oracle is more intuitive and in line with the data sources of traditional society.

Provable [5] is a Web API-based Oracle service. It provides data request services for multiple Blockchain intending to create a trusted data gateway between the Blockchain and the off-chain world. The centralized oracle Town Crier [6] delivers Ethereum-based data feedback. It focuses on providing reliable data for smart contracts and Blockchain applications through Inter Software Guard Extension. However, the types of APIs and data streams that Town Crier may give are limited. It is also vulnerable to a single point of failure due to its centralized structure.

2.2 *Decentralized Oracle*

The design of the decentralized oracle mechanism is in line with the Blockchain's distributed ideology, which focuses on enhancing the fault tolerance of the entire prophecy machine system by having a large number of trusted nodes working together to provide data services. The decentralized oracle does not enhance the trustworthiness of the oracle through technology, but rather through economic incentives and multiple signatures to achieve data trustworthiness.

ASTRAEA [7] is a decentralized oracle that can resist malicious voting attacks by using different roles (voters and verifier) for aggregating the result. Meanwhile, it organizes all the propositions into the form of a list and randomly selects a proposition from the list for voters during the voting process, thus avoiding the problem of voters wanting to make a targeted attack on a certain proposition. By doing so the correctness of data is guaranteed. Chainlink [8] is an Ethereum-based decentralized oracle. It focuses on providing reliable data to users by accessing trustworthy data resources through a specified API. Augur [9], a decentralized Blockchain-based prediction marketplace, is another worth mentioning. Augur is heavily influenced by Truthcoin [10], which prohibits the malicious behavior of participants by introducing a reporting and dispute system. Witnet [11] is a reputation-based decentralized oracle. The reputation mechanism on Witnet rewards the successful majority consensus witnesses while punishing the erroneous witnesses by deducing their reputation. Aeternity [12] is an open-source decentralized application platform that utilizes a decentralized oracle system, which can be used to request and access off-chain data from various providers.

3 Our Proposal Oracle

In this paper, we propose a new scheme in the reputation mechanism and incentive mechanism work together to deal with the robust performance design of a decentralized oracle network. In our proposal, the adaptive reputation algorithm can aggregate the result from more professional oracles, which prevents attackers' oracles from joining a network at a low cost. Meanwhile, the incentive mechanism introduces gambling and reputation to calculate their prize, which forces them to give a guarantee of good behavior to receive more rewards. Our proposal is adaptive to most situations and reduces human participation (Table 1).

Table 1 Explanation of symbols

Symbol	Meaning
$\gamma_{i,j}$	The reputation of i-th oracle after j-th task
β	The coefficient of attenuation
$\alpha_{i,j}$	The reputation value i-th oracle acquires in the j-th task
$score_{i,j}$	The score of i-th oracle in the j-th task
$stake_{i,j}$	The amount of stake from i-th oracle in the j-th task
$prize_{i,j}$	The amount of prize from i-th oracle gets in the j-th task
$pool_j$	The amount of prize pool in the j-th task
K	There are K oracles in the network
K_W	There are K correct oracles in the j-th task
$result_{i,j}$	The result from i-th oracle in the j-th task
$result_j$	The result of j-th task after aggregating
$P_{x,j}$	The sum of the proportion of oracles in $D_{x,j}$
$D_{x,j}$	The class with oracles' $result_{i,j}$ are x in the j-th task

3.1 Reputation Mechanism

The reputation mechanism is related to the aggregation of the result and the calculation of the reward. So it will encourage oracles to provide reliable service consistently with less downtime.

Every oracle in the decentralized oracle network owns a value of reputation, which is changing according to the performance of the oracle in previous tasks. The reputation takes oracles' history performance into consideration because we tend to trust the most reliable oracles that provide the correct result in the past. But even the most honest oracles will experience downtime. So in this paper, the calculation considers time as a dimension of reputation and assigns high reputation values to oracles that have performed well recently.

When a new oracle first joins the decentralized oracle network, the reputation of the oracle is set to 0, which means that the oracle should accumulate its reputation through the correct answers in the later tasks. The reputation of one reputation is changed after a task. Oracles providing the correct answer in the task can get a 100 reputation score in this task, while oracles providing wrong answers get a 0 reputation score. Meanwhile, those oracles that do not participate in this task will be considered to be negligent and also get a 0 reputation score in this task. The reputation in this task is multiplied by a coefficient and then added to the historical reputation which is multiplied by another coefficient. The historical reputation declines by multiplying a coefficient and adds to the reputation given in a new task to form a new reputation using in the next task.

$$\gamma_{i,j} = \beta * \gamma_{i,j-1} + (1 - \beta) * \alpha_{i,j} \tag{1}$$

$$\alpha_{i,j} = \begin{cases} 100 & True \\ 0 & False \end{cases} \quad (2)$$

One of the advantages of using this model is that there is no need to keep reputation scores for each past task. In order to incorporate the smart contract calculation capabilities in a realistic manner, a simple linear equation is employed rather than a sophisticated non-linear equation. Here the $\alpha_{i,j}$ is 100 if the i-th oracle provides the correct result in the j-th task. But if the result is incorrect, it should be 0. From the formula, a high reputation score oracle will drop down quickly if it has downtime and could not provide a correct result.

3.2 Incentive Mechanism

The incentive mechanism is a supplement to the reputation mechanism. The more oracles the network has, the more robustness it has. To attract more oracles, our decentralized oracle network would have an adaptive incentive mechanism to organize the prize, which is an economic reward like Bitcoin or the token in the Blockchain, or money in real. Another reason for introducing an incentive mechanism is that if our oracle network relies solely on the reputation mechanism, then the penalty an oracle receives for incorrect data response is only the deduction of reputation value, then a malicious oracle can attack the network of oracles by responding some requests correctly and some incorrectly.

At first, gambling is a fascinating method not only for the oracles giving their guarantee to their result but also the showing their confidence toward their result, which makes sense to trust to whom give the higher stake. But considering the situation that those oracles owning high reputation scores and daring to trust their result would give very high stake and then give the most prize. We limit the amount of the stake, the maximum of the stake should not overtake the amount of the payment from users. The prize pool $pool_j$ consists of users' payments and all oracles' stakes. When the answer to the task is generated, the money in the prize pool is distributed to the winners who provide the correct answer to this task. The amount of $prize_{i,j}$ for i-th oracle depends on the proportion of $score_{i,j}$ among all winners.

In this paper, our proposal believes the honest oracles who have high reputation scores should have the extra rewards, which encourages oracles to reduce their downtime and work consistently.

We define a score related to oracles' reputation and the stake. And the scores affect their importance in the result aggregation and prize distribution. Considering the attackers can own a large number of oracles and join the network before a specific task, then all those oracles give a wrong answer to mislead the answer aggregation. We set a threshold, which is 50 in this paper. Only those oracles whose reputations are greater than this threshold can play a defining role in the answer aggregation. However, in the situation that the network just start, none of the oracles has a reputation that is greater than the threshold. So we design the function $f(x)$ to tackle the cold boot process.

$$score_{i,j} = \frac{f(\gamma_{i,j-1})}{\sum_{i=1}^{K} f(\gamma_{i,j-1})} * stake_{i,j} \tag{3}$$

$$f(x) = \begin{cases} 51 + 10(x - 50) & x > 50 \\ 1 + x & x \leq 50 \end{cases} \tag{4}$$

$$prize_{i,j} = \frac{score_{i,j}}{\sum\limits_{i=1}^{K_W} score_{i,j}} * pool_j \tag{5}$$

3.3 Result Aggregation

We propose that any value including continuities can be viewed as discontinuities when aggregating the result. When receiving different results from the oracles, the management contract can divide those oracles with the same result into one class, then calculates the proportion of the sum of the score of each oracle in each class. The highest score class's result will be the result in this request.

$$P_{x,j} = \sum \frac{score_{i,j}}{\sum\limits_{i=1}^{K} score_{i,j}} \quad i \in D_{x,j} \tag{6}$$

The result x of the class with the highest proportion $P_{x,j}$ is the final result. However, if there are two classes with the highest proportion means oracles do not reach a consensus. So the solution to this situation is that the management contract repeats the whole process again.

3.4 Whole Process

Step I: Users can invoke the interface provided by management contract to send requests to the management contract to get data and also provide the remuneration for this request. The management contract will record the information about the request such as user's address, amount of the rewards.

Algorithm 1: dataRequest

```
Data: User's data request
Result: Response to the request
initialization;
if Oracles are not available then
    Abandon the transaction;
else
    Log the current request;
    getAllOracleResult();// get results from all oracles
    Aggregate the results to get final result;
    if Not reach consensus then Abandon the transaction;
    else Return the final result;
end
```

Step II: According to the request problem id, management contract will send message to the oracles, which subscribe to this problem to get result.

Step III: Management contract gets the result and the stakes form oracles.

Algorithm 2: getAlloracleResult

```
Data: User's data request
Result: Oracles' data response
initialization;
for i ← 1 to n do
    Send request to Oracle_i;
    tempAddr ← Oracle_i.address;
    tempResult ← Oracle_i.result;
    tempStake ← Oracle_i.stake;
    if tempStake > 0 then
        oracles[tempAddr].result ← tempResult;
        oracles[tempAddr].stake ← tempSkate;
    else
        do nothing
    end
end
```

Step IV: Management contract calculates the $score_{i,j}$ of each oracles. Then it aggregates the result from all oracles and get a final result.

Algorithm 3: aggregateResult

```
Data: Oracles' results to data request
Result: Aggregated result
initialization;
flag ← false;
for i ← 1 to n do
    calcuOracleWeight(Oracle_i);
end
foreach result of results set do
    calcuResultSumWeight(result);
end
foreach result of results set do
    if result.sumWeight > 0.5 then
        request.result ← result;
        flag ← true;
    else
        do nothing
    end
end
```

Step V: Management contract returns the final result to user's smart contract.

Step VI: Management contract distributes the $pool_j$ to oracles with correct result.

Step VII: Management contract updates the reputation of all oracles.

Algorithm 4: updateReputation

```
Data: Oracles' reputation value acquire in the j-th task
Result: Oracles' new reputation score
initialization;
for i ← 1 to n do
    Oracle_i.reputation ←
      β × Oracle_i.tempReputation + (1 − β) × Oracle_i.reputation
end
```

The following Fig. 1 shows the flowchart in our design.

4 System Analysis

4.1 Reliability Analysis

For an oracle, its reputation is enhanced if its response data are consistent with the final aggregated data. The improvement in reputation score will lead to a greater weighting of the oracle in future data aggregation and a larger share of the bonus.

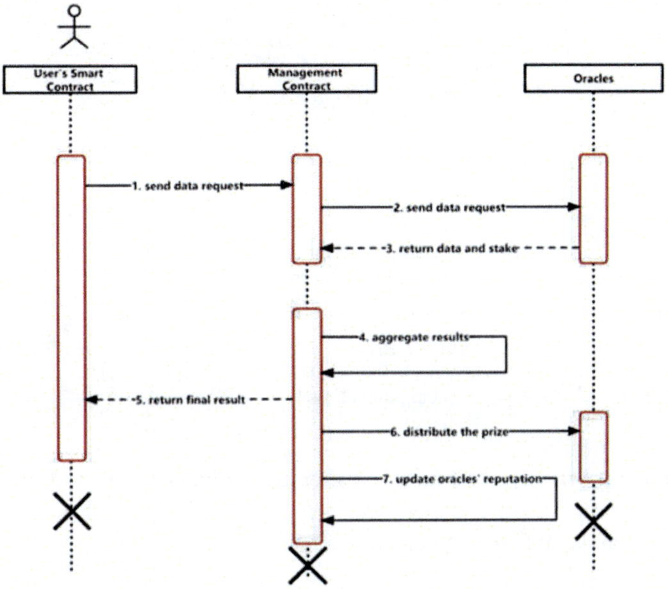

Fig. 1 Data request service process

And for malicious oracles, false data reports deduct all its deposits and reduce its reputation score. Such a mechanism creates an effective closed-loop that motivates the oracles to perform correct data reporting.

Figure 2 shows our design can resist the oracles experiencing downtime well. And the situation in that attackers use oracles to send incorrect answers to the network is the same as in Fig. 2. For an oracle that does not experience errors and downtime, its long-term infallible performance will continuously enhance its reputation score as shown in asterisk in Fig. 2. And for those oracles that have the probability of making incorrect data responses, their reputation score will be diminished by their mistakes. From the graph, we can learn that the higher the probability of a prophecy machine making a mistake, the smaller the mean of its reputation value and the larger the variance. For a reputable oracle, a sudden large number of errors can cause his reputation to drop significantly, which can seriously affect his future earnings.

4.2 *Resistance to Attacks*

If a malicious participant wants to control the final consensus outcome, then it needs to control more than 50% of the oracles, assuming that they all have the same reputation score. In this paper, since the deposit of incorrect oracles will be paid to correct oracles, when the benefits that participants can obtain from reporting the correct data are much more than bribes, it is more likely that the oracles will fend off attackers

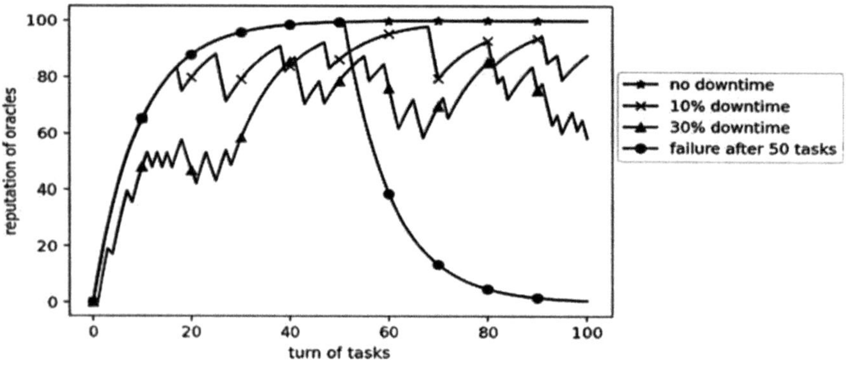

Fig. 2 Reputation of oracle experiencing downtime

and report data correctly. And for the malicious participant, it would be extremely expensive to control the oracle network, which can be far more than the benefits it can obtain from its malicious behavior.

5 Conclusion and Future Work

5.1 Reliability Analysis

Now, many answers of oracles are non-discrete values or even vectors instead of simply Boolean or discrete types. So we need to develop an algorithm to deal with the problems in the aggregation of non-discrete values or vector values from different oracles. Especially, with the development of machine learning, many companies provide services that predict answers from requesters' input. For example, some outputs in the NLP tasks are vectors, so how to aggregate the answers from different oracles and how to manage the reputation of oracles with different answers are two main problems to be solved in the future.

5.2 Conclusion

In this paper, we propose a reliable decentralized oracle system that is based on reputation and incentive mechanism. A simulation-based experiment is conducted to evaluate our reputation mechanism. It shows that the malicious behavior of the oracle will bring down its reputation value significantly and cause it to take much less weight in aggregating results in future tasks. In future work, we will try to extend our scheme to accommodate all types of data requests, not just Boolean types.

References

1. Nakamoto, S.: Bitcoin: a peer-to-peer electronic cash system, 1–2 (2008)
2. Xu, X., Pautasso, C., Zhu, L., Gramoli, V., Ponomarev, A., Tran, A.B., Chen, S.: The blockchain as a software connector. In: 2016 13th Working IEEE/IFIP Conference on Software Architecture (WICSA), pp. 182–191. Venice (2016)
3. Moudoud, H., Cherkaoui, S., Khoukhi, L.: An iot blockchain architecture using oracles and smart contracts: the use-case of a food supply chain. In: 2019 IEEE 30th Annual International Symposium on Personal, Indoor and Mobile Radio Communications, pp. 1–6. Istanbul (2019)
4. Al Breiki, H., Al Qassem, L., Salah, K., Rehman, M.H.U., Sevtinovic, D.: Decentralized access control for iot data using blockchain and trusted oracles. In: 2019 IEEE International Conference on Industrial Internet (ICII), pp. 248–257. Orlando (2019)
5. Provable Dev Community: Provable Documentation (2019). https://docs.provable.xyz. Last Accessed 13 Oct 2022
6. Zhang, F., Cecchetti, E., Croman, K., Juels, A.: Town crier: an authenticated data feed for smart contracts. In: Proceedings of the 2016 ACM CIGSAC Conference on Computer and Communications Security, pp. 270–282. Vienna (2016)
7. Adler, J., Berryhill, R., Veneris, A., Poulos, Z., Veira, N., Kastania, A.: Astraea: a decentralized blockchain oracle. In: 2018 IEEE International Conference on Internet of Things (IThings), pp. 1145–1152. Halifax (2018)
8. Ellis, S., Juels, A., Nazarov, S.: Chainlink: a decentralized oracle network. https://www.academia.edu/41100629/ChainLink_A_Decentralized_Oracle_Network. Last Accessed 13 Oct 2022
9. Peterson, J., Krug, J.: Augur: A decentralized, open-source platform for prediction market. https://arxiv.org/abs/1501.01042. Last Accessed 13 Oct 2022
10. Sztorc, P.: Truthcoin, peer-to-peer oracle system and prediction marketplace. https://github.com/psztorc/Truthcoin. Last Accessed 13 Oct 2022
11. De Pedro, A.S., Levi, D., Cuende, L.I.: Witnet: a decentralized Oracle network protocol. http://arxiv.org/abs/1711.09756. Last Accessed 13 Oct 2022
12. Hess, Z., Malahov, Y., Pettersson, J.: Aeternity blockchain: the trustless, decentralized and purely functional oracle machine. https://aeternity.com/aeternity-blockchainwhitepaper.pdf. Last Accessed 13 Oct 2022

Super-Resolution Reconstruction Based on Kernel Regression Method

Guohong Liang, Guangrong Liu, and Junqing Feng

Abstract Kernel regression method is one of the non-parametric nonlinear regression estimation methods. The univariate regression function is estimated by one-dimensional kernel function. Based on Taylor expansion, the value of regression coefficient is obtained by optimization method. Using the same method, two-dimensional kernel regression is studied. Finally, the adaptive control kernel is summarized. On the basis of the non-parametric estimation kernel regression model, the two-dimensional kernel regression function is extended to three-dimensional and each pixel in the video sequence is represented as a three-dimensional Taylor expansion. The displayed coefficients are obtained by the locally weighted least squares method, and the weights of the kernel regression are used to capture the spatiotemporal local motion information, avoiding the explicit sub-pixel precision motion estimation. Experiments in the standard test video database show that the algorithm has better reconstruction effect and larger scope of application, and can be used for videos with local and complex motion. The 3D-SKR super-resolution reconstruction algorithm solves the sub-pixel accuracy registration problem of the previous super-resolution reconstruction algorithms and extends the application range of sequence image super-resolution reconstruction to any sequence image, which is of great significance. The new algorithm framework proposed in this paper improves the robustness of the original algorithm and eliminates the influence of outliers on the reconstruction results. The algorithm proposed in this paper can adaptively capture the structural information of the image, has a good suppression effect on noise, and can effectively eliminate the artifacts caused by inaccurate motion estimation.

G. Liang (✉) · G. Liu · J. Feng
Foundamentals Department, Air Force Engineering University, Xi'an 710051, China
e-mail: liangguohong321@163.com

K. Nakamatsu et al. (eds.), *Advanced Intelligent Technologies for Information and Communication*, Smart Innovation, Systems and Technologies 365,
https://doi.org/10.1007/978-981-99-5203-8_25

1 Introduction

Regression analysis is a statistical analysis method that examines the changing law of random variables on the basis of analyzing the correlation between random variables through observational data [1–3]. It selects a regression model with a better fitting effect based on the specific performance of the correlation relationship or uses a scatter plot. In layman's terms, it is to establish a mathematical expression between random variables, so as to determine whether the change of one or several variables affects another specific. The degree of influence of variables provides a basis for people to predict and control. Regression analysis has a wide range of applications, including finance, economics, business, meteorology, law, medicine, biology, physics, engineering, chemistry, education, history, sports, military, psychology and sociology, etc. Regression analysis includes the following steps: problem statement; selection of relevant variables; data collection; model setting; selection of fitting method; model fitting; model demonstration; According to different assumptions, regression analysis can be divided into parametric regression, non-parametric regression kernel and semi-parametric regression. Kernel regression method is a non-parametric nonlinear regression in mathematical statistics. The weight is assigned to the data that is relatively close to the estimated point, and the kernel regression theory is estimated by the kernel regression interpolation function. At present, it is wide-ranging used in image processing and has achieved remarkable denoising and interpolation effects [4–6].

Generally, the regression model is:

$$y_i = R(t_i) + \varepsilon_i, i = 1, 2, 3, \ldots n$$

where ε_i is a random variable that satisfies: $E(\varepsilon_i) = 0$, $D(\varepsilon_i) = \sigma^2 < \infty$, ε_i and $\varepsilon_i (i \neq j)$ is irrelevant. R(t) is the conditional expectation in Y condition, namely:

$$\mathrm{R(t)} = \mathrm{E}(Y|T = t) = \int yp(y|t)dy = \frac{\int yp(t, y)dy}{p(t)}$$

where T and Y are the random variable and the joint probability density function is $p(x, y)$, and the marginal probability density of T is p(t).

Nadaraya-Watson kernel estimate:

$$\widehat{R}_{NW}(t; h) = \frac{\frac{1}{nh}\sum_{i=1}^{n} K\left(\frac{t-t_i}{h}\right)y_i}{\frac{1}{nh}\sum_{i=1}^{n} K\left(\frac{t-t_i}{h}\right)}$$

where h is the bandwidth, window width or smoothing parameter, its size depends on the data and determines the degree of participation of each data point; K(·) is the kernel function, which is the weight function in the weighted average, the range and shape control the utilization of the data points used in the estimation R(t) and the degree of number [7–10]. Satisfy: K(t) ≥ 0, $\int K(t)dt = 1$.

Common kernel functions are:
Gaussian kernel function: $\frac{1}{\sqrt{2\pi}}exp\left(-\frac{1}{2}t^2\right)$,
Uniform kernel function: $\frac{1}{2}I(|I| \leq 1)$,
Trigonometric kernel function: $(1-|t|)I(|I| \leq 1)$,
Ipanichkov: $\frac{3}{4}\left(1-t^2\right)I(|I| \leq 1)$,
Cosine kernel function: $\frac{\pi}{4}\cos\left(\frac{\pi}{2}t\right)I(|I| \leq 1)$,
Quadratic kernel function: $\frac{15}{16}\left(1-t^2\right)I(|I| \leq 1)$,
The three-weight kernel function is: $\frac{35}{32}\left(1-t^2\right)I(|I| \leq 1)$,
$\widehat{p}(t)\frac{1}{nh}\sum_{i=1}^{n}K\left(\frac{t-t_i}{h}\right)$ is an estimate of the probability density function $p(t)$.

2 Classic Kernel Regression Method

Let the two-dimensional regression model be:

$$y_i = R(t_i) + \varepsilon_i, t_i = (\mu_i, \sigma_i), i = 1, 2, 3, \ldots n$$

where the regression function $R(t_i)$ is the true value at the coordinate point μ_i, σ_i is the observed approximation, ε_i is the independent and identically distributed error with zero mean, n is the number of sampling points in the neighborhood [10–14].

Given that the form of the regression function $R(t_i)$ is uncertain, in order to estimate its value at any point t under the given data. Assuming that the point t to be found is near the known data sampling point t_i, its N order Taylor Expansion is

$$\begin{aligned}
R(t_i) &\approx R(t) + \{\nabla R(t)\}^T(t_i - t) \\
&\quad + \frac{1}{2!}(t_i - t)^T\left\{\nabla^2 R(t)\right\}vec\left\{(t_i - t)(t_i - t)^T\right\} + \cdots \\
&= R(t) + \{\nabla R(t)\}^T(t_i - t) + \frac{1}{2!}vec^T\left\{\nabla^2 R(t)\right\}vec\left\{(t_i - t)(t_i - t)^T\right\} + \cdots \\
&= \omega_0 + \omega_1^T(t_i - t) + \omega_2^T vec\left\{(t_i - t)(t_i - t)^T\right\} + \cdots
\end{aligned}$$

where ∇ and ∇^2 are the gradient operator Hessian operator,

$$\omega_0 = R(t)$$

$$\omega_1 = \nabla R(t) = \left[\frac{\partial f(t)}{\partial \mu}\ \frac{\partial}{\partial \sigma}\right]^T \triangleq \left[\omega_{11}\ \omega_{12}\right],$$

A semi-vectorized operator that converts the lower triangular part of a symmetric matrix (converts a symmetric matrix to a vector in a certain order), for example:

$$vec\begin{bmatrix} a & b \\ b & c \end{bmatrix} = [a, b, c]^T$$

then,

$$\omega_2 = \nabla^2 R(t) = \frac{1}{2}\left(\frac{\partial^2 R(t)}{\partial \mu^2} \ \frac{\partial \mu^2}{\partial \mu \partial \sigma} \ \frac{\nabla^2 R(t)}{\partial \sigma^2} \right)^T \triangleq \frac{1}{2}\left[\omega_{21} \ \omega_{22} \ \omega_{23} \right].$$

If the Taylor expansion is regarded as a local approximation of the regression function, then the estimated value of the parameter ω_0 is the estimation of the regression function $R(t)$ based on the data. The parameter $\{\omega_i\}_{i=1}^N$ gives the n order differential information of the regression function $R(t)$.

Equation is based on the local approximation, and the sampling points in the local neighborhood will be given corresponding weights, and the weight of the sampling points closer to the value to be estimated is higher than the weight of the sample points farther from the value to be estimated t. Therefore, on the basis of the least squares method, the following problems are obtained for the optimization:

$$\lim_{\{\omega_0\}_{i=0}^N} \sum_{i=1}^{M} \left[\sigma_i - \omega_0 - \omega_1^T (t_i - t) - \omega_2^T vec\left\{ (t_i - t)(t_i - t)^T \right\} - \cdots \right]^2 K_H(t_i - t)$$

where M is the known number in the neighborhood window.

$$K_H(t_i - t) = \frac{1}{\det(H_i)} K\left(H^{-1} t\right)$$

where H_i is the smoothing matrix parameter, $K_H(x)$ is the kernel function, and satisfies:

$$\int t K(t) dt = 0, \int t t^T K(t) dt = h^2 I$$

That is, it is symmetrical about the y axis and takes its maximum value at zero. As long as the functions that meet the above requirements can be used as kernel functions, different kernel functions have different effects on the accuracy of the kernel regression calculation. The kernel function is a Gaussian function.

3 Kernel Regression Method with Control Kernel

In order to be suitable for various types of data and make the regression model have the ability to effectively deal with nonlinear problems, the gray value will be introduced on the basis of the above-mentioned kernel regression method to determine the weight by position, that is, the air force distance and the gray distance will be considered, this method is called adaptive kernel regression method. The optimization problem becomes

$$\lim_{\{\omega_0\}_{i=0}^{N}} \sum_{i=1}^{M} \left[\sigma_i - \omega_0 - \omega_1^T(t_i - t) - \omega_2^T vec\{(t_i - t)(t_i - t)^T\} - \cdots\right]^2$$

$$K_{H_{adapt}}(t_i - t, y_i - y)$$

The most intuitive processing of adaptive kernel function $K_{H_{adapt}}$ is

$$K_{H_{adapt}}(t_i - t, y_i - y) = K_H(t_i - t)K_{H_1}(y_i - y)$$

where H and H_1 are the spatial bandwidth and grayscale bandwidth.

In fact, $K_{H_1}(y_i - y)$ is a function of the local gradient of the gray level in its neighborhood, use the value of this function to weight the data. Consider the estimation in favor of the edge, forming a kernel function that controls the shape of the elliptical outline along the direction of the local edge structure.

$$K_{H_{adapt}}(t_i - t, y_i - y) = K_{H_i^s}(t_i - t)$$

The smooth matrix of adaptive kernel regression is $H_i^S = hC_i^{-1/2}$, the kernel function is:

$$\begin{aligned} K_{H_i^s}(t_i - t) &= \sqrt{\frac{\det(C_i)}{\left(2\pi h^2\right)^3}} exp\left\{-\frac{1}{2h^2}\left\|C_i^{1/2}(T - T_i)\right\|_2^2\right\} \\ &= \sqrt{\frac{\det(C_i)}{\left(2\pi h^2\right)^3}} exp\left\{-\frac{1}{2h^2}(T_i - T)^T C_i (T_i - T)\right\} \end{aligned}$$

where C_i is the symmetric covariance matrix, and the covariance matrix is estimated by singular value decomposition.

4 Adaptive Estimation of Super-Resolution Reconstruction Algorithms

The adaptively estimated super-resolution reconstruction sequence is proposed in this paper [15, 16].

Input: $\{t_i\}_{i=1}^{N}$ is the input low-resolution video sequence, where N is the sampling point and t_i is the total number of sampling points in the local space–time of the center.

Step 1 Set the global smoothing parameter h and the spatial support size of the kernel function, and use the second-order three-dimensional classical adjustment kernel regression to calculate the gradient $\omega_1^{(0)} = [Z_{x1}(\cdot), Z_{x2}(\cdot), Z_{x3}(\cdot)]^T$ of the sampling point at the sampling point position $\{t_i\}_{i=1}^N$.

Step 2 According to the gradient value obtained in the first step, calculate their covariance matrix C_i and smoothing matrix H_i^S for all sample points $\{t_i\}_{i=1}^N$.

Step 3 Set the global smoothing parameters and space and time support dimensions of the three-dimensional adjustment kernel regression. According to the covariance matrix obtained in the second step, the second-order three-dimensional adjustment kernel regression is used to calculate the adjustment kernel function and the unknown pixel value $R(t)$of the pixel of interest, and update its value. Gradient $\omega_1^{(1)}$.

Step 4 Set the number of iterations, and iteratively update the covariance matrix C_i and smoothing matrix H_i^S according to the gradient $\omega_1^{(1)}$ to get better orientation information.

Step 5 According to the more accurate direction information obtained, the three-dimensional adjustment kernel regression is used to calculate the spatial distance of the sampling points and the weight $W_i(\cdot)$ of the adjustment kernel to determine the position of the low-resolution image on the high-resolution grid, and then output the high-resolution image through weighted interpolation rate image.

5 Experiment

Since the 3D-SKR algorithm realizes super-resolution reconstruction without registration, it is not comparable with some existing classical super-resolution reconstruction algorithms (POCS, MAP, regularization, etc.). Therefore, the experiment verifies the effectiveness of the improved algorithm in this paper by comparing the original algorithm, the original algorithm after median filtering and the improved algorithm in this paper.

After blurring and down-sampling the standard test videos "foreman" and "akiyo", the first 30 frames of each video are taken and the video frames are blurred with a 3 × 3 uniform point spread function with a downsampling factor of 2. The following figures are the 1st frame, the 6th frame and the 16th frame of the original algorithm, the original median algorithm and the algorithm in this paper (Fig. 1).

It can be intuitively seen from the images of this group of experimental results that the algorithm in this paper can better eliminate the influence of outliers on the original 3D-SKR algorithm. The reconstruction results of the algorithm proposed in this paper have a good deblurring effect and the reconstruction results are better, the texture details are rich, and the noise can be suppressed.

Fig. 1 Reconstruction results of the three algorithms

6 Conclusion

The 3D-SKR super-resolution reconstruction algorithm solves the sub-pixel accuracy registration problem of the previous super-resolution reconstruction algorithms and extends the application range of sequence image super-resolution reconstruction to any sequence image, which is of great significance. However, its algorithm is very sensitive to outliers. The new algorithm framework proposed in this paper improves the robustness of the original algorithm and eliminates the influence of outliers on the reconstruction results. However, the algorithm still has defects, such as the high computational complexity of 3D data, how to solve this problem still needs further research. For videos with local motion or violent motion, it is difficult for traditional super-resolution reconstruction algorithms to achieve sub-pixel accuracy registration results. In this paper, the linear combination of spatial and temporal local adjacent sampling points is used to represent the pixels to be reconstructed. By adjusting the weight of kernel function to capture motion information, a video super-resolution reconstruction algorithm based on adaptive motion estimation without explicit motion estimation is proposed. Experiments on standard test videos show that the algorithm proposed in this paper can adaptively capture the structural information of the image, has a good inhibitory effect on noise, and can effectively eliminate the artifacts caused by inaccurate motion estimation.

References

1. Zhengrong, Z., et al.: Interpolation method for edge preserving Kernel regression. Comput. Eng. **37**(19), 194–196 (2011)
2. Bingbing, T., et al.: Image super-resolution reconstruction based on kernel regression regular term. Radio Eng. **39**(4), 17–19 (2009)
3. Fengjie, X., et al.: Research on welding X-ray image segmentation based on adaptive kernel regression kernel rough set. J. Changshu Institute Technol. **32**(5), 46–48 (2018)

4. Donghuan, J., et al.: Image enlargement algorithm based on weighted ENO. Comput. Eng. **35**(6), 223–224 (2009)
5. Casciola, G., et al.: Edge-driven image interpolation using adaptive anisotropic radial basis function. J. Math. Imag. Vision **36**(2), 125–126 (2010)
6. Takeda, H., et al.: Kernel regression for image processing and reconstruction. IEEE Trans. Image Process. **16**(2), 348–366 (2007)
7. Wang, X., et al.: An image enhancement method based on robust estimation. Comput. Appl. **26**(7), 1612–1613 (2006)
8. Huaiyu, C., et al.: Large-scale image based on normalized gradient phase correlation fast registration algorithm. Pattern Recogn. Artif. Intell. **28**(8), 695–701 (2015)
9. Guohao, Z., et al.: An adaptive 3D kernel regression based remote sensing spatiotemporal fusion method. J. Wuhan Univ. (Information Science Edition) **43**(4), 562–570 (2018)
10. Xiaoxu, S., et al.: Remote sensing image using improved RPCALike the cloud algorithm. Comput. Eng. Des. **39**(6), 1653–1659 (2018)
11. Feng, W., et al.: Remote sensing image based on dual-tree complex wavelet transform like the cloud and fog system design. Appl. Opt. **39**(1), 64–67 (2018)
12. Chaowei, L., et al.: Remote sensing image thin cloud removal based on wavelet analysis research on division algorithm. Digital Technol. Appl. **6**, 137–140 (2017)
13. Wanyun, L., et al.: Variable order dictionary learning AO-DL resource number three remote sensing image cloud removal. J. Surv. Mapping **46**(5), 624–630 (2017)
14. Jiang, X.-F., et al.: Analysis of quality of removing cloud for monochromatic remote sensing image. LESER Technology **33**(3), 331–333 (2009).
15. Mei, Y.: Video super-resolution reconstruction based on adaptivemotion estimation. Comput. Appl. Softw. **33**(10), 155 (2016)
16. Chengcheng, C., Peikang, W.: Sequence image super-resolution reconstruction based on robust 3D-SKR. Electron. Measur. Technol. **35**(6), 86 (2012)

On the Impact of Artificial Intelligence Application on Public Administration

Zhang Wanliang

Abstract In the process of rapid social and economic development, the level of science and technology of human beings is constantly improving, and it is developing more and more in the direction of intelligence, modernization and convenience. The emergence of artificial intelligence has greatly changed people's daily life, working style and work content, so that the development of various industries and progress in various fields have new directions, new ideas and new choices. In this case, combining artificial intelligence technology, changing the methods and approaches of public management, improving the quality and level of public management technology, artificial intelligence is the current trend of social development. Its application in various fields has completely changed all walks of life and promoted the development and progress of the whole society. The introduction of artificial intelligence into public management can liberate manpower, save costs, optimize management and organizational forms, promote the modernization of governance, innovate the mode of talent training, greatly improve work efficiency, and promote the scientific, efficient and intelligent development of public management. The application of information technology in public management can better promote the progress and development of public management. This paper puts forward reasonable guidance strategies to enhance national comprehensive competitiveness.

1 The Close Relationship of Artificial Intelligence in Public Administration

Artificial intelligence should be an important part of social development. As the technology continues to evolve, it has provided huge changes to the way people produce and live. The application of artificial intelligence in public relations management also makes it play a very good role. The development of this technology is of greater significance to the improvement of national competitiveness.

Z. Wanliang (✉)
Guangdong University of Science and Technology, Guangdong 523083, China
e-mail: 273965296@qq.com

K. Nakamatsu et al. (eds.), *Advanced Intelligent Technologies for Information and Communication*, Smart Innovation, Systems and Technologies 365,
https://doi.org/10.1007/978-981-99-5203-8_26

With the continuous development of artificial intelligence, there are more characteristics in line with the development of The Times and human needs. It is closely related to people's lives. Artificial intelligence technology is of great significance in people's daily life and work. However, in the process of development, we must realize that artificial intelligence cannot replace human intelligence, and the execution system of artificial intelligence will be fundamentally different from the thinking mode of human beings. Because artificial intelligence is a mechanical and physical process without consciousness, human thought patterns are acquired on the basis of constant changes and developments in physiology and psychology [1]. Secondly, human beings have certain social attributes, and human nature is constantly formed in the process of social production and practice. However, AI itself is not social, it doesn't evolve. Most importantly, artificial intelligence can only be the implementation of human consciousness, without its own subjective initiative and creative consciousness.

2 The Embodiment of Artificial Intelligence in Public Management Service

2.1 The Relationship Between Public Administration and Artificial Intelligence

Public management refers to the public organization of management subject for the purpose of reflecting public welfare. Activities that integrate social public resources through economic, political and other means to enhance the actual governance ability of management subjects, so as to improve performance and quality, and better provide corresponding welfare for the public. From the definition of PR management, there are many definitions in practice. First of all, the functions and responsibilities of artificial intelligence in public management take public interests as the main goal, and its main function is to assist the coordinated development of various sectors of society. Secondly, the subject of public management is diversified and multi-level, mainly including public organizations and other social organizations. Common management has many subcategories under each category [2]. The object of public management is the main service object related to social public affairs and shows an expanding trend. The main function of public relations is to fulfill public responsibilities and coordinate and control the coordinated development of public affairs accordingly. In this process, the means used by artificial intelligence should also change accordingly with the overall environment of social development. The application layer of AI research and development enterprises is mainly based on the basic layer and the technology layer. Realize the integration with traditional industries, realize the application of different scenarios, For users to bring more convenient, high-quality experience, such as robots, driverless cars, smart home, intelligent medical, intelligent agriculture and so on. Annual application of AI research and development enterprises is shown in Fig. 1.

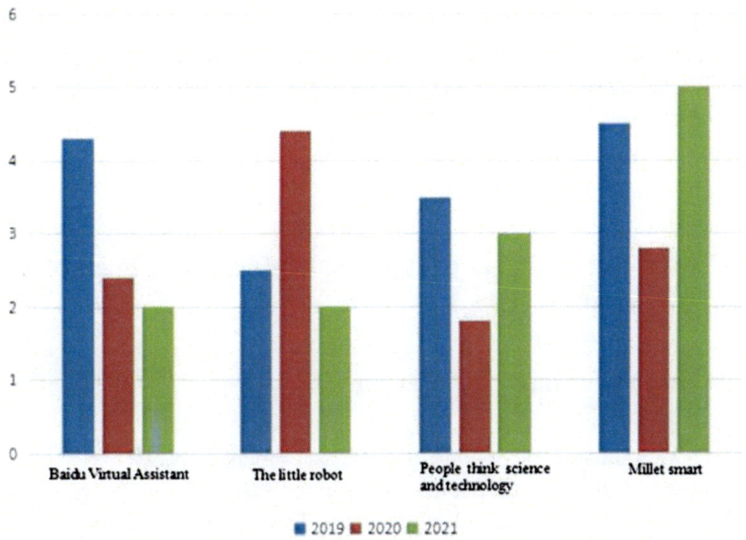

Fig. 1 Annual application of AI research and development enterprises

2.2 Application of Artificial Intelligence in Public Administration

Artificial intelligence is not very common in daily life and work, but it has filled people's lives. In terms of the social economy, AI exists both at the source of production and at the consumer. Through artificial intelligence, consumers' needs will be fed back to enterprises, so that enterprises can develop better services and products. At this stage, electronic payment has become a very important part of people's life. People no longer need to carry a lot of cash, but pay via Alipay or wechat. In the logistics industry, sorting robots also fall into this category [3]. They can complete more than 200,000 workloads per day, effectively solve the problem of poor logistics capacity and greatly reduce labor economic costs. For example, bike-sharing has been on people's radar since 2016. People of all ages install bike-sharing-related apps. Because of its many advantages, shared bikes have become the first choice for people to travel. In bike-sharing, there is the application of artificial intelligence. Through the artificial intelligence platform, virtual riding and parking can be predicted, so as to integrate weather, time and other variables, analyze their supply and demand, so that bike-sharing enterprises can obtain higher management efficiency. Thus, in people's life, artificial intelligence has entered all aspects, greatly changed people's way of life.

Education management is also an important direction of AI application. Judging from the current application of artificial intelligence in the field of education, especially in the field of distance education, this has achieved good results. Teachers can do speech recognition, you can also take photos and query questions, this makes distance education work get very good development. Teachers can provide targeted education to students according to their actual situation. With AI, education can be more flexible and all students can get personalized education so that both teachers and teachers' efficiency can be significantly improved. Artificial intelligence is bound to achieve greater development in the field of education [4].

3 The Application Effect of Artificial Intelligence in Public Administration

3.1 Research Methods and Design

This paper adopts the method of questionnaire experiment and presents experimental materials in the form of questionnaire survey in the experimental design to find the causal relationship between social phenomena. The experiment takes 1500 undergraduates of public administration as samples and divides them into three groups to explore the public's trust in the participation of artificial intelligence technology in two different types of public decision-making. There are three main types of public decision-making: artificial intelligence technology, artificial intelligence technology + public manager (hybrid), and public manager.

3.2 Research Methods and Design

In this experiment, three experimental groups will be formed to represent the three common decision-making methods in the current public decision-making. The dependent variable in this experiment is trust. Do you trust the above three good students? Respondents can choose a number from 0 (no trust at all) to 6 (no trust at all), and the final measurement results in three groups of trust. Analysis of variance was performed on the data, and the results are shown in Table 1. It can be seen that the information degree of artificial intelligence technology and hybrid decision-making methods is significantly higher than that of public managers.

Artificial intelligence will change the way some employees work, which may make them feel a sense of crisis. But those involved need to be aware that AI is an auxiliary tool [5]. Therefore, people need to correctly understand this situation, in order to better play the actual role of artificial intelligence in public management. In the process of the continuous development of society, public managers are required to learn and master new technology. In the process of AI development, they need to improve the efficiency of the use of AI, some clear rules but extremely demanding work to hand over to AI, so that they can engage in more precise work. Artificial intelligence is a new technology developed and popularized. It can transmit information

Table 1 Analysis of variance

Summary				
Group	Observed reading	Sum	Average	Variance
AIT	500	3295	823.75	927.5833333
Public administrator	500	3101	775.25	258.9166667
Mix	500	3417	854.25	599.5833333

T-test: two-sample heteroscedastic hypothesis

	AIT	Public administrator
Average	823.75	775.25
Variance	927.5833333	258.9166667
Observed reading	500	500
Assumed mean difference	0	
df	5	
t Stat	2.816033799	
P(T < = t) single tail	0.018641148	
t Single tail critical	2.015048373	
P(T < = t) Double tail	0.037282296	
t Double tail critical	2.570581836	
	AIT	Mix
Average	823.75	854.25
Variance	927.5833333	599.5833333
Observed reading	500	500
Assumed mean difference	0	
Df	6	
t Stat	−1.560941463	
P(T < = t) single tail	0.084779716	
t Single tail critical	1.943180281	
P(T < = t) Double tail	0.169559433	
t Double tail critical	2.446911851	
	Public administrator	Mix
Average	775.25	854.25
Variance	258.9166667	599.5833333
Observed reading	500	500
Assumed mean difference	0	
df	5	

(continued)

Table 1 (continued)

T-test: two-sample heteroscedastic hypothesis		
t Stat	−5.392460061	
P(T < = t) single tail	0.001480141	
t Single tail critical	2.015048373	
P(T < = t) Double tail	0.002960283	
t Double tail critical	2.570581836	

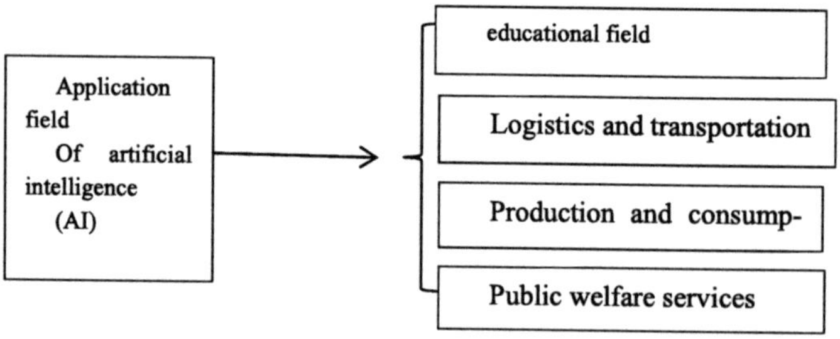

Fig. 2 Application field of artificial intelligence in public management

quickly and change the way people work and communicate. In the process of continuous construction of artificial intelligence technology, its role in public management will continue to improve, and greater progress can be made in public relations. On this basis, China needs to further strengthen the in-depth research on this technology, so that it can play a better role and create more power for China's construction as shown in the Fig. 2.

4 Application of Artificial Intelligence in Public Life

Artificial intelligence has been quietly popularized in people's daily life. Artificial intelligence complements various public services such as shared bikes, shared cars and shared charging banks. The emergence of these new public service bikes has brought great convenience to people's daily life and work. With the application of artificial intelligence in Internet technologies such as big data, shared bikes can be where people need them in a timely and accurate manner. Compared with traditional modes of transportation, the emergence of artificial intelligence technology has made the entire transportation system cleaner and more environmentally friendly. Through artificial intelligence technology, management systems including geography, weather, time and crowd distribution have been established. In addition to tourism, the application of AI technology has also been popularized in various fields such as food, entertainment and accommodation. It has had a huge impact on people's way of life. It has been gradually integrated into the production and consumption fields. On the one hand, through artificial intelligence technology, the production source can receive the demand feedback of the consumer terminal in a more timely and accurate manner, so that manufacturers can more accurately grasp the individual needs and consumer psychology of consumers, and launch more targeted products, popular high-quality products and services [6]. On the other hand, due to the progress of artificial intelligence technology, people begin to gradually get rid of the bondage and restriction of payment mode in the traditional economy. With the emergence of intelligent electronic payment, people's life is becoming more and more efficient and free. The emergence of software has greatly facilitated the consumption life of the public, and the intelligent payment mode is gradually popularized in important public management and service fields such as medical care, education and logistics [7]. The operating efficiency of the system reduces the work pressure and workload of public administration departments and alleviates the labor shortage problem in the rapid development of various industries and fields to a large extent.

5 The Future of Artificial Intelligence is Expanding

Artificial intelligence technology has been gradually integrated into production and consumption fields. On the one hand, through artificial intelligence technology, the production source can receive the demand feedback of consumer terminals in a more timely and accurate manner, so that manufacturers can more accurately grasp the personalized demand and consumer psychology of consumers, and launch more popular high-quality targeted products and services [8]. On the other hand, thanks to advances in artificial intelligence technology, people are gradually getting rid of the payment constraints of traditional economic models. In smart electronic payment, people's life is more and more efficient and more free. The emergence of various intelligent payment software has greatly facilitated the consumption of public life. In addition to the popularization of intelligent and payment methods in clothing, food, housing and transportation. It has also gradually promoted the use of important public management and services such as health, education and logistics, greatly improving the efficiency of public administration, reducing the pressure and workload of public administration [9].

Application status of artificial intelligence in education. Education is one of the most important public services in the development of a country [10]. Therefore, the development of education is bound to keep pace with The Times and develop in combination with artificial intelligence technology. At the same time, the application of artificial intelligence technology can select targeted teaching methods and formulate solutions for students at different levels from the database, providing new possibilities for the implementation of teaching methods. It makes students' learning more targeted, greatly improves the efficiency of teaching activities, and promotes the deeper reform and development in the field of education management.

6 The Future of Artificial Intelligence

In the process of the combination of artificial intelligence and social public management, artificial intelligence will inevitably replace some traditional basic types of work requiring artificial. But on the whole, artificial intelligence is still only an auxiliary intelligent tool, and the specific operation of artificial intelligence still needs the participation of human beings [11]. Therefore, the part replaced by artificial should actively improve its own technology and ability, improve its control level of artificial intelligence, more actively accept and adapt to the application of artificial intelligence technology in the field of social public service, and regard artificial intelligence as a tool and assistant of public management, rather than as a competitor. Even in the Internet technology revolution, computer and artificial intelligence technology is only an auxiliary role of human beings rather than a substitute role. Therefore, public managers in the new era must strengthen their ability to learn and accept new ideas and technologies and improve their professional skills and levels, so as to show their irreplaceable advantages and abilities in the rapid popularization of it. The development of it also affects the changes in the internal structure of China's digital economy. Statistics on the proportion of digital industry from 2019 to 2021 are shown in Fig. 3.

In addition, in the process of the rapid development of artificial intelligence, new job opportunities will inevitably be created. In these new jobs, humans still have an advantage that machines cannot replace [12]. The original public management structure is gradually changing to the structure of "intelligent machine of management subject". This means that public management organizations and public managers should learn to cooperate with intelligent machines as soon as possible and adapt to the changes of management structure. Public managers should foresee the cooperation and work allocation among ordinary employees, machines and intelligent machines more quickly and effectively, and give full play to the maximum value and utility of different individuals in different positions. For senior managers, artificial intelligence can complete more trivial work and simplify complex work, so that managers can devote more time and energy to more needed areas.

Fig. 3 Annual proportion of digital industry

7 Conclusion

With the popularization of artificial intelligence technology, public managers should foresee the collaboration and work allocation among ordinary employees, machines and intelligent machines more quickly and efficiently, so as to give full play to the maximum value and utility of different individuals. Different jobs. For top managers, AI can complete more trivial tasks, simplify complex tasks, and allow managers to devote more time and energy to more needed areas. Provide basis for management decision. Artificial intelligence technology can quickly collect more relevant data and information for specific fields, specific services and products, and provide more reliable decision-making basis for public management organizations to make relevant management decisions. With the proliferation and expansion of human social media and social channels, the data output capacity of human society is also increasing. In the face of massive data, artificial intelligence can quickly screen out useful information to help public managers' complete basic consulting, reply and other services. In a long time to come, artificial intelligence technology will be more comprehensive and more rapidly penetrated into public administration, education, management, resources, construction and other fields. There will be many artificial intelligence fields, and gradually build multiple fields and links deep into the public management system to achieve more convenient and efficient public management and services. At the same time, improve the ability of society to deal with complex problems, risk early warning, major accidents and other important public management work, promote all-round social development.

References

1. Liang, C.: Analysis on the application of artificial intelligence in public administration. Youth Soc. **14**, 291 (2019)
2. Zhang, F.: Application of artificial intelligence in public administration. Today's Wealth (China Intellectual Property) **8**, 24 (2019)
3. Like, J., Nie, C.: International trend analysis of artificial intelligence technology applied to public administration and its enlightenment to China. Today Sci. Park **11**, 21–30 (2019)
4. Hu, X., Chen, Y.: Risk analysis of artificial intelligence application in public administration. Sci. Technol. Think Tank **1**, 66–70 (2020)
5. Wang, Z.: Research on the Reform of Government Governance Mode Based on Artificial Intelligence. Xiangtan University (2020)
6. Li, Y.: On the application of artificial intelligence in public administration. Today's Wealth **24**, 198–199 (2020)
7. Zhao, E.: Application status and methods of artificial intelligence in public administration. Modern Enterprise **05**, 8–9 (2021)
8. Huang, Y.: Application research of artificial intelligence in public administration. Sci. Technol. Inf. **20**(7), 37–39 (2022)
9. Hui, H., He, H., Wang, T.: Application of blended teaching in public administration courses under the background of artificial intelligence. J. Phys.: Conf. Series **1575**(1), 012174 (2020)
10. Young, M.M., Bullock, J.B., Lecy Jesse, D.: Artificial discretion as a tool of governance: a framework for understanding the impact of artificial intelligence on public administration

11. Perspectives on public management and governance. **2**(4), 301–313 (2019)
12. Kouziokas, G.N.: The application of artificial intelligence in public administration for forecasting high crime risk transportation areas in urban environment. Transport. Res. Procedia **24**, 467–473 (2017)

Research on Civil Aircraft Longitudinal Stability and Maneuverability Flight Test Technology Based on Virtual Flight

Haiyun Zhu

Abstract The primary goal of aircraft design is to obtain satisfactory flight performance and excellent flight characteristics. The excellent flight characteristics depend on the aircraft's maneuverability, stability and flight quality. For civil aircraft, the corresponding virtual flight test model is established, and the influence of wing folding angle on the longitudinal static stability of the whole aircraft is calculated. Using the longitudinal small disturbance motion Hall of the aircraft, the long-period and short-period modes of wing deployment and folding are obtained, respectively, and their dynamic stability is discussed. The engineering estimation method is combined with the wind tunnel test method. The moment coefficients, elevator control derivatives and elevator leveling deflection angles of the whole airplane during the folding process of the inner wing are calculated. Through the analysis of the calculation results and the flight test of the demonstrator, the problems of longitudinal Maneuverability after wing folding are found, and the corresponding improvement measures are put forward.

1 Introduction

For civil aircraft, ground operation tasks such as taxiing, take-off and landing occupy a very important position in the entire operation mission envelope. Statistics show that the proportion of civil aircraft accidents during take-off and landing exceeds 70% of the total number of accidents. The aircraft's maneuverability and stability are related to its flight safety and must be highly valued [1]. The airworthiness standards put forward clear requirements for the stability and Maneuverability of the aircraft, and meeting the relevant airworthiness requirements is a principle that must be followed in the design of aircraft and landing gear [2]. The idea of virtual flight was put forward by Burdun and Mavris in the late 1990s. Specifically, in the design stage, based on the

H. Zhu (✉)
Flight Test Center, Commercial Aircraft Corporation of China, Ltd. (COMAC), Shanghai 201323, P.R. China
e-mail: zhuhaiyun@comac.cc

K. Nakamatsu et al. (eds.), *Advanced Intelligent Technologies for Information and Communication*, Smart Innovation, Systems and Technologies 365,
https://doi.org/10.1007/978-981-99-5203-8_27

mathematical modeling of aircraft flight dynamics and pilot control, engineers and technicians use digital computers to conduct man–machine closed-loop simulation calculation for specific assessment flight tasks and investigate whether the safety of aircraft flight and operation and control characteristics can meet the design requirements according to the results [3]. Conventional flight simulation methods mostly calculate the open-loop response of aircraft under given input, while virtual flight fully considers the characteristics of pilot's control behavior after introducing pilot control model, and can complete the simulation of closed-loop flight tasks specified by airworthiness standards [4]. Compared with virtual flight in the strict sense, virtual flight does not require hardware support and the participation of real pilots. It is more economical, and at the same time relaxes the requirements for online real-time calculation, thereby allowing more precise and accurate aircraft system movement construction. The model is especially suitable for the evaluation of the operational stability and airworthiness compliance of the aircraft design scheme [5].

During the flight of aircraft in the atmosphere, it is often subject to various unpredictable movements, such as atmospheric disturbance, engine thrust pulsation, pilot's unconscious moving rod and so on. These disturbances will change the flight state of the aircraft. Therefore, it is necessary to study and learn the ability of the aircraft to automatically restore the original state after being disturbed, that is, the stability of the aircraft [6]. Generally speaking, the flight state and the equilibrium state of the aircraft before disturbance are called trim state. Therefore, the stability problem is to study whether the aircraft can generate torque to return to the original trim state when the aircraft deviates from the equilibrium state due to external disturbance [7]. The primary goal of aircraft design is to obtain satisfactory flight performance and excellent flight characteristics. The excellent flight characteristics depend on the maneuverability, stability and flight quality of the aircraft [8]. For a given aircraft, it is the inherent characteristic of the aircraft but it is not obtained by calculation, nor by test flight after days on the aircraft, but designed and manufactured [9]. The aircraft conceptual design stage is one of the most important design stages, and the design analysis of maneuverability, stability and flying qualities should be one of the most important tasks [10]. Therefore, this paper puts forward the research on longitudinal stability and maneuverability flight test technology of civil aircraft based on virtual flight, so as to ensure that the aircraft scheme has excellent flight characteristics.

2 Overview of Aircraft Stability and Maneuverability

The stability of an aircraft is an important parameter in aircraft design to measure flight quality. It indicates whether the aircraft has the ability to return to its original state after being disturbed. If the aircraft is disturbed and can return to the initial state without the pilot performing any control, the aircraft is said to be stable, otherwise the aircraft is said to be unstable [11].

The stability of the aircraft includes longitudinal stability, which reflects the stability characteristics of the aircraft in the pitch direction. Heading stability reflects the directional stability characteristics of the aircraft. And lateral stability, reflecting the rolling stability characteristics of the aircraft [12]. In order to facilitate the analysis of problems, the stability of aircraft is usually divided into static stability and dynamic stability. Their essential difference is that static stability refers to whether the aircraft in an original flight state has a tendency to return to the original reference flight state at the initial moment after the disturbance stops. Dynamic stability refers to whether the disturbance motion finally returns to the original reference flight state after the disturbance stops [13].

The stability of the aircraft is particularly important to flight safety. If the aircraft is stable, the pilot does not need to intervene in the aircraft when it encounters disturbances, and the aircraft will automatically return to a balanced state: if the aircraft is unstable, in the event of disturbances, Even if it is a slight disturbance, the pilot must control the aircraft to maintain a balanced state, otherwise the aircraft will get farther and farther from the initial state. Unstable aircraft not only greatly increases the pilot's control burden and makes the pilot in a tense state anytime and anywhere but also the mutual interference between the pilot's control of the aircraft and the aircraft's own motion is easy to induce the aircraft oscillation and cause flight accidents.

Maneuverability of aircraft can also be called the handling quality of aircraft, which refers to the response characteristics of aircraft to handling. Manipulation means that the pilot changes the flight state of the aircraft through the driving mechanism. According to the different movement directions, the control of aircraft can be divided into longitudinal control, transverse control and heading control [14].

The quality of aircraft control is a subjective issue related to pilots, but there are still some basic standards to measure the quality of aircraft control. Maneuvering quality is often expressed by the maneuverability index of the ratio of input and output. These ratios should not be too small or too large. If the ratio is too small, the control input is small and the output is large. This kind of aircraft is too sensitive to manipulation, not only difficult to accurately control but also easy to stall or structural damage due to excessive response. If the ratio is too large, the control input is large, the output is small, and the aircraft is slow to respond to the control, which is easy to make the pilot make wrong judgment, and may also cause large amplitude oscillation of the aircraft, which also leads to stall or structural damage. If the aircraft does not need the pilot's complex control actions during maneuvering flight, the driving rod force and rod displacement are appropriate, and the response of the aircraft is not fast or excessively delayed, it is considered that the aircraft has good maneuverability.

3 Longitudinal Stability and Maneuverability

3.1 Longitudinal Stability Analysis

The general layout of the research model proposed in this paper is the flying wing layout, which adopts the S8025 airfoil. The engine inlet is located at the back of the fuselage and the trailing edge of the W-shaped wing. The whole aircraft is divided into five sections, the middle section is the fuselage, the connecting fuselage is the folding section wing, and the outer section wing is outside the folding section. The folding section can be folded upward by 120°, while the outer wing always remains level, and the upper dihedral angle of the wing is 0°. The outer wing is equipped with elevons, which are used to control the pitch and roll of the aircraft, and its deflection angle is −20° ~ 30°. The main overall parameters are shown in Table 1.

At medium and small angles of attack, the variation range of the aerodynamic focus position of the aircraft is small, so it is generally considered that its position is unchanged in analysis and engineering. However, at high angle of attack, due to the flow separation in a large area on the aircraft wing surface, the aerodynamic focus position of the aircraft will move forward, resulting in the upward bending of the longitudinal moment curve of the aircraft, as shown in Fig. 1, making $m_z^\alpha > 0$ aircraft lose the longitudinal static stability according to overload.

Longitudinal static stability depends on two conditions, namely $C_{mo} > 0$ and $C_{m.C_L} < 0$. $C_{mo} \approx 0$ airfoil can be used for swept wing layout, and C_{mo} of S8025 airfoil is about 0.006, so it meets $C_{mo} > 0$ conditions. $C_{m.C_L}$ is the derivative of the pitching moment coefficient to the lift coefficient, which depends on the ratio of the distance from the center of gravity of the civil aircraft to the aerodynamic focus and the average aerodynamic chord length b_A of the wing, which reflects the pitch static stability margin. Its calculation formula is as follows:

$$C_{m.C_L} = \overline{X}_T - \overline{X}_F \quad (1)$$

Table 1 Overall parameters of the model

Overall parameters	Wing unfolded	Wing folding
Wingspan/m	1.9	1.4
Wing area/m^2	0.879	0.526
Wing load/kg.m^{-2}	4.36	7.27
Full length/m	1.12	1.12
Total weight/kg	3.8	3.8
Push-to-weight ratio	0.7	0.7
Sweep angle of 1/4 chord line/(°)	33.8	34.8
Shoot root ratio	0.142	0.181

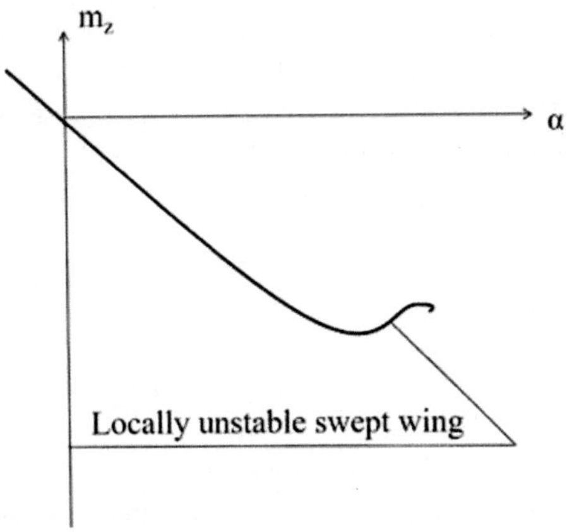

Fig. 1 Longitudinal moment curve of aircraft

$$C_{m.C_L} = \frac{C_m - C_{mo}}{C_L} \tag{2}$$

where $\overline{X}_T = X_G/b_A$ is the position of the center of gravity of civil aircraft on the average aerodynamic chord. $\overline{X}_F = X_F/b_A$ is the position of the aircraft aerodynamic focus on the average aerodynamic chord.

When calculating, determine the center of gravity position X_G of the whole machine, which is unchanged. b_A, C_m, C_{mo}, C_L with different folding angles of the wing is calculated by Vortex Latitude Method (VLM), and $C_{m.C_L}$ with different folding angles can be obtained by substituting the above formula. The variation curve of static stability margin is shown in Fig. 2.

It can be seen from Fig. 2 that as the folding angle of the wing increases, the focus position moves back significantly. With the center of gravity position fixed, the static stability margin increases from 5.0% to 20.1%. After the wings are fully

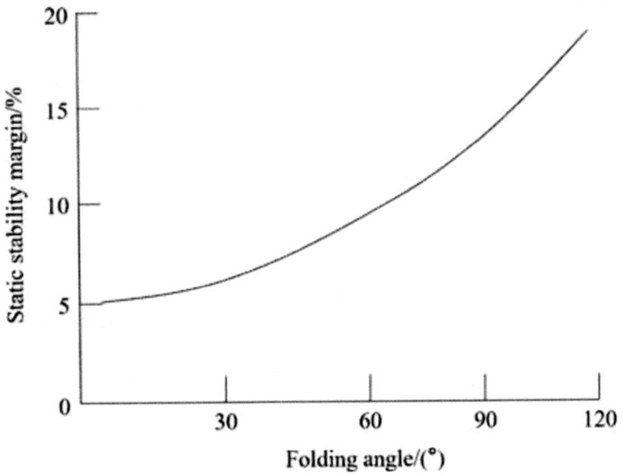

Fig. 2 The influence of the wing folding angle on the margin of static stability

folded, the static stability margin is too large, which will make the aircraft control sluggish, and increase the trim resistance of the civil aircraft, which affects the longitudinal Maneuverability of the civil aircraft. In the case of a small angle of attack, the assumption of small disturbance is used to separate the aircraft's lateral and longitudinal movement. The longitudinal dynamic stability of a civil aircraft is represented by the typical mode of disturbance motion. Using the civil aircraft longitudinal small disturbance motion equation, ignoring the influence of altitude changes on the force and moment, the characteristic equation is obtained:

$$\left(S^2 + 2\zeta_{sp}\omega_{sp}S + \omega_{sp}\right)\left(S^2 + 2\zeta_p\omega_pS + \omega^2\right) = 0 \tag{3}$$

Generally, two pairs of conjugate complex roots can be obtained from the characteristic equation: a pair of large conjugate complex roots describe the rapid motion, corresponding to the short period motion of the aircraft. The other pair of small common complex roots corresponds to long-period motion (ups and downs). According to the established model, the long-period and short-period modes of civil aircraft wing folding and unfolding can be calculated, and the stability of longitudinal small disturbance motion can be judged by rouse Horwitz criterion.

When the wing is deployed, the short-period undamped oscillation frequency $\omega_{sp} = 19.84$, damping ratio $\zeta_{sp} = 0.5874$, oscillation period $T_{sp} = 0.391s$ and half-life $T_{1/2.sp} = 0.0595s$. Long-period undamped oscillation frequency $\omega_p = 0.693$, damping ratio $\zeta_p = 0.07171$, oscillation period $T_p = 9.1s$, half-life $T_{1/2.p} = 12.4s$. Two pairs of characteristic roots are:

$$\begin{aligned} \lambda_{1,2} &= -11.654 \pm 16.056i \\ \lambda_{3,4} &= -0.0497 \pm 0.6912i \end{aligned} \tag{4}$$

R > 0 is calculated by the Rouss-Horwitz discriminant, indicating that the small disturbance motion is stable when the wing is deployed. When the wings are folded, $\omega_{sp} = 29.68$, $\zeta_{sp} = 0.4853$, $T_{sp} = 0.242s$, $T_{1/2,sp} = 0.0481s$. $\omega_p = 0.531$, $\zeta_p = 0.07709$, $T_p = 12.3s$, $T_{1/2,p} = 17.5s$. The two pairs of characteristic roots are:

$$\begin{aligned} \lambda_{1,2} &= -14.405 \pm 25.950i \\ \lambda_{3,4} &= -0.0395 \pm 0.5115i \end{aligned} \tag{5}$$

R > 0 is calculated by rouse Horwitz discriminant, which shows that the small disturbance motion is stable when the wing is folded. Comparing the long-term and short-term modes of the two states, it can be seen that when the wing is folded, the convergence speed of short-term oscillation is faster, the long-term sinking and floating motion are gentle, and the aircraft has better dynamic stability.

3.2 Maneuverability Analysis

Because of the flying wing layout, the pitching control adopts the elevon, which is used as the elevator when the rudder surface deflects symmetrically upward or downward at the same time. For aircraft with flying wing layout, there is no better theoretical calculation method to calculate the control characteristics of ailerons. Therefore, this section adopts the method of engineering estimation combined with wind tunnel test to study the longitudinal control characteristics of the model. In the engineering estimation method, firstly, the position of the center of gravity is used as the moment reference point, and the VLM estimation method is used to calculate the influence of different wing folding angles on the pitching moment coefficient C_m of the whole aircraft. Then regard it as the trailing edge simple flap to calculate its steering derivative C_m. The address increase caused by the deflection elevator is:

$$\Delta C_m\left(\overline{x}_{erf} - 0.25\right)C_{Lf} + K_\Lambda\left(\frac{A}{1.5}\Delta CL_{ref}\tan\Delta 1/4\right)+ \left[\left(\frac{\Delta C'm}{\Delta CL_{ref}}\right)\Delta CL_{ref}\left(\frac{c'}{c}\right)^2\right] - 0.25K_bC1\left[\left(\frac{c'}{c}\right)^2 - \frac{c'}{c}\right]+ K_bC_m\left[\left(\frac{c'}{c}\right)^2 - 1\right] \tag{6}$$

The derivative of the pitching moment coefficient to the elevator deflection angle is $C_{m\delta e} = \Delta C_m/30$.

According to the above analysis, two methods can be adopted to improve its maneuverability:

(1) Move back the center of gravity of the aircraft, and design the deployment state of the wing to be statically unstable. After the wing is folded, the focus moves back and the aircraft becomes statically stable, which is controlled by the automatic driving system.
(2) The center of gravity position adjustment mechanism is added in the aircraft to reduce the moment coefficient after the wing is folded, and to improve the problem that the angle of the elevator is too large due to the increase of the moment coefficient.

4 Flight Test Verification

Modern civil aircraft adopt a large number of new technologies and methods, and the cross-linking between systems is complicated. During the flight, the coupling effect of various complex weather, special environmental conditions and other unforeseen factors increase the complexity of virtual flight. Flight test involves many disciplines

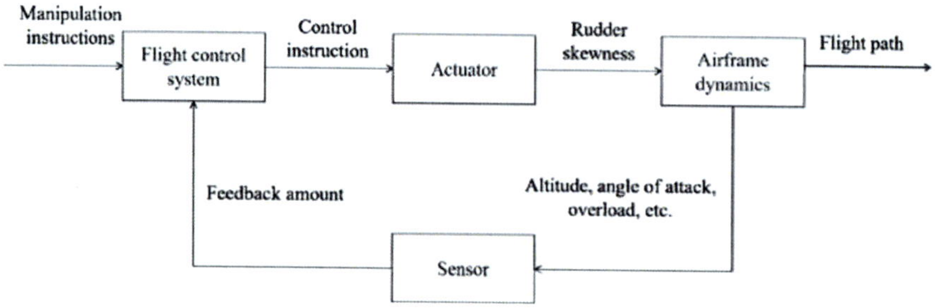

Fig. 3 Aircraft virtual flight frame model

such as aerodynamics, structure, strength, machinery, electronics, computer and so on. It needs a complete set of test methods, test methods and test modification, which forms a very complex working system.

In order to verify the stability and maneuverability of the aircraft in this layout, a virtual flight verification is made according to the overall parameters. The verification is controlled by the digital virtual technology simulation system, and the wing can be folded and unfolded completely according to the design requirements. The parameters of the ailerons are the same as the design values to ensure the validity of the verification.

The general structure diagram of the aircraft virtual flight model with flight control system is shown in Fig. 3.

After the virtual flight model of the aircraft is established, the virtual calculation software is compiled. After determining the control input, the dynamic response of the aircraft can be virtually calculated to establish the basis for trajectory calculation, maneuverability and stability analysis. The structure diagram is shown in Fig. 4.

In the flight test verification, after the inner wing is folded, the bow torque increases, which shows that the aircraft tends to enter the dive. At this time, it is necessary to increase the elevator trim angle to ensure level flight. The movement of the rocker controlled by the digital virtual technology simulation system reflects the deflection of the elevator. Through observation, the variation of the trim angle of the elevator folded by the inner wing in actual flight is in good agreement with the calculated value. After the inner wing is folded, due to the increase of the absolute value of longitudinal moment coefficient, the trim rudder deflection angle increases, which leads to the decrease of available rudder deflection angle, and the pitching motion is not flexible enough during operation. The virtual flight test shows that the results obtained by VLM method and wind tunnel test method are correct.

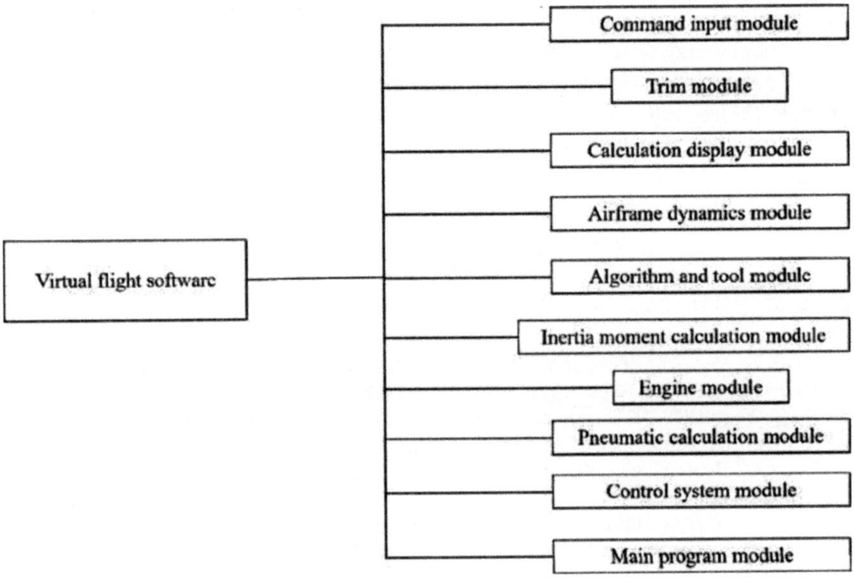

Fig. 4 System module structure diagram

5 Conclusions

Aircraft flight safety has a huge negative impact on aircraft stability and maneuverability and brings serious hazards to flight safety. If the safety impact is underestimated, equipment defects or insufficient capabilities will lead to fatal accidents. Therefore, the research on aircraft stability and maneuverability plays an important role in actual flight, aircraft design and airworthiness certification. Finally, an evaluation method for civil aircraft longitudinal stability and Maneuverability flight test technology based on virtual flight is established. The minimum pitch attitude angle occurs in the high-speed taxiing section during takeoff and landing, that is, the moment before the front wheel at the end of takeoff taxiing leaves the ground or after the front wheel is grounded at the initial section of landing taxiing. The main wheel brake will significantly increase the load of the front wheel and make the fuselage have a significant tendency to roll forward. The forward movement of the front landing gear position or the overall forward movement of the landing gear position can weaken the forward tipping tendency of the fuselage in the high-speed taxiing section to a certain extent. After the inner wing is folded, the absolute value of elevator control derivative increases, and longitudinal control should be more sensitive. However, the absolute value of moment coefficient increases faster, which leads to the increase of elevator trim rudder deflection angle, the decrease of available rudder deflection angle and the increase of trim resistance. Two methods can be adopted to improve longitudinal maneuverability. The flight altitude not only has a great impact on the aircraft's longitudinal stability and maneuverability but also has a certain impact on the aircraft's longitudinal stability at low altitudes and low speeds, which may affect flight safety. Regarding the layout of the flying wing, there is a large error in the method of calculating the steering derivative of the elevator as a simple trailing edge flap. The wind tunnel test should prevail to further promote the progress and development of China's civil aircraft development and test flight technology.

References

1. Lu, X.B., Wu, Z.M., Dou, B.Y., et al.: Research on the design of civil aircraft pitch control force characteristics. Flight Mech. **35**(4), 6–9 (2017)
2. Wang, Y., Liu, Q.L.: Research on the longitudinal static stability flight test method of a certain aircraft. Civil Aircraft Des. Res. 2016(3), 66–68.
3. Qu, X.Y.: Research on civil aircraft flight control simulation system. Software Guide **15**(6), 107–109 (2016).
4. Xue, J.: Analysis on the configuration management technology of civil aircraft flight simulator data package. Sci. Technol. Commun. **18**, 111–112 (2016)
5. Xi, C., Liu, J.: Research on longitudinal stability flight test of large seaplane. China Sci. Technol. Inf. **8**, 24–25 (2018)
6. Li, J., Ye, Y., Juanni, S.: Longitudinal stability and maneuverability of aircraft after icing. Inf. Wkly **26**, 125–125 (2018)
7. Qu, L., Li, Y.H., Yuan, G.Q., et al.: Longitudinal nonlinear stability region analysis of icing aircraft based on phase plane method. Acta Aeronautica Sinica **37**(3), 865–872 (2016).
8. Tu, Z.J., Wang, L.X., Chen, J.P.: Evaluation of the climb gradient of a civil aircraft based on digital virtual flight. J. Beijing Univ. Aeronaut. Astronaut. **43**(12), 2530–2538 (2017).
9. Qu, X.Y.: Research on civil aircraft flight simulation software. Software Guide **15**(7), 115–117 (2016)
10. Zuo, L.X., You, M.: Research on aerodynamic layout and maneuverability characteristics of hypersonic aircraft. Aeronaut. Sci. Technol. **31**, No. 228(11), 50–56 (2020)
11. Ma, Z., He, R.: On the control and simulation of helicopter longitudinal attitude tracking stability. Comput. Simul. **36**(1), 61–64 (2019)
12. Xu, P.H., Cao, Y.H.: Research on the longitudinal stability of autogyro. Aeronaut. Sci. Technol. **28**(1), 42–47 (2017)
13. Dong, R., Zhang, K.P., Wang, X.F., Wang, L.P., Li, J.T.: The Maneuverability of the automatic landing longitudinal flight control system can be optimized. Sci. Technol. Rev. **38,** No. 603(21), 146–151 (2020)
14. Luo, D.C, Zhang, M., Wang, W.X., et al.: Longitudinal stability evaluation method of ultra-low-altitude airdrop carrier considering ground effect. Sci. Technol. Innov. Appl. **24**, 11–14 (2019)

Application of Computer Simulation Technology in Hydraulic Resistance Test of Fire Hose

Tao Li, Jin Cao, and Guijun Zhu

Abstract In the modern technological innovation development, in order to better meet the demand of urban fire control safety, research scholars according to the current problems of fire hoses, mainly discuss how to improve the nozzle jet distance and fire extinguishing effect, and combining the theory of fluid mechanics and dynamics of space, based on computer simulation technology simulation fire hoses water resistant and experimental operation, Therefore, the corresponding improvement plan is made clear. Therefore, on the basis of understanding the development status of fire water gun, according to the application principle of computer simulation technology, it is clear that under working pressure, the maximum stress of fire water gun appears in the claw area, and the actual strain part will be concentrated throughout the claw, so it is very easy to damage. Through experimental analysis, it is found that stress concentration can be effectively controlled in the operation of increasing the rounded corner and width of the catch, and the safety performance of the fire water gun can be comprehensively improved during the use of the fire water gun.

1 Introduction

In the steady development of social economy, the number of urban residents is increasing, and the social material living conditions are getting better and better, and the fire phenomenon is becoming more and more complicated. Especially in the construction and promotion of high-rise and multi-storey buildings, the need to deal with the fire scale is becoming larger and larger, and put forward higher requirements for fire protection technology. Nowadays, the urban construction chooses the type of fire extinguishing system is more, such as carbon dioxide extinguishing system, low expansion foam extinguishing system and foam spray extinguishing system, sprinkler system, etc., but almost all the fire departments will choose water club fire

T. Li (✉) · J. Cao · G. Zhu
Chongqing Academy of Metrology and Quality Inspection, Chongqing, China
e-mail: ltxiaocong@163.com

K. Nakamatsu et al. (eds.), *Advanced Intelligent Technologies for Information and Communication*, Smart Innovation, Systems and Technologies 365,
https://doi.org/10.1007/978-981-99-5203-8_28

extinguishing system, the reason is that the cost of water resources is low, won't cause pollution to the urban environment, and easy to store and take in the daily work [1–3]. It can be seen that the water gun is the core content of the sprinkler system in the fire fighting work, and the overall improvement of the application performance is the main topic discussed in the field of fire science and technology. According to historical records, the earliest fire brigade in the world appeared in the seventeenth century. In a painting that is relatively well preserved at present, there is a primitive water gun fixed on the upper part of a machine, which is about 1 m long and belongs to copper products, from which pressure water can be spewed out. Then, in the nineteenth century, Dutch inventor Johann Von Herscher wanted to separate the primordized hoses attached to the pump from the pump body, using leather hoses so that the primordized hoses could move freely and deal with more fires. After the twentieth century, fire pumps are quickly applied to multiple fields, firefighters can choose nozzle has a lot of kinds, and in the late 50s has carried on the elaboration to the development of fire hoses, the rest of the various types of fire phenomenon developed special nozzle, such as automatic injection nozzle are mainly used for processing electrical fires. At the same time, the application material of the fire water gun has changed with the innovation of science and technology, gradually changing from the traditional bronze or brass material to light alloy products, so that the staff can operate more conveniently. From the point of view of the current firefighting work, water gun, as a basic part of fire equipment products, plays a key role in the rescue and disaster relief period, so the specific performance of product material and application technology should be guaranteed in the development of products, to avoid quality problems during the use. According to the analysis of the technical index of the current fire water gun in China, it is found that the water pressure test, as one of the most common parameters, needs to keep the pressure unchanged for at least 2 min when the maximum working pressure is 1.5 times. However, it is found that there are many unqualified phenomena in the current use of fire water gun, which directly affects the application quality of water gun interface and claw [4–6]. Therefore, this paper chooses the computer simulation technology, establishes the simulation system according to the computer platform, and makes a deep exploration of the water pressure resistance test of the fire nozzle after defining the basic experimental conditions. In essence, computer simulation is also seen as computer simulation, which mainly uses digital computers and related technologies to simulate and analyze complex real systems and operating states. The application of computer simulation technology in the hydraulic test of fire water gun can not only present the real and perfect experimental process, but also grasp more valuable parameter data, so as to ensure the safety and effectiveness of the research and development design of fire water gun.

2 Method

2.1 Computer Simulation Technology

Computer simulation technology contains three elements, the first refers to the system, the second refers to the model, and finally refers to the computer. The core of this technology is to construct a real computer simulation mathematical model according to the principle of systems engineering, and to complete the experimental analysis in computer based on the model instead of the real system. The practical operation process is as follows: first, build the system model. In the observation and research of the actual system, the secondary factors and undetectable variables should be ignored, and the system should be described in a mathematical or physical way, so as to obtain the indirect approximate model of the actual system [7–9]. The functional parameters of the model and the actual system have correspondence. Second, build a simulation model. Only when the simulation model is obtained, the corresponding computer program can be obtained and the system simulation analysis can be completed. Third, simulation experiment. During the running of simulation model, the relationship between system variables should be clarified and the whole process of system model variables should be observed carefully. In this process, after several runs and parameter optimization, the simulation model results can be analyzed, and the experimental conclusion can be obtained.

According to the analysis of the computer simulation flow chart shown in Fig. 1, it is found that the basic idea of finite element method should be used to solve the practical problems in the study of the water pressure resistance test of fire water gun.

This kind of thinking mainly contains two contents. On the one hand, it is necessary to construct multiple combinations of a certain continuum, li Shanchang, and the relationship between them is equivalent. Cells are connected only by nodes. And the internal ball carrying quantity can be calculated according to the difference of function relation. Because the element shape is very simple, it is more convenient to use the equilibrium relation to get the equation of node quantity. On the other hand, all the equations should be integrated together to form a total system of equations. After the boundary conditions are clear, the equations can be calculated and solved. In the finite element method, the displacement method is used to define coordinate variables, and the corresponding displacement function is shown as follows:

$$y = \sum_{i=1}^{n} a_i \psi_i$$

Fig. 1 Flow chart of computer simulation

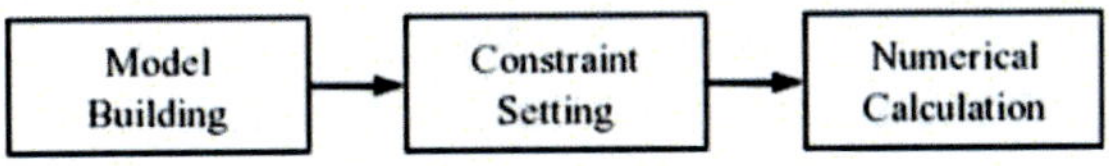

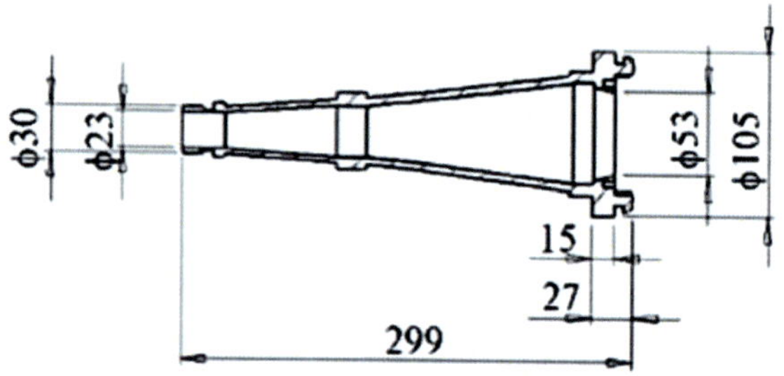

Fig. 2 Section model diagram

In the above formula, ai represents the undetermined coefficient, and ψ_i represents some function related to the coordinates.

The definition of unit set is to reconnect the balance conditions, boundary conditions, and all elements of the structural force according to the original structure, thus obtaining the overall finite element equation, as shown below:

$$Kq = f$$

In the above formula, K represents the stiffness matrix of the overall structure, Q represents the node displacement array, and F represents the load array.

2.2 Model Building

In this paper, the fire dc water gun Q Z3.5/7.5 model is regarded as the main target. The specific section model is shown as follows. The whole water gun includes the wall and k-D type pipe tooth interface, which is conical from the entrance to the exit. During the use period, both the exit and the entrance belong to the through hole. According to the experimental requirements of GB 8181-2005, the water gun hydraulic test should seal the exit (Fig. 2).

2.3 Finite Element Analysis

First, materials. In this paper, the material of fire water gun is aluminum alloy, which has high corrosion resistance, air tightness, technology, and other characteristics in practical application. The specific density can reach 2.65 g/cm^3, elastic modulus can reach 70 GPa, and Poisson's ratio is 0.3. After solution treatment, the actual tensile strength can reach 235 MPa and brinell hardness can reach 70 HBS.

Second, load constraint. Using linear static analysis of the sealing experiment and strength experiment of fire water gun, 1.6 mpa, and 2.4 mpa two kinds of load for deep analysis, water gun will be subjected to uniform circumferential radial pressure, the actual direction will be perpendicular to the wall surface. Therefore, in order to better study the interface connection form of fire water gun experiment, it is necessary to make clear that the restraint part is installed on the buckle of K-D type interface, so as to ensure that the cavity is fixed and not rotated.

Third, calculate. The water gun is constructed by tetrahedral solid grid, and the solid unit is solved after the above analysis. This paper mainly chooses EEEplus algorithm.

3 Result Analysis

3.1 *Specific Content*

According to the simulation calculation of computer simulation technology software, as shown in Fig. 3, the simulation data results of various fire water guns are finally obtained as shown in Table 1.

According to the above analysis, it is found that in computer simulation application software, different areas will show different profits, among which the darker area represents too little stress, and the lighter area represents too much force. In the state of 1.6mpa and 2.4mpa, the maximum stress and strain of the fire nozzle appear in the entrance and the latch area, in other words, under the working pressure, the two places are most likely to be damaged.

At the same time, combined with the analysis of the actual experimental results, it is found that more than 90% of the problems occur in the claw area, which is consistent with the actual phenomenon. The analysis of the area of the clasp shows that the maximum stress is close to the yield strength of the material if the area

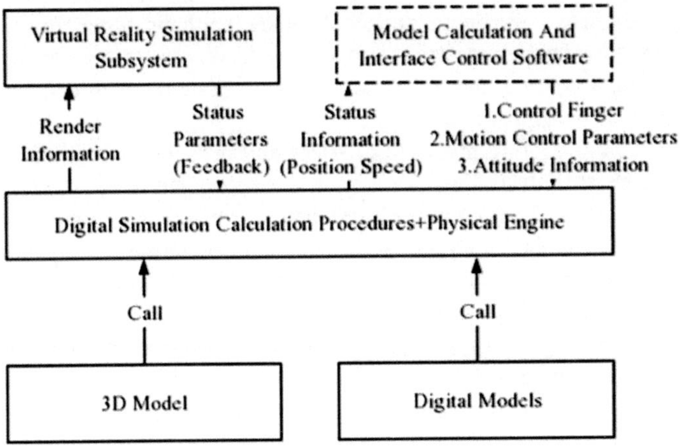

Fig. 3 Computer simulation technical structure diagram

Table 1 Experimental results

Pump type	Traffic		Pressure		Speed	The input power		The weight of the	Overall dimensions
	l/ min	G. P.M H. (USA)	har	P,s,i		hp	kw		
WS102	21	5.55	100	1450	1450	5,5	4.04	14,5	321 × 224 × 176
WS131	15	3.96	130	1885	1450	5,5	4.04	14,5	321 × 224 × 176
WS151	15	3.96	150	2175	1450	5,5	4.04	14,5	321 × 224 × 176
WS171	13	3.43	170	2465	1450	5,5	4.04	14,5	321 × 224 × 176

is reduced to 87mm3. In other words, on the basis of protecting the strength of the material, if the area of the kill is not less than 87mm3, then the basic design requirements can be met. In addition, if the contact position during installation is not fit, or the contact area is too small, resulting in stress concentration, or the selection of materials during manufacturing defects, fracture phenomenon can occur.

3.2 *Optimization and Improvement*

According to the above research and analysis, it is found that on the basis of the stress concentration area running through the whole clasp, if there is a defect in the applied material of the clasp area, then the failure problem will inevitably occur. Therefore, in order to improve the application effect of fire water gun, it is necessary to strengthen the control of selected materials and pay attention to the study and solution of casting defects and material structure, so as to improve the strength of practical application. The specific improvement method involves the following points: on the one hand, the designer will adjust the right Angle of the claw and increase the rounded Angle R1, while other parameters remain unchanged; On the other hand, when the width of the claw is increased, other parameters are not changed. According to the material stress analysis results, it is found that the stress concentration area is effectively cleared, and only small areas appear at both ends.

In extinguishing cooling principle of buses and bus, for example, according to the following chart analysis shown in Fig. 4, to be installed above the engine automatically and destroy the fire valve closed, can also be combined with the specific requirements to install in other areas of the car, the inlet can and multiple or a nozzle connection, the will use fast plug and liquid inlet pipe connection, The piping will be installed in a variety of ways above the car and will be connected to an on-off valve on the extinguishing agent container. In Fig. 4, 1 represents the carriage, 2 represents the water tank, 3 represents the nitrogen tank, 4 represents the pressure relief valve, 5 represents the soda mixing valve, 6 represents the solenoid valve, 7 represents the lever, 8 represents the jet hole, 9 represents the controller, and 10 represents the connection release button [10].

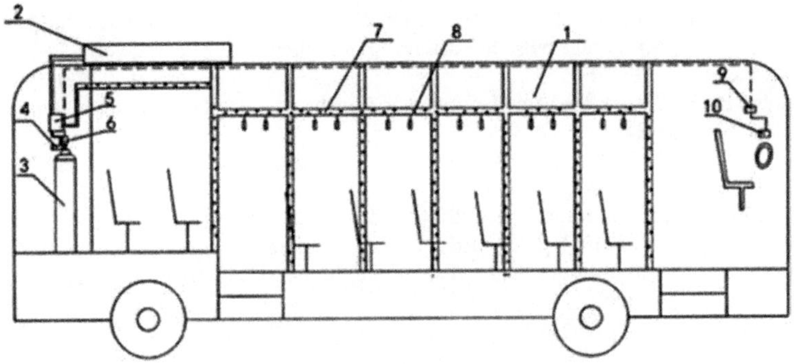

Fig. 4 Working principle of fire extinguishing and cooling of buses and school buses

4 Conclusion

To sum up, computer simulation technology is used to conduct a deep study on the water pressure resistance test of fire water gun, so as to clarify the sealing and strength conditions of the current urban fire water gun. Based on the linear static analysis, the stress and strain information of the water gun is obtained, and a comparative study is made on the application of the actual fire water gun. The final results show that there is maximum stress in the water gun interface during the use and pressure test. If there are unqualified factors in the casting material, it is bound to be damaged during the application, and serious problems such as fracture will occur. Therefore, in the development of modern technology innovation, according to the computer simulation technology, scientific selection of production materials, effectively adjust the casting process, so that not only can reduce the stress concentration area, but also can comprehensively improve the quality of fire water gun products.

Acknowledgements cstc2020jxjl120011(Formulation of calibration specification and development of calibration device for fire bolt buckle hydraulic testing machine cstc2020jxjl120011)

References

1. Zhang, J., Chen, S.G., Zhang, P., Zhang, L., Gong, B. J. Shenyang Univ. Chem. Technol. **34**(3), 6 (2020)
2. Zhang, Z.X., Wang, D., Wang, Q.H. et al.: J. Mil. Equip. Eng. **43**(2), 6 2022
3. Gai, Y.X., Zhang, N.: Application of computer simulation technology in water pressure test of fire water gun. Fire Technol. Prod. Inf. (10), 4 (2013)
4. Xu, W.W., Zheng, C.S., Ju, Y.: Research and application of intelligent multi-purpose fire water gun system. Water Fire (4), 5 (2018)
5. Wen, J.L., Wang, J.F., Sha, Y. et al.: J. Drainage Irrig. Mach. Eng. **020**(004), 21–23,26 (2002)

6. Cui, H., Zhang, Y.L., Shu, S.: Combination and application of wheeled forest fire truck and high voltage energy storage pulse fire extinguishing water gun. J. Central South Univ. Forest. Technol. **38**(3), 6 (2018)
7. Ren, K., Pu, J.Y.: Parametric simulation method for flame control of new fire training device. Fire Sci. Technol. **32**(8), 5 (2013)
8. Li, C.Y., Tang, X.Y.: Structure design and simulation analysis of micro-intelligent fire truck. Mach. Des. Res. **37**(3), 6 (2021)
9. Hao, W.R., Kan, J.M.: Forest Eng. **33**(5), 6 (2017)
10. Li, C.Y.: Research on fire fighting tactics of large-span and large-space warehouse based on PyroSim Simulation. J. Wuhan Univ. Technol. Inf. Manage. Eng. **40**(1), 5 (2018)

Research on Construction of Higher Vocational Teaching Quality Evaluation System Based on BP Neural Network

Kai Liu

Abstract With the development of education today, how to improve the teaching of higher education is an important link in the education cycle. From the experience of higher vocational education in recent years, teaching effect will be affected by teaching environment, teachers' ability, students' talent, and many other factors. Based on the understanding of the status quo of higher education and the basic concept of BP neural network, this paper studies how to use BP neural network to create evaluation and test the effectiveness of BP neural network in higher education. The experimental results show that BP neural network plays an important role in higher level training evaluation, and the overall evaluation is feasible.

1 Introduction

According to the analysis of the current higher vocational teaching quality evaluation system, it can be seen that there are a large number of non-quantitative factors, and there is a complex nonlinear relationship between the input and output of the system, so it is impossible to build a scientific and reasonable mathematical model. At the same time, around the university teaching quality evaluation chose a large number of simple methods, such as comprehensive score method, realistic method, evaluation, relative evaluation, and absolute evaluation method, etc., [1] because these methods are too subjective, or directly by a simple mathematical operation evaluation of teaching effect, ignored the nonlinear relation between evaluation index and the teaching effect, So the end result is hard to show the real quality of teaching. For example, the single factor evaluation method, multivariate statistical analysis method, method of operational research, fuzzy mathematical evaluation method, etc., these modern number though statistical method is widely applied in teaching quality evaluation, and obtained the application effect of different level, but these methods also exist many drawbacks, for example when determining the secondary indexes,

K. Liu (✉)
Shandong Institute of Commerce and Technology, Shandong, China
e-mail: 1263655454@qq.com

K. Nakamatsu et al. (eds.), *Advanced Intelligent Technologies for Information and Communication*, Smart Innovation, Systems and Technologies 365,
https://doi.org/10.1007/978-981-99-5203-8_29

weights numerical problems. The evaluation only based on expert experience will increase the subjectivity and arbitrariness of evaluation results, resulting in errors between evaluation results and actual values [2, 3].

Combined with the experience of teaching concept in recent years, higher vocational colleges should first provide teaching evaluation when changing traditional teaching, which puts forward higher requirements for training talents. This paper starts with classroom teaching, combines the actual situation, understands the effective teaching and management objectives, obtains the evaluation result, and provides a good foundation for the administrative training work. Based on the background of the artificial intelligence technology is widely used, artificial neural network as the simulation of the brain a good way of information transmission, in the education, teaching has been widely used in quality assessment, which is given priority to BP neural network, the BP neural network algorithm can help researchers understand more deeply the characteristics of deep learning model. The specific model is shown in Fig. 1 below. In 1986, Rumelhart, McClelland et al. put forward the concept of BP neural network in their research. Minsky and Papert et al. claimed in their study that in order to solve such problems, the network must have a layer of understanding, but did not study the rules for learning neurons during hiding. Combined with the perceptron proposed in this paper, the learning strategy analysis shows that the adjustment of the actual weight value should be determined by combining the difference between the output demand and the actual output price. Note that there is no obvious output value for the hidden process of nodes, so relational learning cannot be used to adjust the weight value. According to the empirical study, the main idea of BP neural network is that there are two processes in the learning process: signal forward propagation and error back propagation [4, 5].

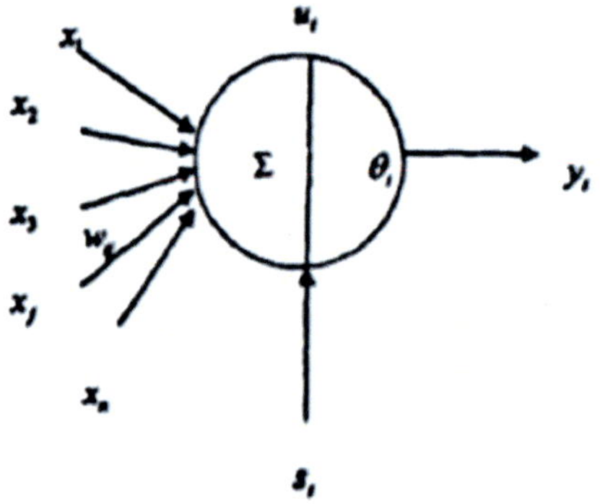

Fig. 1 Structure diagram of artificial neuron

2 Method

2.1 Artificial Neural Network

Artificial neuron, as the simulation result and abstract expression of biological neuron, belongs to the basic unit constituting artificial neural network. The specific structure is shown in Fig. 1.

From the perspective of neural networks, by analyzing the connection structure, it can be seen that it is divided into two kinds of structures. The first is the feedforward type, which requires all neurons to go directly to the next layer after receiving an input, with no input for all functions. Nodes of this form are input units and computation units. Each member of the terminal can input as many points as it wants, but the final output can only be one. The other is feedback mode, where internal content is available in the test unit, including input and output. Based on the analysis of the BP neural network model diagram shown in Fig. 3, it can be seen that, according to the application algorithm of three or more layers of neurons, although the upper and lower layers can be well connected, the neuronal layers of each layer cannot be connected. The structure of the algorithm includes input layer, half layer, and output layer. In a learning model sent to the network, the activity of neurons passes from the input layer to the output layer and receives the neurons in the next layer (Fig. 2).

2.2 Specific Operation Steps of Higher Vocational Education Teaching Quality Evaluation System Based on BP Neural Network

From the point of higher vocational teaching quality evaluation, the BP neural network technology advantage can be in unclear causes on the basis of the data, process modeling of nonlinear analysis, real-time optimization, learning classification, the nonlinear mapping characteristics, therefore can provide the nonlinear classification, model analysis and so on with new channels, This is in line with the current education quality evaluation system construction requirements. This paper

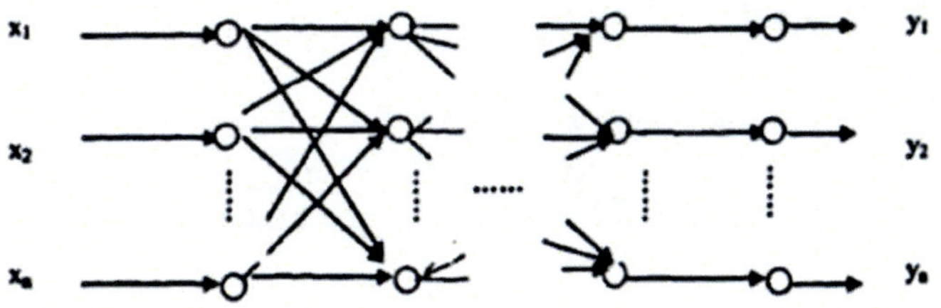

Fig. 2 Structure diagram of BP neural network

uses the neural network toolbox to construct a neural network model for empirical teaching quality evaluation, and transforms expert experience into specific weight values between neurons through sample learning, so as to build a more scientific evaluation model [6, 7].

This study focuses on the evaluation of classroom teaching. After evaluating the teaching quality of many colleges and universities, it should be noted that although there are many studies on teaching evaluation at home and abroad, the indicators created by the evaluation are not the same. According to the needs of today's higher education, the teaching methods for measuring exam quality are reproduced, as shown in Table 1: Set the score as the evaluation result, excellent 8 —— 10 good 6 —— 84 —— 6 bad 2 —— 4.

The teaching quality evaluation process of higher vocational colleges with BP neural network as the core is shown in Fig. 3.

Before entering the network design, the planning process is first carried out, and then the output types of the network are defined, which are represented by values in the interval [0, 1]. Second, the number of nodes in the output layer. Effective teaching evaluation is based on the quality assessment of classroom teaching and student feedback. Generally speaking, it can be represented by a real number that changes continuously within a certain range, so the number of neurons in the output layer is 1. Third, the number of nodes in the hidden layer. According to the above description, the specific design includes the following elements: first, the output model and the target vector. The input to the neural network is the extracted quantity. When removing the material to be valued, carefully consider whether there is a good relationship with the quality of teaching. According to the main points of the teaching process, determine an 18-dimensional vector, that is, the process has 18 neurons. Generally speaking, the number of neurons in the hidden layer is determined by the integration degree of the network. Since hidden neurons are only included in the mean, the actual number is not strictly defined. The identifying feature is the smallest network that can match a given pattern in the absence of other known sources. At the same time, the number of hidden neurons is too small to display the network, let alone recognize patterns that have not been seen before. In other words, the system is very fault tolerant. Therefore, how to determine the correct number of layers is an important problem. Finally, select Change [8–10].

3 Result Analysis

Based on the above research, 20 teachers are selected for informatization to improve the speed of network connection. Based on the analysis of the quality of higher learning performance evaluation, the BP neural network ensures the network structure. The test results are shown in Table 2.

In this paper, the alternating construction function of the input layer and the middle layer is used to establish the BP network model by software, and 15 groups of training samples are randomly selected for analysis, as well as the transition from the middle

Table 1 Higher vocational teaching quality evaluation system

The serial number	Evaluation of project	Evaluation content	Index code
1	Basic quality	Neat appearance, simple and generous, civilized behavior	X_1
2		Putonghua (foreign language) standard, rigorous words, language is appealing	X_2
3		Board writing standard, reasonable design, neat handwriting	X_3
4	Teaching attitude	Attend classes on time and follow rules and regulations	X_4
5		Full lesson preparation, teaching equipment, and instruments in place	X_5
6		Respect students, strict requirements for students, student-oriented	X_6
7		The relationship between teachers and students is harmonious, and the teaching information feedback is emphasized	X_7
8	The teaching design	Knowledge target, ability target is reasonable, reflect occupation standard	X_8
9		Teaching design combined with working process, teaching fluency	X_9
10	The teaching content	The content is rich, substantial, and appropriate in depth	X_{10}
11		Key points are highlighted and difficult points are properly handled	X_{11}
12		Pay attention to the real adjustment, the teacher's accurate demonstration of skills	X_{12}
13		Combine theory with practice and pay attention to knowledge application	X_{13}
14	The teaching method	Teach students according to their aptitude and appropriate method, good at enlightening, and guiding students, teachers, and students' interaction	X_{14}
15		Cultivate students' thinking, teach scientific learning and operation methods	X_{15}
16		Timely use of multimedia and other modern teaching methods	X_{16}
17	The teaching effect	High learning enthusiasm, concentration, and interest	X_{17}
18		Students can fully grasp the content of the lecture	X_{18}

layer to the output layer. The target learning error reaches 0.0001, and the maximum number of training steps is 1000 times. A total of 53 experiments were trained, and the final training results are shown in Table 3.

After the neural network training, the trained network system is verified and analyzed by using untrained samples. Finally, it is found that the output values of the test results are basically consistent with the expected output data. The network

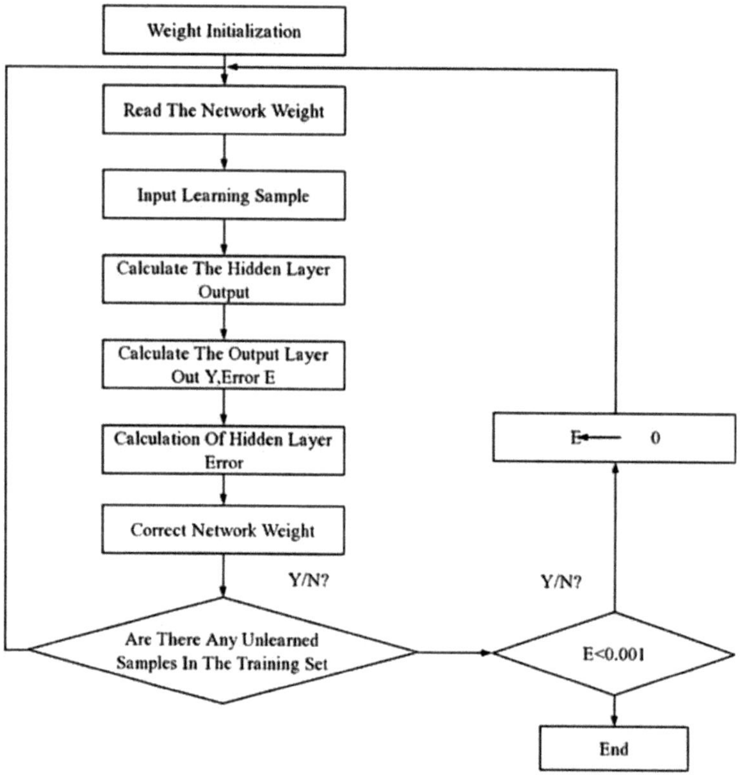

Fig. 3 Training flow chart

Table 2 Sample results of data pretreatment

The serial number	X_1	X_2	X_3	X_4	X_5	X_6	X_7	X_8	X_9	X_{10}
1	0.725	1.000	0.541	0.490	0.245	0.755	0.205	0.000	0.296	0.990
2	0.966	0.926	0.698	0.611	0.00	1.000	0.336	0.349	0.416	0.993
3	0.787	1.000	0.639	0.648	0.271	0.762	0.000	0.549	0.410	0.705
…	…	…	…	…	…	…	…	…	…	…
18	1.000	0.720	0.398	0.258	0	0.925	0.559	0.226	0.301	0.699
19	0.932	0.985	0.432	0.462	0.303	0.932	0.076	0	0.447	1.000
20	1.000	0.733	0.622	0.533	0.244	0.889	0.144	0.111	0	0.989

test grade and the actual grade are also identical, which proves that the BP neural network system is feasible to evaluate the teaching quality index of teachers, and can be comprehensively promoted in the development of modern education.

Table 3 Training results of BP neural network

The serial number	Desired output	The test results	Test level
1	(001)	(0.0000,0.0159,0.9840)	Good
2	(010)	(0.0007,0.9905,0.0066)	Good
3	(010)	(0.0002,0.9991,0.0001)	Good
4	(010)	(0, 0.9962, 0.0092)	Good
5	(011)	(0, 0.9772, 0.9639)	General
6	(010)	(0, 0.9912, 0.0005)	Good
7	(010)	(0,0.9864, 0.0117)	Good
8	(001)	(0, 0.0134, 0.9196)	Good
9	(010)	(0, 0.9771,0.0307)	Good
10	(010)	(0, 0.9955,0.0103)	Good
11	(011)	(0, 0.9767, 0.9866)	General
12	(010)	(0.0002,0.9987,0.0032)	Good
13	(001)	(0, 0.0598, 1.0000)	Good
14	(010)	(0,0.9593, 0.0392)	Good
15	(011)	(0,09,823,0.9554)	General

3.1 Specific Countermeasures

The advantages of some algorithms are reflected in the following points: first, the ability of nonlinear mapping. The essence of the algorithm is the mapping function from output to input, which can accurately approximate the nonlinear continuous function and effectively solve the complex mechanism problems in the system. Second, independent learning and adaptive ability. Combined with the above experimental research results, the application of BP neural network algorithm in higher vocational teaching quality evaluation system can help educators and managers to improve the evaluation rate and obtain more valuable information. In the training process, the algorithm can extract the scientific rules between the input data and the output data through autonomous learning, and record them in the weight of the network. Third, the ability to summarize. The algorithm can apply the learned knowledge to practice; Fourth, fault tolerance. After a few neurons of the algorithm are affected, the global training result will not be damaged, and the whole system can still run normally.

4 Conclusion

For example, combined with the basic theory and analysis, the BP neural network algorithm as the basis for the evaluation of higher performance training system is deeply studied in this paper. Quality is a hot issue in practice. It can not only solve the problem of the content of the scale, but also correct the different data in the process of measuring students, which is in line with the needs of today's higher education management.

References

1. Zhang, Y., Zhu, Y., Xu, H.: Research on online teaching quality evaluation of higher vocational colleges based on PCA-BP neural network. J. Taiyuan City Polytech., (11), 4 (2021)
2. Xu, X., Wei, J., Ding, Y., et al.: Research on performance evaluation of asset management in higher vocational colleges based on BP neural network. J. Jinhua Polytech. Coll. (2), 1–7 (2021)
3. Zhang, L., Fang, Z., Wang, L., et al.: Research and application of curriculum quality evaluation System based on BP Neural Network. J. Chifeng Univ.: Nat. Sci. Ed., **36**(1), 3(2020)
4. Ning, J.: Research on agricultural enterprise credit evaluation system based on bp neural network—a case study of Anda city in Heilongjiang province. Shanxi Agric. Econ., (11), 2(2021)
5. Yuan, C. W.: Construction and application of teaching quality evaluation index system in higher vocational education. J. Hubei Corresp. Univ., **033**(002), 25–26, 31(2020)
6. Guo, X., Fan, B., Li, C.: Research on effective evaluation mechanism of education quality of communication specialty in higher vocational colleges. China New Commun., (1), 2(2020)
7. Cen, J.: Research on the quality evaluation system of packaging professional group practice teaching based on CIPP model. China Packag., **42**(2), 5(2022)
8. Li, Y.: Research on higher vocational network teaching quality evaluation system based on big data analysis. Sci. Technol. Econ. Guid., **28** 730(32), 167–168(2020)
9. Zhou, W., Bin, Xie, Y. D.: Construction of practical teaching quality evaluation system of higher vocational packaging specialty based on KPI. China Packag., (5), 4(2021)
10. Yuan, C. W.: Construction and application research of teaching quality evaluation index system in higher vocational education. J. Hubei Open Vocat. Coll. (2), 3 (2020)

Neuro-Adaptive Fault-Tolerant Control with Prescribed Performance for Nonlinear Systems in Normal Form

Wei Tan

Abstract It is significant to cope with the uncertainty of nonlinear systems in practical work. However, if the system is unknown, and even the controller loses part of its effectiveness and has undetectable disturbance, the system's issues become very tricky. We propose a neuro-adaptive control scheme, which can guarantee UUB stability and adopts the backstepping design technique to construct control law. It can ensure a control scheme with prescribed performance for the completely unknown normal form system. Our control algorithm benefits from introducing a class of new coordinate transformation methods, which generalized t-correlation function and x-correlation function, the mean value theorem, and its related lemma. The proposed control algorithm's effectiveness and benefits are verified by theoretical analysis and numerical simulation.

1 Introduction

In this paper, we mainly cope with how to design the controller to manage the uncertainty in the nonlinear system and actuator and the undetectable actuator and external disturbance. In addition, how to design prescribed performance controls that are not constrained by initial conditions. With the process of control theory, control algorithms for unknown, uncertain nonlinear systems have developed significantly [1, 2]. However, in many pieces of research, the system is entirely or partially known [3, 4], i.e., the control design of the utterly unknown system is still a difficulty. With the increase of requirements, users no longer only would like to meet the stability of the system but also want to be able to preset the performance index of the control process, and many excellent results have emerged [5–7]. But the prescribed performance control of many systems requires very conservative preconditions [8–10]. If the initial states change constantly, the preset performance control algorithm needs to be continuously switched, which often reduces the working efficiency of

W. Tan (✉)
College of Automation, Chongqing University, Chongqing, China
e-mail: 318571296@qq.com

K. Nakamatsu et al. (eds.), *Advanced Intelligent Technologies for Information and Communication*, Smart Innovation, Systems and Technologies 365,
https://doi.org/10.1007/978-981-99-5203-8_30

the controller. In addition, many prescribed performance control algorithms are not designed to take into account the loss of some of the effectiveness of the controller [2, 10], that is, they need to meet the condition that the controller works perfectly and without interference. By introducing a class of coordinate transformation, we present a neural-adaptive fault-tolerant control method, which imposes preset performance constraints on the output state for any initial condition, in this work. Unlike most prescribed performance control methods, which have conservative conditions for initial conditions, the proposed method can choose any initial state value considering fault-tolerance. It can deal with the uncertainties of completely unknown systems well. In this paper, the significant achievements can be summarized as follows:

(a) A neuro-adaptive fault-tolerant prescribed performance control method with any initial state selection is proposed, effectively preset the control performance such as tracking accuracy and convergence time.
(b) The proposed algorithm is based on a radial basis function neural network (RBFNN), which can get an approximation function of the completely unknown smooth function and further expands the application range of the control scheme.
(c) The proposed control algorithm does not like the traditional fault-tolerant control with the conservative preconditions for the control gain and the internal disturbance of the actuator, which has good robustness.

2 Problem Formulation

2.1 System Model and Problem Statement

Taking account of the following n-order normal form nonlinear systems

$$\begin{cases} \dot{x}_i = x_{i+1}, i = 1, 2, ..., n-1 \\ \dot{x}_n = f(\overline{X}, u_a) \\ y = x_1 \end{cases} \quad (1)$$

In model (1), the $x_i \in R(1 \leq i \leq n)$ are the system state and $\overline{X} = [x_1, x_2, ..., x_n]^T \in R^n$, $u_a \in R$ and $y \in R$ are the actual control signal of the system and output of the system, respectively. $f(\overline{X}, u_a)$ is completely unknown, smooth and continuously differentiable function, which is bounded and denotes the lumped uncertainties and the external disturbances. In order to simplify the subsequent derivation process, some function parameters will be omitted in the absence of confusion.

Such a scenario that may occur with unanticipated actuator faults is taken into account in the model additionally, which means the actual control input is different from the actual controller u in that

$$u_a = \rho(t_\rho, t)u + u_s(t_s, t) \quad (2)$$

where $0 \leq \rho(\cdot) \leq 1$, is called health indicator [7], represents the effective status of the actuator, and $u_s(\cdot)$ is the uncontrollable part of the control signal. t_ρ and t_s, which is entirely unknown, respectively denote the time instant at that actuator lost part of its utility and unknown undetectable fault in the actuator occurs. To ensure controllability, we take account of the case that $0 < \rho(\cdot) \leq 1$ in this work, where the actuator loses part of its effectiveness but not, i.e., although the actuator fails part of its utility, the actuator is still active that the actuator u_a can always be affected by the control input u. For the normal form nonlinear system (1), define

$$g = \frac{\partial f\left(\overline{X}, u_a\right)}{\partial u_a} \tag{3}$$

where g is often referred to as the virtual control coefficient, which denotes the control direction.

The objective of this paper is fourfold:

(1) Developing a neuro-adaptive control scheme to achieve trajectory tracking for the normal form nonlinear systems.
(2) All the signals in the systems are ultimately uniformly bounded.
(3) The proposed method can maintain the output constraints and achieve arbitrary convergence time allowed by physics conditions.
(4) The proposed control algorithm is always effective when the internal interference of the actuator is bounded by a controller that does not lose efficacy totally, and the most severe failure is known.

For work to unfold further, the following assumptions are made.

Assumption 1: $g(\cdot)$ is the unknown and time-varying virtual control gain but is never zero, which means there are two positive unknown constants g_l and g_r such that $0 < g_l < |g(\cdot)| < g_r < \infty$, and sign of $g(\cdot)$ is deterministic. In this paper, in order not to lose generality, we assume that $sgn(g) = +1$.

Assumption 2: The system state variables can be used for control design. y_d is the known desired trajectory, and we assume that its derivative up to nth are known, smooth. What's more, ${y_d}^{(n)}$, the nth derivative of y_d, is bounded, i.e. exists an unknown constant y_m so that $\left|{y_d}^{(n)}\right| \leq y_m < \infty, \forall t \geq t_0$.

Assumption 3: $\rho(\cdot)$ and $u_s(\cdot)$ are unknown and unpredictable, possibly fast time-varying. And they are also bounded i.e. exist two unknown constants such that $0 < \rho_m < |\rho(\cdot)| \leq 1$ and $|u_s| \leq d < \infty$.

***Remark 1*:** Assumption 1 and 2 is commonly applied in the tracking control problem of normal form system [7], as for Assumption 3. Unlike many fault-tolerant control jobs [7, 11, 12] that require time-varying interference to change slowly enough, Assumption 3 has no such limitation and is, therefore, more widely used.

To continue, the following lemmas are essential.

Lemma 1 [1]: *Assume there is a continuous function* $\mathrm{w(x, u)} : \mathrm{R^n} \times \mathrm{R} \to \mathrm{R}$, *which is differentiable* $\forall(x, u) \in R^n \times R$, *and exists a constant c such that* $\frac{\partial \mathrm{w(x,u)}}{\partial u} \geq c > 0$, $\forall(x, u) \in R^n \times R$. *And then, there exists a continuous and smooth function* $u^* = u(x)$, *meets that* $\mathrm{w(x, u^*)} = 0$.

2.2 Description of Neural Networks

RBFNN is widely applied in control engineering as an instrument for modeling unknown nonlinear functions due to its nice capabilities in function approximation with some compact sets in [13–15]. In this work, the following RBF neural network is adopted for approximating the unknown function $\mathbb{N}(Z) \in R$:

$$\mathbb{N}_{NN}(Z) = P^T \varphi(Z) \tag{4}$$

where the input vector $Z \in \Omega_Z \subset R^m$, $P \in R^n$ is the weight vector and $\varphi(Z) = [\xi_1(Z), \ldots, \xi_n(Z)] \in R^n$, where the quantity of NN node denotes n, and $\varphi(Z)$ is the Gaussian function that is commonly used. These can be presented as follow

$$\varphi_i(Z) = exp\left[\frac{-(Z - c_i)^T (Z - c_i)}{b_i{}^2}\right], i = 1, ..., n \tag{5}$$

where $c_i = [c_{i1}, c_{i2}, ..., c_{im}]^T$ is the center of receptive field and the width of Gaussian function denotes b_i. In [13–15], it has deduced a conclusion that the NN can approximate $\mathbb{N}(Z)$ over a compact set $\Omega_Z \subset R^m$ within any arbitrary precision as

$$\mathbb{N}(Z) = P^{*T} \varphi(Z) + \delta(Z) \tag{6}$$

where P^* is the optimal constant weight matrix of RBFNN, and the approximation error denotes $\delta(Z) \in R$, which satisfies $|\delta(Z)| \leq \delta_m < \infty$ for all $Z \in \Omega_Z$ where δ_m is an unknown constant. The approximation error of neural network can decrease as the quantity of NN node increase. The practical application of neural network shows that if the quantity of NN node is sufficient, and for the approximation error of neural network, it can be reduced to a enough small value in a compact set. The optimal weight P^* is an artificial constant vector and is used here for analysis purposes only, which is represented as

$$\beta(t) : [0, \infty) \to \mathbb{R}^+ \tag{7}$$

2.3 Space Coordinate Transformation

Before introducing the control algorithm, we first introduce a kind of space coordinate transformation [2], which combines two types of valuable functions, and it is significant for designing the control scheme.

Definition 1 [2]: $\beta(t)$ is a generalized time-varying performance function, which conforms to the following features:

- $\beta(t) : [0, \infty) \to \mathbb{R}^+$ is a differentiable function of order n;
- $\beta(0) = 1$ and $\lim\limits_{t\to\infty} \beta(t) < 1$;
- $\beta(t) \in \mathcal{L}_\infty$and $\dot{\beta}(t) \in \mathcal{L}_\infty, \forall t \in [0, +\infty)$.

Remark 2: A lot of functions, actually an infinite number of functions that conform to these characters. For instance,

$$\beta(t) = \begin{cases} (1-\beta_\infty)\left(\frac{T-t}{T}\right)^n + \beta_\infty, 0 \le t < T \\ \beta_\infty, t \ge T; \end{cases} \tag{8}$$

where $\beta(t) \to \beta_\infty$, as $t \to \infty$, $T > 0$ is a constant and the order of system is represented as n. Note it doesn't have to be monotonically for the performance function, which may be beneficial in situations where large errors are caused by strong parameter variations in the system, resulting in large inputs.

Definition 2 [2]: $\kappa(x)$ is the generalized normalized function, which accords with the following properties:

- $\kappa(x) : \mathbb{R} \to (-1, 1)$ is a differentiable function of order n and monotonically increasing;
- $\lim\limits_{x\to\pm\infty} \kappa(x) = \pm 1$ and $\kappa(0) = 0$;
- $\kappa'(x)$ has a definite lower bound which is positive, where $\kappa'(x) = \frac{d\kappa(x)}{dx}$

Remark 3: In this paper, we choose $\kappa(x)$ as follows:

$$\kappa(x) = tanh(x) = \frac{e^x - e^{-x}}{e^x + e^{-x}} \tag{9}$$

then we have:

$$\kappa'(x) = sech^2(x) \tag{10}$$

The inverse function of κ is represented as $\kappa^{-1}(x)$, it is seen that:

$$\begin{aligned} &\kappa'(x) > 0, \kappa_x \triangleq \frac{\kappa}{x} > 0, \\ &\kappa^{-1}(\beta(0)) = \kappa^{-1}(1) = +\infty \end{aligned} \tag{11}$$

By combining $\beta(t)$ and $\kappa(x)$, we can construct the following coordinate transformation to apply properties to z and x as stated in Lemma 2:

$$z(\beta, \kappa) = \frac{\beta(t)\kappa(x)}{\beta^2(t) - \kappa^2(x)} \tag{12}$$

Lemma 2 [2]: For any $\beta(t)$ as defined in Sect. 2 and z as defined in (12), then $-\kappa^{-1}(\beta) < x < \kappa^{-1}(\beta)$ is founded if $t \geq 0$ and $z \in \mathcal{L}_\infty$.

***Remark 4*:** The coordinate transformation in (coordinate transformation) is simple, smooth, and non-singular. Another point to notice is that the coordinate transformation enables errors to reach a predetermined accuracy in a predetermined time.

3 Motivating Example

Taking account of the following first-order system

$$\dot{x} = f(x, u_a) \tag{13}$$

where x is the state, u_a is the actuator, $f(\cdot)$ is completely unknown system function that satisfies the definition in (13).

The tracking error is denoted as follows

$$e = x - y_d \tag{14}$$

where e is the tracking error and y_d is the desired trajectory.

By using the coordinate transformation (coordinate transformation), the system (13) can be converted into z-dynamics

$$\dot{z} = \varpi(e, t)\dot{e} + \Theta(e, t) \tag{15}$$

With

$$\varpi(e, t) = \frac{\left(\beta^2(t) - \kappa^2(e)\right)\beta(t)\kappa'(e) + 2\kappa^2(e)\kappa'(e)\beta(t)}{\left(\beta^2(t) - \kappa^2(e)\right)^2}$$

$$\Theta(e, t) = \frac{\beta(t)\kappa(e)\left(\beta^2(t) - \kappa^2(e)\right) - 2\beta^2(t)\dot{\beta}(t)\kappa(e)}{\left(\beta^2(t) - \kappa^2(e)\right)^2},$$

where ϖ and Θ are known time-varying smooth functions and bounded if $z \in \mathcal{L}\infty$ is provided, and $\varpi > 0$ for $\forall z \in \mathcal{L}\infty$. These afore mentioned conditions guarantee the controllability of system (13). Substitute (13) and (14) into (15), we have

$$\dot{z} = \varpi(f(x, u_a) - \dot{y}_d) + \Theta \tag{16}$$

Define ω as$\omega = \dot{y}_d - \frac{\Theta}{\varpi}$, then we can conclude that $\frac{\partial(f(\overline{X}, u_a)+\omega)}{\partial u_a} > g_a > 0$ by definition (3) where g_a is an unknown constant. According to Lemma 1, for x and ω, , there exists a continuous (smooth) desired unknown function $\alpha^*(x, \omega)$ such that

$$f(x, \alpha^*) + \omega = 0 \tag{17}$$

And then, by using mean value theorem [1], we also notice that there exists a constant $\mu(0 < \mu < 1)$ that satisfies

$$f(x, u_a) = f(x, \alpha^*) + g_\mu(\cdot)(u_a - \alpha^*) \tag{18}$$

where $g_\mu(\cdot) = \frac{\partial f(\overline{X}, u_a)}{\partial u_a}\Big|_{u_a = v}$ with $v = \mu u + (1-\mu)\alpha^*$. Adding and subtracting ω in the right-hand side of the Eq. (18), we have

$$f(x, u_a) = -\omega + g_\mu(\cdot)(u_a - \alpha^*) \tag{19}$$

According to the definition of ω. . Substituting (2), (19) into (15), we have

$$\dot{z} = \varpi g_\mu(\rho u + u_s - \alpha^*) \tag{20}$$

Note that α^* is an unknown continuous (smooth) function, then we can utilize the RBFNN in a compact $Z \in \Omega_Z$ to approximate $\alpha^* : \alpha^* = P^{*T}\varphi(Z) + \delta(Z)$, where $Z = [x, y_d, \dot{y}_d, \beta] \in R^5$, P^* represents the optimal constant weight, and $|\delta(Z)| \le \delta_m$ is approximate error with constant $\delta_m > 0$. From Assumption 2 and the definition of $\beta(t)$, it is seen that y_d, $\dot{y}_d$ and $\beta(t)$ are known and bounded, then it is not difficult to find a compact set such that $Z \in \Omega_Z$. Therefore (19) can be expressed as

$$\dot{z} = \varpi g_\mu(\rho u + u_s - P^{*T}\varphi - \delta) \tag{21}$$

From Assumption 1, there exist two unknown positive constants g_l and g_r such that gr_l, then the candidate Lyapunov functions can be selected as $V_c(z) = \frac{1}{2g_l}z^2$. Taking the derivative of $\frac{1}{2g_l}z^2$ with respect to time yields.

$$\begin{aligned} \frac{1}{g_l} z\dot{z} &= \frac{g_\mu}{g_l}\varpi z(\rho u + u_s - P^{*T}\varphi - \delta) \\ &= \frac{g_\mu}{g_l}\varpi z\rho u + \frac{g_\mu}{g_l}\varpi z u_s - \frac{g_\mu}{g_l}\varpi z(P^{*T}\varphi + \delta) \end{aligned} \tag{22}$$

By using Young's inequality, it follows that

$$\begin{aligned}\tfrac{g_\mu}{g_l}\varpi z u_s &\le \varpi^2 z^2 d^2 + \tfrac{{g_r}^2}{4{g_l}^2}\\ -\tfrac{g_\mu}{g_l}\varpi z(P^{*T}\varphi + \delta) &\le \varpi^2 z^2(\|P_1^*\|^2\|\varphi\|^2 + \delta_m^2) + \tfrac{{g_r}^2}{2{g_l}^2}\end{aligned} \tag{23}$$

then we have

$$-\frac{g_\mu}{g_l}\varpi z(P^{*T}\varphi + \delta) + \frac{g_\mu}{g_l}\varpi z u_s \le z^2 c\phi + \frac{{g_r}^2}{2{g_l}^2} \tag{24}$$

where $c = max\{\|P^*\|^2, {\delta_m}^2, d^2\}$ and $\phi = \left(\|\varphi\|^2 + 1\right)\varpi^2$.

Substituting (24) into (22), we have

$$\frac{1}{g_l}z\dot{z} \le \frac{g_\mu}{g_l}\varpi z\rho u + \varpi^2 z^2 d^2 + z^2 c\phi + \frac{3{g_r}^2}{4{g_l}^2} \tag{25}$$

Note that c is an unknown constant, which can be estimated in an adaptive way, and ϕ can be calculated with available x, y_d, $\dot{y}_d$ and $\beta(t)$. To proceed the work, we construct adaptive law $\dot{\hat{c}}$ as

$$\dot{\hat{c}} = -k_1\hat{c} + k_0\phi z^2, \hat{c}(0) > 0 \tag{26}$$

where $\hat{c}$ is the estimate of c.

Choosing the control law as.

$$u = -\frac{\gamma z}{\rho_m} - \frac{z\hat{c}\phi}{\rho_m\varpi} \tag{27}$$

Let $\tilde{c} = c - \hat{c}$, then (25) within the adaptive law and control law follows that

$$\frac{1}{g_l}z\dot{z} \le -\gamma\varpi z^2 + \tilde{c}\phi z^2 + \frac{3{g_r}^2}{4{g_l}^2} + \varpi^2 z^2 d^2 \tag{28}$$

Choosing Lyapunov function candidate as $V(z, \tilde{c}) = \frac{1}{2g_l}z^2 + \frac{1}{2k_0\rho_m}\tilde{c}^2$ and taking its derivative w.r.t time yields.

and substituting adaptive law into $V(z, \tilde{c})$, we have

$$\begin{aligned}\dot{V}(z, \tilde{c}) &\le -\gamma\varpi z^2 + \tilde{c}\phi z^2 + \frac{3g_r^2}{4g_l^2} + \varpi^2 z^2 d^2 + \frac{1}{k_0}\tilde{c}\left(-\dot{\hat{c}}\right)\\ &\le -\gamma\varpi z^2 + \frac{3g_r^2}{4g_l^2} + \varpi^2 z^2 d^2 + \frac{k_1}{k_0}(c + \tilde{c})(c - \tilde{c})\\ &\le -\gamma\varpi z^2 - \frac{k_1}{k_0}\tilde{c}^2 + \frac{k_1}{k_0}c^2 + \frac{3g_r^2}{4g_l^2} + \varpi^2 z^2 d^2\end{aligned} \tag{29}$$

Let $\ell_1 = max\{2g_l\varpi\gamma, 2k_1\} > 0$ and $\ell_2 = \frac{k_1}{k_0}c^2 + \frac{3g_r^2}{4g_l^2} + \varpi^2 z^2 d^2 > 0$. It is obvious that $\ell_1 \in \mathcal{L}_\infty$, and $\ell_2 \in \mathcal{L}_\infty$, then (29) can be expressed as

$$\dot{V} \leq -\ell_1 V + \ell_2 \tag{30}$$

From (30), we can deduce that there exists a compact set $\Omega = \left\{(z, \tilde{c}) | V \leq \frac{\ell_2}{\ell_1}\right\}$, which is a domain of attraction. Once $(z, \tilde{c}) \notin \Omega$, $\dot{V} < 0$. There consequently exists a finite time T_f meets that $(z, \tilde{c}) \in \Omega$ for $\forall t > T_f$, i.e. $|z| \leq \sqrt{2g_l\left(V - \frac{k_1}{2k_0}\tilde{c}^2\right)} \leq \sqrt{2g_l\left(\frac{\ell_2}{\ell_1} - \frac{k_1}{2k_0}\tilde{c}^2\right)}$. Note c is unknown constant, and $\hat{c}$ is bounded with the adaptive law in (26), then we can conclude that $\hat{c}$ is bounded, so is the transform error z. Therefore, z is ultimately uniformly bounded (UUB). In addition, the function $\beta(t)$ and $\kappa(e)$ are bounded in definition, which means $\beta(t) \in \mathcal{L}_\infty, \kappa(e) \in \mathcal{L}_\infty$, then we can get the conclusion that the tracking error e is also UUB.

Next, we analyze the boundedness of each internal signal in the system. By solving the differential inequality (30), it is derived that, $V \leq V(0)e^{-\ell_1 t} + \frac{\ell_2}{\ell_1} \in \mathcal{L}_\infty$, for $\forall t \geq 0$. It impliesz $\in \mathcal{L}_\infty, \tilde{c} \in \mathcal{L}_\infty$. By recalling that $\beta(t) \in \mathcal{L}_\infty, \kappa(e) \in \mathcal{L}_\infty$, and $y_d \in \mathcal{L}_\infty$, $\dot{y}_d \in \mathcal{L}_\infty$ as we mentioned in Assumption 2. Finally, one can deduce that $x \in \mathcal{L}_\infty$. Therefore we can deduce that all the internal signals in system are continuous and bounded.

Then the neuro-adaptive fault-tolerant control will be extended to higher-order systems in Sect. 4.

4 Main Results

The design process for the first-order system motivates us such that the control algorithm proposed in Sect. 3 can be extended to a higher order in this part, and the object is changed into the model (1). For such a normal form system, we utilize the classical backstepping design technique [6], with particular treatment in the final step, as detailed in what follows:

Step 1: First we apply coordinate transformation to the output state, which is mentioned in (12), and define the virtual error of each order as follows:

$$\begin{cases} z_1 = \frac{\beta(t)\chi(e)}{\beta^2(t)-\chi^2(e)} \\ z_i = x_i - \alpha_{i-1}, (i = 2, 3..., n) \end{cases} \tag{31}$$

where z_i is the virtual error, and z_1 is the conversion error especially, and α_{i-1} is the virtual input, which can also be called virtual control law. Next, we design the controller.

Taking the derivative of z_1 w.r.t the time yields as

$$\dot{z} = \varpi(e,t)\dot{e} + \Theta(e,t) \tag{32}$$

The Lyapunov function candidate is chosen as $V_1 = \frac{1}{2}{z_1}^2$, and take its derivative with respect to the time yields as

$$\dot{V}_1 = z_1\dot{z}_1 = z_1(\varpi(x_2 - \dot{y}_d) + \Theta) \tag{33}$$

To make this derivative negative semidefinite, choosing the first desired virtual control law as

$$\alpha_1 = -\frac{\Theta}{\varpi} - \frac{\gamma_1 z_1}{\varpi} + \dot{y}_d \tag{34}$$

where $\gamma_1 > 0$ is a constant parameter that can be designed by user. Substituting virtual error and the virtual law in (34) into (33), we have

$$\dot{V}_1 = -\gamma_1{z_1}^2 + \varpi z_1 z_2 \tag{35}$$

where $-\gamma_1{z_1}^2$ is negative semidefinite, and $\varpi z_1 z_2$ will be dealt with in the next step.

Step $i(2 \le i \le n-1)$: We choose the Lyapunov function candidate as $V_i = \frac{1}{2}{z_1}^2 + \frac{1}{2}{z_2}^2 + \ldots + \frac{1}{2}{z_i}^2$, and take its derivative w.r.t the time yields as

$$\begin{aligned}\dot{V}_i &= -\gamma_1 z_1^2 - \gamma_2 z_2^2 \ldots - \gamma_{i-1} z_{i-1}^2 \\ &\quad + z_i(kz_{i-1} + x_{i+1} - \dot{\alpha}_{i-1})\end{aligned} \tag{36}$$

with

$$\begin{aligned}\dot{\alpha}_{i-1} = \sum_{j=1}^{i-1}\frac{\partial\alpha_{i-1}}{\partial x_j}x_{j+1} + \sum_{j=0}^{i-1}\frac{\partial\alpha_{i-1}}{\partial\beta^{(j)}}\beta^{(j+1)} \\ + \sum_{j=1}^{i-1}\frac{\partial\alpha_{i-1}}{\partial y_d^{(j)}}y_d^{(j+1)}\end{aligned} \tag{37}$$

where $k = \varpi$ with $i = 2$, and $k = 1$ when $i > 2$. we can construct the i-th desired virtual control law as

$$\begin{aligned}\alpha_i = -\gamma_i z_i - kz_{i-1} + \sum_{j=1}^{i-1}\frac{\partial\alpha_{i-1}}{\partial x_j}x_{j+1} \\ + \sum_{j=0}^{i-1}\frac{\partial\alpha_{i-1}}{\partial\beta^{(j)}}\beta^{(j+1)} + + \sum_{j=1}^{i-1}\frac{\partial\alpha_{i-1}}{\partial y_d^{(j)}}y_d^{(j+1)}\end{aligned} \tag{38}$$

where $\gamma_i > 0$ are constant parameters that can be designed by user. Substituting virtual error and the virtual law in (38) into (37), we have

$$\dot{V}_i = -\gamma_1 {z_1}^2 - \gamma_2 {z_2}^2 \ldots - \gamma_i {z_i}^2 + z_i z_{i-1} \tag{39}$$

where $-\gamma_1 {z_1}^2 - \gamma_2 {z_2}^2 \ldots - \gamma_i {z_i}^2$ is negative semidefinite, and $z_i z_{i-1}$ will be handled in the next step.

Step n: Choosing the Lyapunov function candidate as $V_n = \frac{1}{2}{z_1}^2 + \frac{1}{2}{z_2}^2 + \ldots + \frac{1}{2g_l}{z_n}^2$, and take its derivative w.r.t the time yields as

$$\begin{aligned} \dot{V}_n &= -\gamma_1 z_1^2 - \gamma_2 z_2^2 \ldots - \gamma_{n-1} z_{n-1}^2 \\ &+ z_n\left(z_{n-1} + \frac{1}{g_l}\left(f(\bar{X}, u_a) - \dot{\alpha}_{n-1}\right)\right) \end{aligned} \tag{40}$$

with

$$\begin{aligned} \dot{\alpha}_{n-1} = &\sum_{j=1}^{n-1} \frac{\partial \alpha_{n-1}}{\partial x_j} x_{j+1} + \sum_{j=0}^{n-1} \frac{\partial \alpha_{n-1}}{\partial \beta^{(j)}} \beta^{(j+1)} \\ &+ \sum_{j=1}^{n-1} \frac{\partial \alpha_{n-1}}{\partial y_d^{(j)}} y_d^{(j+1)} \end{aligned} \tag{41}$$

Define ω_1 as $\omega_1 = -\dot{\alpha}_{n-1}$, then we can conclude that $\frac{\partial(f(\overline{X}, u_a) + \omega_1)}{\partial u_a} > g_1 > 0$ by definition (3) where g_1 is unknown constant. According to Lemma 1, for $\overline{X}$ and ω_1, there exists a continuous (smooth) desired unknown function $\alpha_1^*(\overline{X}, \omega_1)$ such that

$$f(\overline{X}, \alpha_1^*) + \omega_1 = 0 \tag{42}$$

then, the mean value theorem [1] is adopted, and it should be noted that there exists a $\mu_1 (0 < \mu_1 < 1)$ such that

$$f(\overline{X}, u_a) = f(\overline{X}, \alpha_1^*) + g_{\mu 1}(\cdot)(u_a - \alpha_1^*) \tag{43}$$

where $g_{\mu 1}(\cdot) = \left.\frac{\partial f(\overline{X}, u_a)}{\partial u_a}\right|_{u_a = v_1}$ with $v_1 = \mu_1 u_a + (1 - \mu_1){\alpha_1}^*$. Adding and subtracting ω_1 in the right-hand side of the Eq. (43), then we have

$$f(\overline{X}, u_a) = -\omega_1 + g_{\mu 1}(\cdot)(u_a - \alpha_1^*) \tag{44}$$

Substituting (44) into $\dot{z}_n$, we have

$$\dot{z}_n = g_{\mu 1}(\rho u + u_s - {\alpha_1}^*) \tag{45}$$

Note that $\alpha_1{}^*$ is an unknown continuous (smooth) function, then we can utilize the RBFNN in a compact $Z_1 \in \Omega_{Z_1}$ to approximate $\alpha_1{}^* : \alpha_1{}^* = P_1{}^{*T}\varphi_1(Z_1) + \delta_1(Z_1)$, where $Z_1 = \left[x_1, ..., x_n, y_d, ..., y_d{}^{(n)}, \beta\right] \in R^{2n+2}$, $P_1{}^*$ represents the ideal constant weight, and $|\delta_1(Z_1)| \le \delta_m$ is approximate error with constant $\delta_m > 0$. From Assumption 2 and the definition of $\beta(t)$, it is seen that $x_1, ..., x_n, y_d, ..., y_d{}^{(n)}, \beta(t)$ are known and bounded, then it is not hard to find a compact set such that $Z \in \Omega_Z$. Therefore (40) can be expressed as

$$\begin{aligned}\dot{V}_n &= -\gamma_1 z_1^2 - \gamma_2 z_2^2 ... - \gamma_{n-1} z_{n-1}^2 \\ &+ z_n\left(z_{n-1} + \frac{g_{\mu 1}}{g_l}\left(\rho u + u_s - P_1^{*T}\varphi_1 - \delta_1\right)\right) \\ &= -\gamma_1 z_1^2 - \gamma_2 z_2^2 ... - \gamma_{n-1} z_{n-1}^2 + z_n z_{n-1} \\ &+ \frac{g_{\mu 1}}{g_l}\rho z_n u + \frac{g_{\mu 1}}{g_l} z_n u_s - \frac{g_{\mu 1}}{g_l} z_n\left(P_1^{*T}\varphi_1 + \delta_1\right)\end{aligned} \tag{46}$$

From Assumption 1, there exist two unknown positive constants g_l and g_r such that $g\mu 1_{rl}$, then by using Young's inequality, it follows that

$$\begin{aligned}\tfrac{g_{\mu 1}}{g_l} z_n u_s &\le z_n{}^2 d^2 + \tfrac{g_r{}^2}{4 g_l{}^2} \\ -\tfrac{g_{\mu 1}}{g_l} z_n\left(P_1{}^{*T}\varphi_1 + \delta_1\right) &\le z_n{}^2(\|P_1^*\|^2\|\varphi_1\|^2 + \delta_m^2) + \tfrac{g_r{}^2}{2 g_l{}^2}\end{aligned} \tag{47}$$

then we have

$$-\frac{g_{\mu 1}}{g_l} z_n\left(P_1{}^{*T}\varphi_1 + \delta_1\right) + \frac{g_{\mu 1}}{g_l} z_n u_s \le z_n{}^2 c_f \phi_1 + \frac{g_r{}^2}{2 g_l{}^2} \tag{48}$$

where $c_f = max\left\{\|P_1{}^*\|^2, \delta_m{}^2\right\} > 0$ and $\phi_1 = \left(\|\varphi\|^2 + 1\right) > 0$.

Substituting (48) into (46), we have

$$\dot{V}_n \le -\sum_{i=1}^{n-1}\gamma_i z_i^2 + \frac{g_{\mu 1}}{g_l}\rho z_n u + z_n^2 c_f \phi_1 + + \frac{3 g_r^2}{4 g_l^2} \tag{49}$$

Note that c_f is an unknown constant, which can be estimated in an adaptive way, and ϕ_1 can be calculated with available $x_1, ..., x_n, y_d, ..., y_d{}^{(n)}, \beta, ..., \beta^{(n)}$. To proceed the work, we construct adaptive law $\dot{\hat{c}}_f$ as

$$\dot{\hat{c}}_f = -k_1\hat{c}_f + k_0\phi_1 z_n{}^2, \hat{c}(0) \ge 0 \tag{50}$$

where $\hat{c}_f$ is the estimate of c_f.

Choosing the true control law as

$$u = -\frac{\gamma_n z_n}{\rho_m} - \frac{z_n \hat{c}_f \phi_1}{\rho_m} \tag{51}$$

Let $\tilde{c}_f = c_f - \hat{c}_f$, then (49) within the adaptive law and control law follows that.

$$\dot{V}_n \le -\sum_{i=1}^{n} \gamma_i {z_i}^2 + {z_n}^2 \tilde{c}_f \phi_1 + \frac{3{g_r}^2}{4{g_l}^2} \tag{52}$$

To handle the estimate error $\hat{c}_f$, choosing Lyapunov function candidate as $V_N(z_1, \ldots, z_n, \tilde{c}) = V_n + \frac{1}{2k_0}\tilde{c}_f^2$, and taking the derivative of $V_N(z_1, \ldots, z_n, \tilde{c})$ w.r.t time yields. Then substitute the adaptive law into $\dot{} V_N(z_1, \ldots, z_n, \tilde{c})$, we have

$$\begin{aligned} V_N &\le -\sum_{i=1}^{n} \gamma_i z_i^2 + z_n^2 \tilde{c}_f \phi_1 + \frac{3g_r^2}{4g_l^2} + z_n^2 d^2 + \frac{1}{k_0}\tilde{c}_f\left(-\dot{\hat{c}}_f\right) \\ &\le -\sum_{i=1}^{n} \gamma_i z_i^2 + \frac{3g_r^2}{4g_l^2} + z_n^2 d^2 + \frac{1}{k_0}\tilde{c}_f \hat{c}_f \\ &\le -\sum_{i=1}^{n} \gamma_i z_i^2 + \frac{3g_r^2}{4g_l^2} + z_n^2 d^2 + \frac{1}{k_0}\tilde{c}_f (c_f - \tilde{c}_f) \\ &\le -\sum_{i=1}^{n} \gamma_i z_i^2 + \frac{3g_r^2}{4g_l^2} + z_n^2 d^2 + \frac{k_1}{k_0}(c_f + \tilde{c}_f)(c_f - \tilde{c}_f) \\ &\le -\sum_{i=1}^{n} \gamma_i z_i^2 - \frac{k_1}{k_0}\tilde{c}_f^2 + \frac{k_1}{k_0}c_f^2 + \frac{3g_r^2}{4g_l^2} + z_n^2 d^2 \end{aligned} \tag{53}$$

Let $\ell_3 = min\{2\rho_m\gamma_1, \ldots, 2\rho_m\gamma_{n-1}, 2g_l\rho_m\gamma_n, 2k_1\rho_m\} > 0$ and $\ell_4 = \frac{k_1}{k_0}{c_f}^2 + \frac{3{g_r}^2}{4{g_l}^2} + {z_n}^2 d^2 > 0$. Then (53) can be expressed as

$$\dot{V}_N \le -\ell_3 V_N + \ell_4 \tag{54}$$

From (54), there is a compact set $\Omega = \left\{(z_1, \ldots, z_n, \tilde{c}_f) \middle| V_N \le \frac{\ell_4}{\ell_3}\right\}$, which is a domain of attraction. Once $(z_1, \ldots, z_n, \tilde{c}_f) \notin \Omega$, $\dot{} \dot{V} < 0$. There consequently exists a finite time T_0 satisfies that $(z_1, \ldots, z_n, \tilde{c}_f) \in \Omega$ for $\forall t > T_0$, i.e. $\left|\sum_{i=1}^{n} z_i\right| \le \sqrt{a\left(V_N - \frac{k_1}{2k_0}\tilde{c}_f^2\right)} \le \sqrt{a\left(\frac{\ell_4}{\ell_3} - \frac{k_1}{2k_0}\tilde{c}_f^2\right)}$, where $a = min\{0.5, 0.5g_l\}$. Note c_f is unknown constant, and $\hat{c}_f$ is bounded with the adaptive law in (26), then we can conclude that $\hat{c}_f$ is bounded, so is $z_1, \ldots, z_n$. Therefore, $z_1, \ldots, z_n$ is ultimately uniformly bounded (UUB). In addition, the function $\beta(t)$ and $\kappa(e)$ are bounded in definition, which means $\beta(t) \in \mathcal{L}\infty$, $\kappa(e) \in \mathcal{L}\infty$, then we can get the conclusion that the tracking error e is also UUB.

Next, we analyze the boundedness of each internal signal in the system. By solving the differential inequality (54), it is derived that, $V_N \leq V_N(0)e^{-\ell_3 t} + \frac{\ell_4}{\ell_3} \in \mathcal{L}_\infty$, for $\forall t \geq 0$. It implies $z_1, ..., z_n \in \mathcal{L}_\infty, \tilde{c} \in \mathcal{L}_\infty$. By recalling that $\beta(t) \in \mathcal{L}_\infty, \kappa(e) \in \mathcal{L}\infty$, and $y_d{}^{(i)}(1 \leq i \leq n) \in \mathcal{L}_\infty$ as we mentioned in Assumption 2. Then, one can conclude that $e \in \mathcal{L}_\infty$. So we have $x_1 \in \mathcal{L}_\infty$ and $\alpha_1 \in \mathcal{L}_\infty$ in further. By a similar recursion, we can conclude that $x_2, ...x_n \in \mathcal{L}_\infty$ and $\alpha_2, ..., \alpha_{n-1}, u \in \mathcal{L}_\infty$. Therefore, we can deduce that all the internal signals in system are continuous and bounded.

Remark 5: Note that $\kappa^{-1}(\beta(0)) = \infty$, i.e. user can choose any initial conditions if the physical conditions permit, and $-\kappa^{-1}(\beta(t)) < e(t) < \kappa^{-1}(\beta(t))$, which means the tracking error is bounded with preset constraints. Moreover, the preset constraints that represent the prescribed performance can be adjusted through adjustable parameters by users for the specific requirements, even the index of convergence time.

Remark 6: The method presented in this paper can deal well with the situation where the system is completely unknown and contains actuator failure and still achieves ultimately uniformly bounded.

5 Simulation Results

To verify the proposed neuro-adaptive fault-tolerant prescribed control, we take account of the following second-order normal form nonlinear systems

$$\begin{cases} \dot{x}_1 = x_2 \\ \dot{x}_2 = f(\overline{X}, u_a) \\ y = x_1 \end{cases} \tag{55}$$

where $\overline{X} = [x_1, x_2]^T$ denote the state variables vector, and u_a and y are the actuator and system output, respectively. Note the $f(\overline{X}, u_a)$is completely unknown system function and it can be selected as $f(\overline{X}, u_a) = 0.1\sin^2(x_1x_2) + u_a$, Where $u_a = \rho(t_\rho, t)u + u_s(t_r, t)$ is the actuator with $t_\rho = t_r = 0$, and here we choose unknown undetectable disturbance u_s in actuator as $u_s = 0.01\sin(t)$ and the unknown health indicator ρ as

$$\rho = \begin{cases} 1 - 0.5t, 0 \leq t \leq 1 \\ 0.3 + 0.2e^{1-t}, t > 1 \end{cases} \tag{56}$$

It is not difficult to verify that Assumptions 1–3 are satisfied. In the simulation, the control objective is to let the output x_1 track the desired trajectory $y_d(t) = 0.5\sin(t)$ as closely as possible and guarantee the boundedness of all signals in the closed-loop systems as well as the following state constraints$-\kappa^{-1}(\beta(t)) < x_1 < \kappa^{-1}(\beta(t))$, where we choose $\beta_\infty = 0.3$ and $T = 3$, i.e. the final preset tracking accuracy $\kappa^{-1}(\beta_\infty) = \frac{1}{2} In\left(\frac{1+0.3}{1-0.3}\right) \approx 0.31$ will be achieve in three seconds, which is the preset time. In addition, the initial conditions are chosen as$[x_1, x_2]^T = [-0.5, 0.5]^T$, $\hat{c}_f(0) = 0$ and the design parameters are given $\gamma_1 = 5$, $\gamma_2 = 3$, $k_1 = 0.01$, $k_2 = 2$. And RBFNN are chose as: Neural network $P_1^{*}\varphi_1$ contains 41 nodes with centers spaced evenly in the interval $[-2, 2] \times [-2, 2] \times [-2, 2] \times [-2, 2] \times [-2, 2] \times [-2, 2] \times [-2, 2] \times [-2, 2]$ and widths being equal to 1.

The simulation results are presented in Figs. 1–3. Figure 1 shows the evolutions of health indicator ρ and unknown undetectable disturbance u_s in the actuator, and Fig. 2 shows the evolutions of control law u and estimate parameter $\hat{c}_f$. Figure 3 shows the evolutions of tracking error e and conversion error z_1, the evolutions of state trajectory x_1 and desired trajectory, and the evolution of state x_2, respectively.

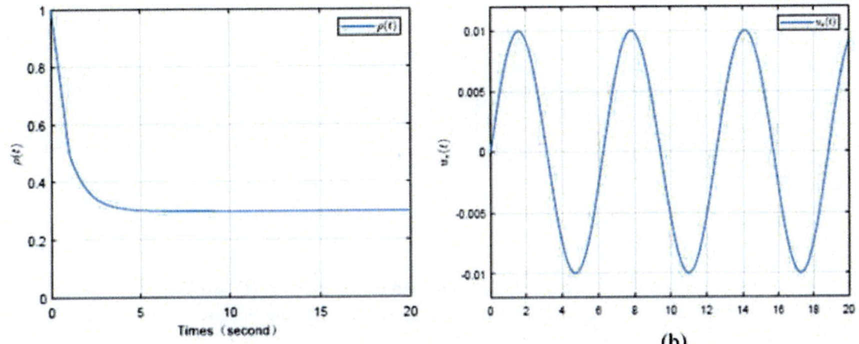

(a) (b)

Fig. 1 The health indicator $\rho(t)$ of controller and the undetectable disturbance $u_s(t)$ in actuator

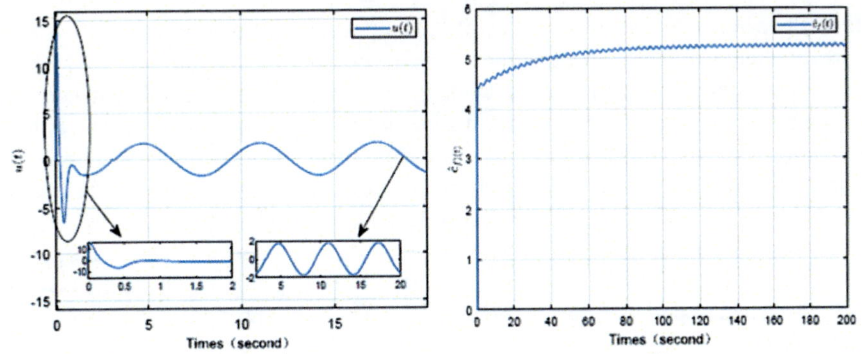

(a) (b)

Fig. 2 The evolutions of the real control law $u(t)$ and estimate parameter $\hat{c}_f(t)$ are shown in (**a**), (**b**), respectively

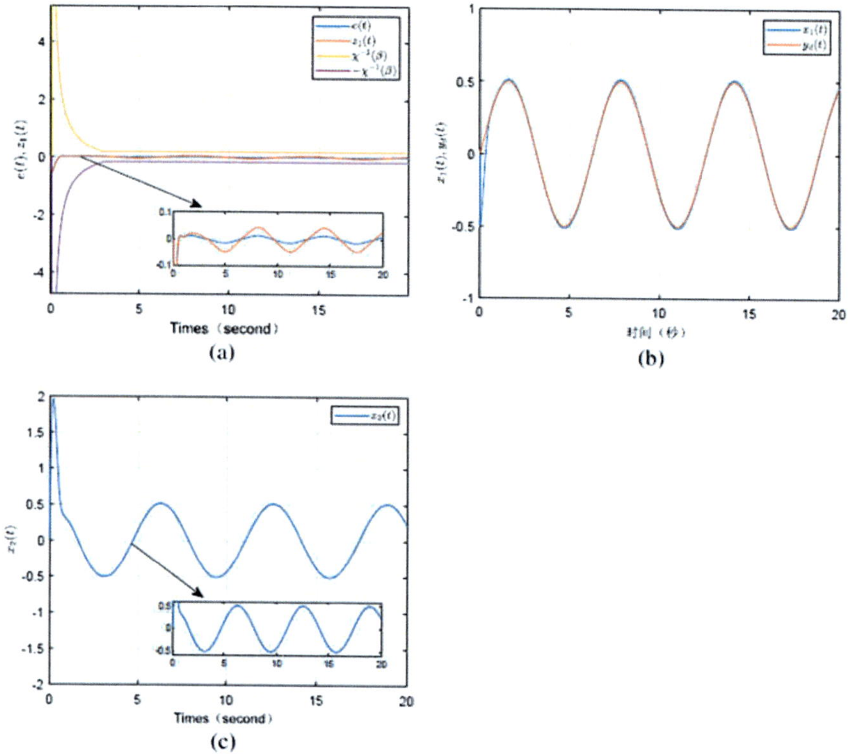

Fig. 3 The evolutions of tracking error $e(t)$ and conversion error $z_1(t)$ are shown in (**a**), and the evolution of output state and desired trajectory are shown in (**b**), and the evolution of state $x_2(t)$ is shown in (**c**)

6 Conclusion

In this paper, a neuro-adaptive fault-tolerant strategy has been developed for normal form nonlinear systems in the presence of output constraints. By introducing a spatial coordinate transformation, we developed a neuro-adaptive fault-tolerant control scheme that possesses the following characteristics: (1) Users can choose any initial conditions, i.e., the conservative initial conditions in the traditional way are removed; and (2)The tracking error can converge to a preset accuracy in a preset time; and (3) Completely unknown smooth functions can be handled gracefully by neural networks. It is still an open, challenging problem when the control gained in this paper becomes time-varying and the direction is unknown. The complexity of the system is higher, such as pure feedback nonlinear system, which represents interesting work in the future.

References

1. Zhao, K., Song, Y.: Neuroadaptive robotic control under time-varying asymmetric motion constraints: A feasibility-condition-free approach. IEEE Trans. Cybern. **50**(1), 15–24 (2018)
2. Ye, H., Song, Y.: Adaptive control with guaranteed transient behavior and zero steady-state error for systems with time-varying parameters. *arXiv preprint* arXiv:2202.06320, (2022)
3. Chen, K., Astolfi, A.: Adaptive control for nonlinear systems with time-varying parameters and control coefficient. IFAC-PapersOnLine **53**(2), 3829–3834 (2020)
4. Chen, K., Astolfi, A.: Adaptive control of linear systems with timevarying parameters. In: 2018 Annual American Control Conference (ACC), pp. 80–85, IEEE, (2018)
5. Song, Y., Zhao, K., Krstic, M.: Adaptive control with exponential regulation in the absence of persistent excitation. IEEE Trans. Autom. Control **62**(5), 2589–2596 (2016)
6. Ye, H., Song, Y.: Backstepping design embedded with time-varying command filters. Express Briefs, IEEE Transactions on Circuits and Systems II (2022)
7. Song, Y., Wang, Y., Wen, C.: Adaptive fault-tolerant pi tracking control with guaranteed transient and steady-state performance. IEEE Trans. Autom. Control **62**(1), 481–487 (2016)
8. Liu, Y.-J., Tong, S.: Barrier lyapunov functions-based adaptive control for a class of nonlinear pure-feedback systems with full state constraints. Automatica **64**, 70–75 (2016)
9. Liu, Y.-J., Tong, S.: Barrier lyapunov functions for nussbaum gain adaptive control of full state constrained nonlinear systems. Automatica **76**, 143–152 (2017)
10. Tang, Z.-L., Ge, S.S., Tee, K.P., He, W.: Robust adaptive neural tracking control for a class of perturbed uncertain nonlinear systems with state constraints. IEEE Trans. Syst., Man, Cybernet: Syst. **46**(12), 1618–1629 (2016)
11. Gao, C., Zhang, C., Liu, X., Wang, H., Wang, L.: Event-triggering based adaptive neural tracking control for a class of pure-feedback systems with finite-time prescribed performance. Neurocomputing **382**, 221–232 (2020)
12. Cai, W., Liao, X., Song, Y.: Indirect robust adaptive fault-tolerant control for attitude tracking of spacecraft. J. Guid. Control. Dyn. **31**(5), 1456–1463 (2008)
13. Wang, W., Wen, C.: Adaptive actuator failure compensation control of uncertain nonlinear systems with guaranteed transient performance. Automatica **46**(12), 2082–2091 (2010)
14. Ge, S.S., Wang, C.: Adaptive nn control of uncertain nonlinear pure-feedback systems. Automatica **38**(4), 671–682 (2002)
15. Zhao, K., Song, Y., Shen, Z.: Neuroadaptive fault-tolerant control of nonlinear systems under output constraints and actuation faults. IEEE Trans. Neural Netw. Learn. Syst. **29**(2), 286–298 (2016)
16. Zhao, K., Song, Y., Ma, T., He, L.: Prescribed performance control of uncertain euler–lagrange systems subject to full-state constraints. IEEE Trans. Neural Netw. Learn. Syst. **29**(8), 3478–3489 (2017)

Research on Improved Genetic Algorithm to Optimize PID Parameters of Second-Order System

Yi Jiangang, Liu Peng, Wu Jiajun, Xu Changsong, Gao Jun, and Guo Xin

Abstract The reliability and stability of the second-order system are mainly based on the PID control parameter tuning, but the traditional parameter tuning method will cause some effects such as long adjustment time, large overshoot, and slow response. Based on the foregoing, an improved genetic algorithm was proposed to optimize the control method of the Proportion Integration Differentiation (PID) parameters of the second-order system. Our research built a typical second-order system—mass-spring-damping single-degree-of-freedom system, constructed the mathematical model of the PID control second-order system, then used MATLAB to perform Simulink simulation experiments and comparative analysis on PID parameter tuning of second-order system based on Artificial Experience (AE), Genetic Algorithm (GA) and Improved Genetic Algorithm (IGA). The results show that the PID control parameters tuned by IGA have fast convergence speed and reach the global optimal state relatively quickly, which makes the control of the second-order system have the advantages of high stability and small overshoot.

1 Introduction

Most control systems in industrial production today are higher-order systems, but under certain specific conditions, some of the non-essential factors can be omitted and they can be approximated as a second-order system [1]. In practical applications, it is accurate enough to analyze some of the higher-order systems as second-order systems, so the study of second-order systems has better practical significance. There are many relevant examples of second-order systems, such as mass-spring-damped mechanical systems, RLC circuits, and so on [2]. The control effect of second-order systems depends on the selection of control parameters, which has been studied by many researchers. In 1942, Ziegler and Nichols proposed the Z-N engineering rectification method starting from the time domain response, but it was not well adapted

Y. Jiangang (✉) · L. Peng · W. Jiajun · X. Changsong · G. Jun · G. Xin
School of Smart Manufacturing, Jianghan University, Wuhan, China
e-mail: Yjg_wh@yeah.net

K. Nakamatsu et al. (eds.), *Advanced Intelligent Technologies for Information and Communication*, Smart Innovation, Systems and Technologies 365,
https://doi.org/10.1007/978-981-99-5203-8_31

[3]. Later, Siemens summarized the "regulator optimal tuning" method [4]. With the deepening of industrialization, this type of engineering rectification method requires engineers to accumulate manual experience by trial and error repeatedly, which has the disadvantage of being time-consuming and laborious. The PID parameters are often selected to ensure stability and accuracy, which are too conservative in terms of rapidity [3]. Today, many research academics have used genetic algorithms to rectify PID parameters [4]. The PID control algorithm is a process control algorithm that consists of three links: proportional, integral, and differential. It is one of the most technically mature and widely used control algorithms for linear control systems, which first appeared in the 1930s and 1940s for applications where the model of the controlled object was not well understood. The essence of PID control is to operate according to the input deviation values, in accordance with the proportional, integral, and differential functions, and the result of the operation is used to control the output. With the deepening of industrialization, this type of engineering calibration method requires engineers to accumulate manual experience through trial and error, which has the disadvantage of being time-consuming and labor-intensive. PID regulators are becoming a fundamental part of process control activities in industry, relying on their flexible structure and ease of parameter adjustment, and have become an integral part of the system. For objects with difficult mathematical model solving and redundant parameters, the choice of PID regulator can also be a good solution to the target problem. Today, as technology is updated and iterated, intelligent optimization algorithms are gradually replacing the functions of the old analog PID regulation, enabling digital PID rectification, whose control process is easier to regulate and perfect than before [6]. Many researchers have used genetic algorithms to rectify PID parameters. The genetic algorithm was first proposed by Holland [7] in 1969 and has been collated and formed by scholars. Hao Qi designed an adaptation function consisting of error squared integral, maximum overshoot, adjustment time, and peak time [8]. Sun Yumeng used penalty function and error integral as the fitness function [9, 10]. At present, there are many adaptation methods, but some of them are not refined enough [11–13], and more single consideration is given to the stability, rapidity, or accuracy of the control system. This paper combines the control problem of a typical mass-spring-damping second-order system, introduces a genetic algorithm, combines the characteristics of PID and genetic algorithm, integrates the three characteristics of the above control system, redesigns the objective function to assign reasonable weights to the importance of the characteristics, and the obtained improved genetic algorithm adjusts and corrects the PID parameters. The results show that the PID parameters are better optimized by using the genetic algorithm, and the results of the simulations will be compared and analyzed by the artificial empirical method (AE), the genetic algorithm (GA), and the improved genetic optimization algorithm (IGA), The effectiveness of the proposed rectification method was verified in conjunction with MATLAB.

2 Mathematical Model of Second-Order System PID Control

2.1 Second-Order System Control Components

A typical second-order control system can be reduced to a PID controller and a second-order system, with the PID controller, used as a series calibration link to dynamically adjust the input through the output to achieve negative feedback control. The control block diagram of the PID regulated second-order system is shown in Fig. 1.

2.2 Typical Second-Order System Model

According to Fig. 1, a typical open-loop transfer function of a two-stage system.

$$G_k(s) = \frac{K}{s(Ts+1)} \tag{1}$$

where T is the time constant of the second-order system and K is the gain. The closed-loop transfer function [9].

$$G_b(s) = \frac{G_k(s)}{G_k(s)+1} = \frac{K}{Ts^2+s+K} = \frac{\omega_n^2}{s^2+2\xi\omega_n s+\omega_n^2} \tag{2}$$

where ω_n is the undamped intrinsic frequency; ξ is the damping ratio. The control object of this paper is a mass-spring-damped single-degree-of-freedom system to study the control tracking accuracy of a typical second-order system. The model of the mass-spring-damped single-degree-of-freedom system is shown in Fig. 2, with initial conditions of m = 10 kg, c = 1 N·s/m, and k = 1 N/m.

According to Newton's second law of motion, the kinetic equations of the mass-spring-damped single-degree-of-freedom system are obtained.

$$m\ddot{x}(t) = f(t) - c\dot{x}(t) - kx(t) \tag{3}$$

After Laplace transformation, we obtain F.

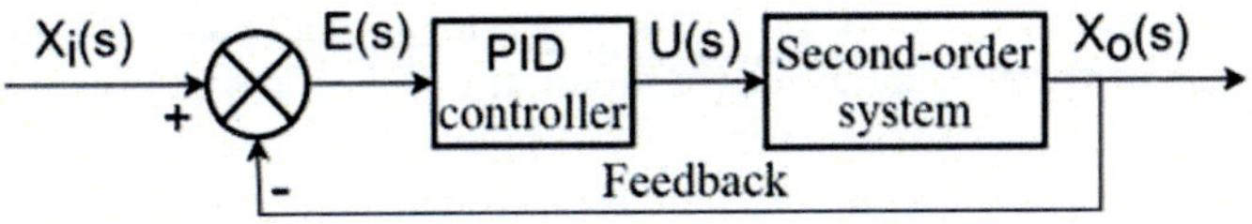

Fig. 1 Second-order system control block diagram of PID regulation

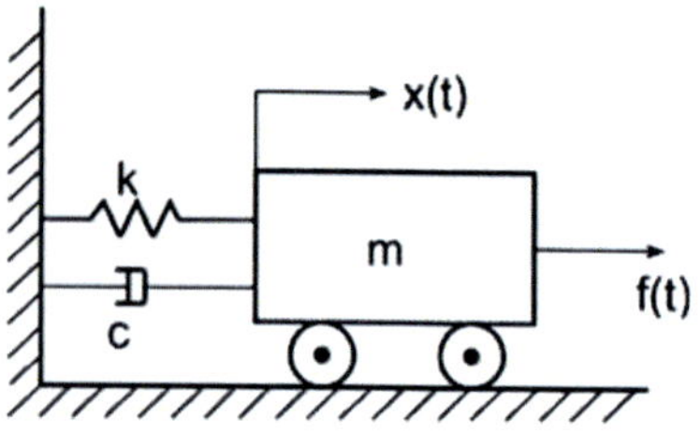

Fig. 2 Mass spring damping single-degree-of-freedom system

$$F(s) = ms^2X(s) + csX(s) + kX(s) \tag{4}$$

Therefore, the transfer function of the system is obtained.

$$G_1(s) = \frac{X(s)}{F(s)} = \frac{1}{ms^2 + cs + k} = \frac{1}{10s^2 + s + 1} = \frac{0.1}{s^2 + 0.1s + 0.1} \tag{5}$$

Calculated without the controller $\xi = 0.16 < 1$, this system is under-damped system, its oscillation characteristics are more strongly expressed, so the system has a very weak following ability, so the need to increase the PID regulator to improve the rapidity and stability of the system.

3 Improved Genetic Algorithm for PID Parameter Tuning

3.1 Theory of Genetic Algorithm

The genetic algorithm is a method designed to simulate the natural evolutionary process and thus search for the optimal solution for the evolution of living organisms in nature [14]. Based on the above conditions of the transfer function Eq. (5) of the control system, the parameters [K_P, K_I, K_D] of the PID controller are used as inputs to form the three groups of genes of the individual, and the genetic algorithm is used to calculate the search for the optimal solution so that the output achieves the optimal performance.

3.2 Improved Genetic Algorithm

The algorithm needs to initialize the genes, individuals, and populations before the individual can perform global search for superiority. The improved genetic algorithm adds system overshoot, rise time, and adjustment time control factors to the objective function, including the weights of the control factors, and adds replication operations to the genetic operations [15]. The flow chart of the improved genetic algorithm is shown in Fig. 3.

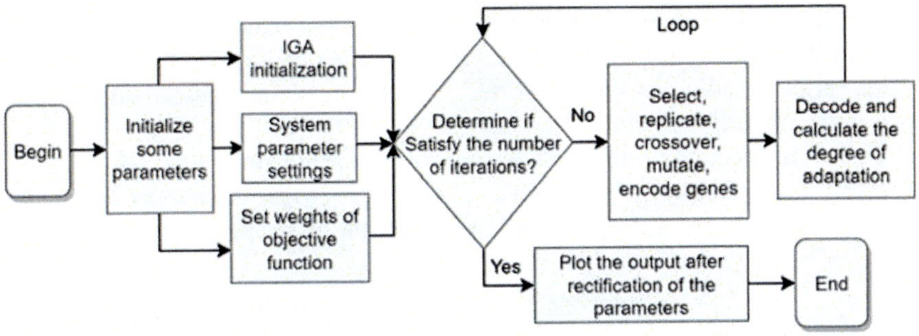

Fig. 3 Flowchart of improved genetic algorithm

3.3 Parameter Adjustment Steps

- There are various ways to encode the genetic algorithm, and here floating-point decimal encoding is used. The parameters to be used as input are [K_P, K_I, K_D], and each decimal code of length 10 represents one of the variables, respectively, and then the three variables are linked in series to form an individual of length X = 3*10 = 30. The size of the random initial population size used is 50, too small a population size will make the algorithm not universal, while too large a size will increase the computational effort to extend the convergence time.
- The fitness function, using the absolute value of the error, the squared term of the control output, and the time integral of the overshoot amount of the three, and then adding the target function as the final determination of the rise time and adjustment time of the system control parameters, can achieve the dynamics of the intermediate transition process. The smaller the final objective function is, the more it can show the characteristics of the control system such as stability, speed, and accuracy. The specific implementation of the control system for the above objectives can be described as:

$$J = \int_0^t (\omega_1 |e(t)| + \omega_2 y^2(t) + \omega_3 pos)dt + \omega_4 t_r + \omega_5 t_s \tag{6}$$

where $e(t)$ is the system error, $y(t)$ is the actual output of the controller, pos is the system overshoot amount, t_r is the rise time, t_s is the adjustment time, where ω_1, ω_2, ω_3, ω_4, ω_5 are the corresponding parameters. The weights are taken as 1/1/1/50/50 respectively. The adjustment time is related to the stability of the system; the system error and rise time are related to the rapidity of the system; the overshoot amount is related to the accuracy of the system; the system output is related to the digestion energy of the system [16]. The objective function is taken as the inverse of the final fitness function, that is $F = \frac{1}{J}$.

- For the natural selection operation, selection was performed using a ranking method in which the population was ranked in descending order of fitness, and the better 50 individuals were selected from the population and the inferior ones were eliminated.
- Replication operations, in which two individuals are randomly selected with probability Pt and the gene of one individual X is copied to the other individual Y, can significantly increase the inheritance ability of the offspring, e.g.

$$X =< 7, 7, 5, 2, 1, 6, 7, 9, 6, 5 \& 9, 7, 8, 3, 4, 0, 5, 2, 1, 0$$
$$\& 5, 7, 5, 8, 0, 6, 3, 3, 5, 9 >$$
$$\updownarrow$$
$$Y =< 7, 4, 5, 3, 8, 7, 3, 3, 2, 3 \& 8, 2, 3, 5, 4, 0, 7, 8, 1, 2$$
$$\& 6, 3, 4, 0, 0, 1, 1, 8, 5, 4 >$$

The & symbol is a separator of three genes, and one gene consists of 10 decimal numbers. A new individual Y is generated by copying the first gene X directly to the second gene.

$$\mathrm{X} =< 7, 7, 5, 2, 1, 6, 7, 9, 6, 5 \& 9, 7, 8, 3, 4, 0, 5, 2, 1, 0$$
$$\& 5, 7, 5, 8, 0, 6, 3, 3, 5, 9 >$$
$$\updownarrow$$
$$\mathrm{Y} =< 7, 7, 5, 2, 1, 6, 7, 9, 6, 5 \& 8, 2, 3, 5, 4, 0, 7, 8, 1, 2$$
$$\& 6, 3, 4, 0, 0, 1, 1, 8, 5, 4 >$$

- The crossover operation, in which two parental X and Y individuals are randomly selected with probability Pc to crossover at some random gene locus to generate two new individuals, can significantly improve the search power of the improved genetic algorithm, and the single-point crossover example used here is as follows.

$$\mathrm{X} =< 7, 7, 5, 2, 1, 6 \$ 7, 9, 6, 5 \& 9, 7, 8, 3, 4, 0, 5, 2, 1, 0$$
$$\& 5, 7, 5, 8, 0, 6, 3, 3, 5, 9 >$$
$$\updownarrow$$
$$\mathrm{Y} =< 7, 4, 5, 3, 8, 7 \$ 3, 3, 2, 3 \& 8, 2, 3, 5, 4, 0, 7, 8, 1, 2$$
$$\& 6, 3, 4, 0, 0, 1, 1, 8, 5, 4 >$$

Here a randomly selected gene crossover site, in this case at the sixth position, is separated by $. The crossover produces new individuals X1 and Y1, as follows.

$$X_1 =< 7,7,5,2,1,6\,\$\,3,3,2,3\,\&\,8,2,3,5,4,0,7,8,1,2$$
$$\&\,6,3,4,0,0,1,1,8,5,4 >$$
$$\updownarrow$$
$$Y_1 =< 7,4,5,3,8,7\,\$\,7,9,6,5\,\&\,9,7,8,3,4,0,5,2,1,0$$
$$\&\,5,7,5,8,0,6,3,3,5,9 >$$

Crossover operations can combine individuals with better performance in the offspring with a certain probability.

- The mutation operation, with a small probability Pm, randomly changes the gene value in a gene locus, thus achieving a mutation effect. For example.

$$X =< 7,7,5,2,1,6,7,9\,\$\,6,5\,\&\,9,7,8,3,4,0,5,2,1,0$$
$$\&\,5,7,5,8,0,6,3,3,5,9 >$$

Individual X mutates at the eighth locus and the gene value changes from 9 before the mutation to 8 after the mutation, resulting in a new individual X after the mutation as follows.

$$X =< 7,7,5,2,1,6,7,8\,\$\,6,5\,\&\,9,7,8,3,4,0,5,2,1,0$$
$$\&\,5,7,5,8,0,6,3,3,5,9 >$$

The introduction of variation allows the improved genetic algorithm to have local stochastic search capability on the one hand, avoiding the local optimum caused by the initial convergence; on the other hand, it can maintain the population diversity and can use variation operators to accelerate the convergence to the optimal solution when the crossover operator is close to the optimal solution.

3.4 PID Control Model

PID controllers are linear controllers commonly used in industry, where the PID controller is mainly based on the linear superposition of the three around the control deviation transformation, and the control deviation e(t) is equal to the difference between the set target value r(t) and the actual output value y(t).

$$e(t) = r(t) - y(t) \tag{7}$$

The proportional (P), integral (I), and differential (D) control deviations are assigned corresponding weights and then linearly superimposed to form the control quantity of the PID controller, which outputs the corresponding control quantity to control the controlled object [16]. The PID controller is expressed as:

$$u(t) = K_p\left[e(t) + \frac{1}{T_I}\int_0^t e(t)dt + T_D\frac{de(t)}{dt}\right] \tag{8}$$

The transfer function form is equal to the system output divided by the system input, which can be inherited from Eq. (8).

$$G_2(s) = \frac{U(s)}{E(s)} = K_p\left(s + \frac{1}{T_I s} + T_D s\right) \tag{9}$$

where K_p is the proportionality factor, T_I is the integration time constant, and T_D is the differential time constant. In control system design and simulation, the transfer function is often written as:

$$G_2(s) = K_P + \frac{K_I}{s} + K_D s = \frac{K_D s^2 + K_P s + K_I}{s} \tag{10}$$

The above equations K_P, K_I, and K_D are proportional, integral, and differential coefficients, respectively. In the PID controller, proportional control determines the strength of the control of the system, and an appropriate KP weight is beneficial to improve the dynamic response speed of the system; integral control is the cumulative effect of deviation over time, and an appropriate KI weight can eliminate the steady-state error in the system; differential control can predict the deviation, and an appropriate KD weight can speed up the response speed of the system, thus improving the dynamic characteristics of the system. The block diagram of the PID correction system after the mass-spring-damped single-degree-of-freedom system is connected in series with the PID controller is shown in Fig. 4.

The closed-loop transfer function of the whole control system can be derived as:

$$G(s) = \frac{G_1(s)G_2(s)}{1 + G_1(s)G_2(s)} = \frac{K_D s^2 + K_P s + K_I}{K_D s^2 + K_P s + K_I + 10s^3 + s^2 + s} \tag{11}$$

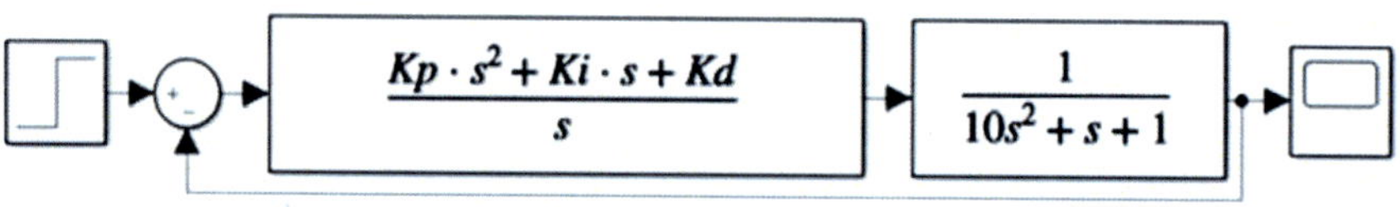

Fig. 4 Block diagram of PID correction system

4 Simulation Experiments

Experiment for the previous control system to design the corresponding MATLAB program and Simulation system block diagram, the former to obtain the genetic algorithm and improved genetic algorithm PID parameters, and then articulated to the latter system block diagram. The number of populations in the MATLAB genetic algorithm program is 50, and the number of genetic generations is 100. The unit step response was used as input and the iterative intelligent search for optimization was combined with the evaluation function described above to obtain the three parameters of the Improved Genetic Algorithm (IGA), Genetic Algorithm (GA) rectified PID. Comparing the three parameters of the artificial experience (AE) design, the parameters in the three PID controllers were obtained as shown in Table 1.

4.1 Simulink Model

Simulink system block diagram control system input using unit step response test experiments, after three control methods (improved genetic algorithm IGA-PID, genetic algorithm GA-PID, artificial empirical method AE-PID) calculated to show the output images and error images of different control methods over time, so as to verify the effectiveness of the improved genetic algorithm optimized PID parameters for the mass-spring-damping single-degree-of-freedom system, as shown in Fig. 5 Simulink control block diagram.

4.2 Simulation Analysis

In Simulink, the system inputs a step from 0 to 1 m signal in the 0th s, a total of 0 ~ 2 s simulation, by running the Simulink system block diagram to get the experimental results, where the system unit step error and response as shown in Fig. 6. The system performance index as shown in Table 2, from the simulation diagram can be seen compared to the manual adjustment method, according to the genetic algorithm adjusted parameters Compared with the manual adjustment method, the parameters adjusted by the genetic algorithm are more accurate, the overshoot M_p is reduced to

Table 1 PID parameter

Algorithm	PID parameter		
	K_P	K_I	K_D
IGA	20	5.8	120
GA	25	8	100
AE	20	2	50

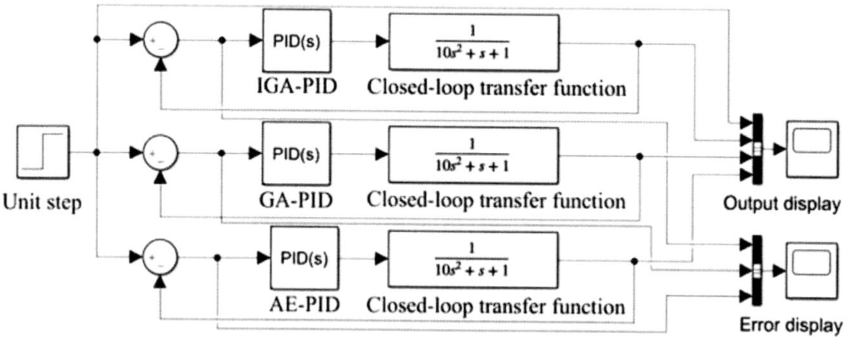

Fig. 5 Control block diagram

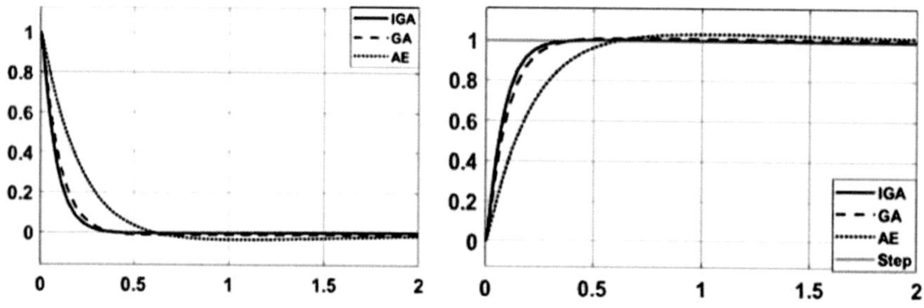

Fig. 6 Unit step error and response

Table 2 System performance indicators

PID parameter	*AE*	*GA*	*IGA*
M_p	0.05%	0.03%	0.02%
t_r	0.61s	0.35s	0.31s
t_p	1.2s	1s	0.8s

0.02%, the rise time t_r is shortened to 0.31 s, and the regulation time t_p is shortened to 0.8 s. The PID parameters optimized by the improved genetic algorithm have good performance.

5 Conclusion

To address the conservativeness and difficulty of manually adjusting PID parameters and the problem of easily falling into local optimum using traditional genetic algorithms, a method of adjusting PID parameters of a typical mass-spring-damped single-degree-of-freedom second-order system with an improved genetic algorithm is proposed. The objective function of the improved genetic algorithm is constructed for three aspects of the second-order system: stability, accuracy, and rapidity. Simulation experiments were conducted in Simulink using a unit step response. The experimental results show that the PID parameter control of the mass-spring-damped single-degree-of-freedom second-order system optimized by the improved genetic algorithm has better stability and accuracy, and the PID control parameters rectified by the improved genetic algorithm can significantly improve the steady-state characteristics of the second-order system compared with the other two methods. However, improving the genetic algorithm requires the transfer function and related parameters of the discriminatory system to be set in advance is not ideal, and changing the point is the next step to be improved.

References

1. Zhu, Y.Q.: Solar tracking system and research on BLDC control. Yangzhou University, (2011)
2. Chen, W.G.: The research and application of intelligent system based on particle swarm optimization algorithm. Hunan University, (2006)
3. Ziegler, J.G., Nichols, N.B.: Optimum setting for automatic controllers. J. Dyn. Syst. Meas. Control., 759–768, (1993)
4. Ruan, Y., Yang, Y., Chen, B.S.: Electric drive automatic control system: Motion control system, 5th ed. Beijing: Mechanical Industry Press, (2016)
5. Li, Y.C., Kang, Y.B.: Servo parameter adjustment method of CNC machine tool. Sci. Technol. Innov. Her. **17**(2), 66–68 (2020)
6. Mo, C., Tai, K., Deng, S., Duan, L.: PID control parameter optimization strategy based on intelligent algorithm. Inf. Technol. Inf., (09), 222–223+227 (2021)
7. Andrew, A.M.: Systems: An introductory analysis with applications to biology, control, and artificial intelligence, by John H. Holland MIT Press (Bradford Books), Cambridge, Mass. 1992, xiv+211 pp. (Paperback 13.50, cloth 26.95). Robotics, **11**(5), 489–489 (1993)
8. Hao, Q., Guan, L.W., Wang, L.P.: Control parameter tuning of parallel machine tool motor servo system based on genetic algorithm. J. Tsinghua Univ. (Sci. Technol.) **50**(11), 1801–1806 (2010)
9. Sun, Y.M., Zhang, X.X.: Tuning parameters and application of auto disturbance rejection controller by improved genetic algorithm LJ. Autom. Instrum., **35**(3),13–17,45 (2020)
10. Gao, X.Q., Huang, D.D., Ding, S.M.: PID parameter optimization based on penalty function and genetic algorithm. J. Jilin Inst. Chem. Technol. **38**(03), 57–60 (2021)
11. Wang, L.P., Kong, X.Y., Yu, G.: Motor servo control parameter tuning for parallel and hybrid machine tools based on a genetic algorithm. J. Tsinghua Univ. (Sci. Technol.) **61**(10), 1106–1114 (2021)
12. Jia, Y.T., Tian, J.: Control optimization of hydraulic servo system based on genetic algorithm. Agric. Equip. & Veh. Eng. **59**(05), 116–118 (2021)
13. Duan, G.R., Liu, G.P., Thompson, S.: A new stability condition for second-order linear systems//2001 European Control Conference (ECC). IEEE, (2001)
14. Gao, C., Peng, J.T.: Parameter tuning of PID controller based on genetic algorithms. Mod. Inf. Technol. **3**(02), 171–172 (2019)
15. Niu, J.M.: Controllers design for ball and beam system based on adaptive genetic algorithm. Northeastern University, (2010)
16. Shao, H.L.: Research on PID control parameter tuning based on genetic algorithm. Wirel. Internet Technol. **21**, 111–112 (2016)

Uncalibrated Visual Servoing Using Dynamic Broyden and Least-Squares Methods

Mingyou Chen, Liucun Zhu, Junqi Luo, Haofeng Deng, Hongwei Wu, and Daopeng Liu

Abstract Visual servo systems usually require a calibration operation before performing their tasks. This involves extra time cost, and the calibration results may fail as the internal and external environment changes. To address this problem, this paper proposes a calibration-free visual servo implementation way based on the least-squares optimized dynamic Broyden method. The dynamic Broyden method estimates the image Jacobian matrix online with unknown depth information of the target, and uses least-squares to optimize the dynamic Broyden method by minimization. The simulation results show that the effectiveness of the method can be achieved under the condition that the camera parameters and hand-eye relationship are unknown.

1 Introduction

In 1979, Hill and Park first proposed the concept of visual servoing [1]. Until the late 1990s, it has rapidly developed into a hot topic in the field of robotics. According to the different feedback signals, the structure of visual servo control system includes position-based visual servoing (PBVS), image-based visual servoing (IBVS), and hybrid visual servoing (HBVS) [2]. The IBVS system forms a closed-loop feedback control by comparing the actual and expected image error information online [3], but it needs to estimate the depth information of the Jacobian matrix of the image [4]. The IBVS system can effectively keep the robot or the target object model always in the field of view of the camera, but cannot ignore the related problems such as singularity

M. Chen · H. Wu · D. Liu
College of Mechanical and Marine Engineering, Beibu Gulf University, Guangxi, China

L. Zhu (✉)
Institute of Advanced Science and Technology, Beibu Gulf University, Guangxi, China
e-mail: lczhu@bbgu.edu.cn

J. Luo · H. Deng
School of Mechanical Engineering, JiangSu University, Zhenjiang, China

K. Nakamatsu et al. (eds.), *Advanced Intelligent Technologies for Information and Communication*, Smart Innovation, Systems and Technologies 365,
https://doi.org/10.1007/978-981-99-5203-8_32

and local minima of the Jacobian matrix J_r. Nowadays, the research of uncalibrated visual servo system has become the key direction of domestic and foreign academic research [5].

The traditional calibration method needs to complete the calibration in Cartesian space. Roger et al. proposed a two-step hand-eye calibration method, which is divided into two parts: rotation and translation. First, the rotation calibration part is solved, and then the translation calibration part is solved using the result. However, due to inaccurate grasping, hand-eye calibration must be frequent [6]. Strobl et al. proposed a calibration error measurement method based on the special Euclidean group SE(3), which transformed the hand-eye calibration problem into a nonlinear optimization problem, and discussed and solved the abstract optimization problem [7]. Zhao proposed a hand-eye calibration method based on convex optimization. The nonlinear optimization problem is solved by a convex optimization problem on the L_∞-norm. The calculation speed is fast and the local optimal problem is effectively solved, however, in practical applications, the calculation error and time-consuming need to be considered [8]. Hosoda et al. first proposed the Broyden method in 1994, which performs online identification and calculation of the image Jacobian matrix in real time, that established the theoretical foundation of calibration-free visual servoing [9]. Qian et al. proposed to convert the Jacobian matrix problem into the state quantity problem of the Kalman filter estimation algorithm, this method does not introduce other redundant motions, through online regression, the Jacobian matrix of the image at the current moment is estimated according to the previous moment, but this method is not suitable for the non-Gaussian environment with unknown noise [10]. Zhong et al. proposed an online identification method of image Jacobian matrix based on ANNKF, which improved the robustness of the algorithm to interference noise [11]. Miljkovic et al. used Q-learning and SARSA training to obtain mapping relations between the space of image and robot motion, but no similar control parameters appeared in the research results, and the velocity signal appeared a large jitter phenomenon [12].

On the basis of summarizing the current research on robot visual servoing, this essay completes the simulation experiment in the environment of Matlab 2020a, the iterative least square method is used to process the image Jacobian matrix of image-based visual servo system (IBVS). Finally, the simulation experiment is completed, and its stability and robustness are verified and analyzed.

2 Visual Servo Control

2.1 Camera Model

As shown in Fig. 1, it represents the imaging of light passing through a thin lens, and the positive direction Z axis represents the direction of the optical axis. Its perspective law is expressed as:

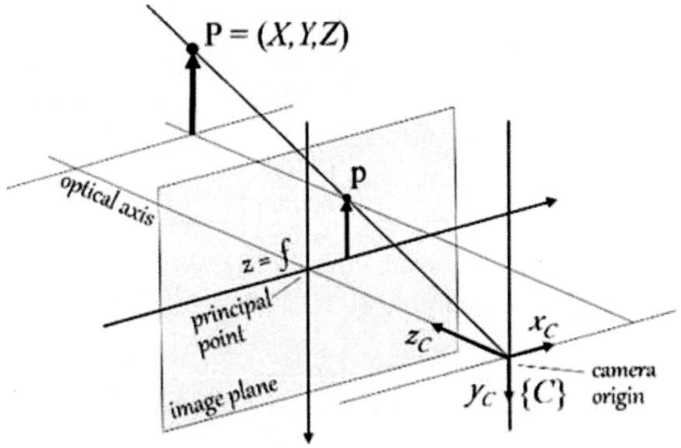

Fig. 1 Perspective projection model

$$\frac{1}{z_o} + \frac{1}{z_i} = \frac{1}{f} \tag{1}$$

where z_o and z_i represent the distance between the object and the image to the lens respectively, and f represents the lens focal length. When $z_o > f$, the corresponding inverted image will be formed on the image plane of $z_o < -f$. For the camera, since the image plane is always on the surface of the sensor chip, in order to obtain the distance z_i (for object $z_i = f$ at infinity) from the image plane to the lens, it is only necessary to move the camera along the optical axis to perform the focusing operation.

According to the perspective projection model, the light rays converge to the origin {C}, and projected on the image plane at $z = f$. According to similar triangles, it can be proved that point $P = (X, Y, Z)$ in the world coordinate system can be projected to $p = (x, y)$ on the image plane, and we have:

$$x = f\frac{X}{Z}, y = f\frac{Y}{Z}. \tag{2}$$

2.2 *Image-Based Visual Servoing*

As shown in Fig. 2, the error amount of the IBVS system is calculated on the image plane. Firstly calculate the feature error between the current image and the expected image, and then with the help of the visual controller, Then, the conversion relationship between the two-dimensional space of the image and the three-dimensional space of the motion is established, and the control amount of the control motion is obtained, and finally, the control of the motion trajectory is realized. In the whole control process of the uncalibrated visual servo system, because all calculations are carried out in the image plane, we do not need to calibrate the camera.

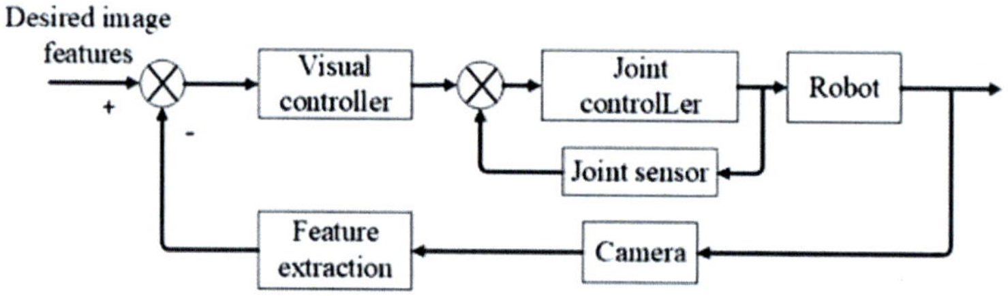

Fig. 2 IBVS control structure

3 Estimation of Image Jacobian Matrix

3.1 Image Jacobian

The image Jacobian matrix essentially describes the correspondence between image information and spatial information [13]. At present, feature parameters such as point feature, line feature, image moment, and centroid have been successfully applied in practical visual servo control. It is defined as follows:

$$\dot{f} = J_r(r) \cdot \dot{r} \tag{3}$$

$$J_r(r) = \left[\frac{\partial y}{\partial r}\right] = \begin{bmatrix} \frac{\partial y_1}{\partial r_1} & \cdots & \frac{\partial y_1}{\partial r_n} \\ \vdots & \ddots & \vdots \\ \frac{\partial y_m}{\partial r_1} & \cdots & \frac{\partial y_m}{\partial r_n} \end{bmatrix} \tag{4}$$

In the above formula, $f = [f_1, f_2, \cdots, f_m]^T$ and $\dot{f} \in R$ represents the change speed of image features, $r = [r_1, r_2, \cdots, r_m]^T$ and $\dot{r} \in R$ represents the change speed of the pose at the end of the robot, and $J_r(r) \in R^{m \times n}$ represents the Jacobian matrix of the image.

When the control variable is set to the joint movement speed of the robotic arm, according to the kinematics expression of the robot arm, we can obtain the speed relationship between the joint $\dot{\theta}$ of the robot arm and the image feature vary $\dot{f}$ as follows:

$$\dot{f} = J_r(r) \cdot \dot{r} = J_r(r) \cdot J_q(\theta) \cdot \dot{\theta} \tag{5}$$

In the above formula, J_q represents the Jacobian matrix of the robot arm.

From this, the varying relationship between the moving joints of the robotic arm and the image features can be established, and its differential expression is as follows:

$$\dot{f} = J \cdot \dot{\theta} \tag{6}$$

Among them, $J = J_r \cdot J_q$ is the Jacobian matrix of the whole system.

Suppose that coordinate $P_c = (x_c, y_c, z_c)^T$ in the camera coordinate system represents point P at the end of the robot arm. According to the camera parameter model standard, we can know that the coordinate corresponding to the point in the image plane coordinate system is $(u, v)^T$. According to the imaging principle

$$\begin{cases} u = \frac{\lambda}{d_x} \cdot \frac{x_c}{z_c} + u_0 \\ v = \frac{\lambda}{d_y} \cdot \frac{y_c}{z_c} + v_0 \end{cases} \tag{7}$$

In the above formula, λ represents the camera focal length parameter.

Denote the speed of the end effector of the robotic arm as $\dot{r} = [v, \omega]^T$, the linear velocity as $v = [v_x, v_y, v_z]^T$, and the angular velocity as $\omega = [\omega_x, \omega_y, \omega_z]^T$, so the coordinates of any point in the space relative to the camera are:

$$\dot{P}_c = -\omega \times P_c - v \tag{8}$$

Then there

$$\begin{cases} \dot{x}_c = -v_x - \omega_y z_c + \omega_z y_c \\ y_c = -v_y - \omega_z x_c + \omega_x z_c \\ \dot{z}_c = -v_z - \omega_x y_c + \omega_y x_c \end{cases} \tag{9}$$

It is useful for the derivation of image horizontal coordinates.

$$\begin{cases} \dot{u} = \frac{\lambda}{d_x}\left(\frac{\dot{x}_c}{z_c} - \frac{x_c \dot{z}_c}{z_c^2}\right) \\ \dot{v} = \frac{\lambda}{d_y}\left(\frac{\dot{y}_c}{z_c} - \frac{y_c \dot{z}_c}{z_c^2}\right) \end{cases} \tag{10}$$

Then there

$$\begin{bmatrix} \dot{u} \\ \dot{v} \end{bmatrix} = \begin{bmatrix} -\frac{\lambda}{z_c} & 0 & \frac{u}{z_c} & \frac{u \cdot v}{\lambda} & -\frac{\lambda^2+u^2}{\lambda} & v \\ 0 & -\frac{\lambda}{z_c} & \frac{v}{z_c} & \frac{\lambda^2+v^2}{\lambda} & -\frac{u \cdot v}{\lambda} & -u \end{bmatrix} \cdot \begin{bmatrix} v_x \\ v_y \\ v_z \\ \omega_x \\ \omega_y \\ \omega_z \end{bmatrix} \tag{11}$$

The characteristic change rate of the system is $f = [\dot{u}, \dot{v}]^T$, and the moving speed of the end effector of the robotic arm is $\dot{r} = [v_x, v_y, v_z, \omega_x, \omega_y, \omega_z]^T$. The J_r of the system can be obtained as:

$$J_r = \begin{bmatrix} -\frac{\lambda}{z_c} & 0 & \frac{u}{z_c} & \frac{u \cdot v}{\lambda} & -\frac{\lambda^2+u^2}{\lambda} & v \\ 0 & -\frac{\lambda}{z_c} & \frac{v}{z_c} & \frac{\lambda^2+v^2}{\lambda} & -\frac{u \cdot v}{\lambda} & -u \end{bmatrix} \tag{12}$$

3.2 Estimation of Image Jacobian Matrix by Dynamic Broyden Method

Let $\widehat{J}_r$ represent the estimation of J_r, expand the first-order Taylor series for the deviation function $y(\varphi, t)$, where φ is the joint angle, t is the time, and ignore its higher-order derivative terms, and the radiation model is defined as $m(\varphi, t)$, Available

$$m_r(\varphi, t) = y(\varphi_r, t_r) + \widehat{J}_r(\varphi - \varphi_r) + \frac{\partial y_r}{\partial t}(t - t_r) \tag{13}$$

At time r−1 we have

$$y(\varphi_{r-1}, t_{r-1}) = m_r(\varphi_{r-1}, t_{r-1})$$

$$= y(\varphi_r, t_r) + \widehat{J}_r(\varphi_{r-1} - \varphi_r) + \frac{\partial y_r}{\partial t}(t_{r-1} - t_r) \tag{14}$$

$$y(\varphi_r, t_r) - y(\varphi_{r-1}, t_{r-1})$$

$$= \widehat{J}_r(\varphi_r - \varphi_{r-1}) + \frac{\partial y_r}{\partial t}(t_r - t_{r-1}) \tag{15}$$

Remember

$$\Delta y_r = y(\varphi_r, t_r) - y(\varphi_{r-1}, t_{r-1})$$

$$\Delta \widehat{J} = \widehat{J}_r - \widehat{J}_{r-1}$$

$$\Delta \varphi = \varphi_r - \varphi_{r-1} \tag{16}$$

$$\Delta t = t_r - t_{r-1}$$

Substitute with

$$\widehat{J}_r = \widehat{J}_{r-1} + \frac{\Delta y_r - \widehat{J}_{r-1}\Delta\varphi - \frac{\partial y_r}{\partial t}\Delta t \Delta\varphi^T}{\Delta\varphi^T \Delta\varphi} \tag{17}$$

Considering that the object is stationary and the velocity is 0, $\frac{\partial y_r}{\partial t}\Delta t = 0$, so the estimation method is also applicable to the static situation.

3.3 *Iterative Least-Squares*

Regarding the image Jacobian matrix formula derived from formula (17), there will be a divergence phenomenon in its iterative process. For the reason that enhances rapidity and stability, the iterative least-squares method is adopted to solve this problem [14]. The radiation model cost function for feature variation is

$$C_k = \sum_{i-1}^{n} \rho^{k-i} \|\Delta m_i\|^2 \tag{18}$$

In the formula, the forgetting factor $\rho \in (0, 1]$, its value affects the performance of the system to a large extent. The purpose of using the iterative least-squares is to minimize C_k, that is

$$q_r = \frac{1}{\lambda}(q_{r-1} - \frac{q_{r-1}\Delta\varphi\Delta\varphi^T q_{r-1}}{\rho + \Delta\varphi^T q_r \Delta\varphi}) \tag{19}$$

$$\widehat{J}_r = \widehat{J}_{r-1} + \frac{\Delta y_r \Delta\varphi^T - \widehat{J}_{r-1}\Delta\varphi\Delta\varphi^T q_{r-1}}{\rho + \Delta\varphi^T q_{r-1}\Delta\varphi} \tag{20}$$

In the above formula, q is the covariance positive definite matrix.

4 Simulation Results and Analysis

The simulation experiments in this paper are carried out under the MATLAB 2020a version, using Machine Vision Toolbox and Robotics Toolbox developed by Peter Corke [15]. The focal length of the analog camera is 8 mm, and the pixel coordinates of the principal point are expressed as [512,512], the image resolution is 1024 × 1024, and the gain lambda is set to 1.2. The initial position of the image is four points on a square with a side length of 1.5 m centered at (0.6, 0.6, 2), and the desired target position is a 600*600 square with the center located at the main point and the positions are [212 212]',[212 812]',[812 812]',[812 212]'.

As shown in Fig. 3a and b, it is the result of tracking the motion trajectory of the target image plane when the depth is manually set to 0.9 and 17 respectively. Figure 3c shows the results of using the depth estimator when the depth value is unknown, where in all three plots, the green circle represents the initial position of the target and the red star indicates the desired target position. It can be known from Fig. 3 that Fig. 3a and c can achieve target tracking, and Fig. 3b fails to track, but Fig. 3c is much smoother in the image plane. Figure 3b fails to track.

As shown in Fig. 4a and b, it is the result that the convergence of camera space velocity components when the depth value is unknown and manually set to 0.9 and 17 respectively. Figure 4c shows the result that the depth estimator is used. It can be

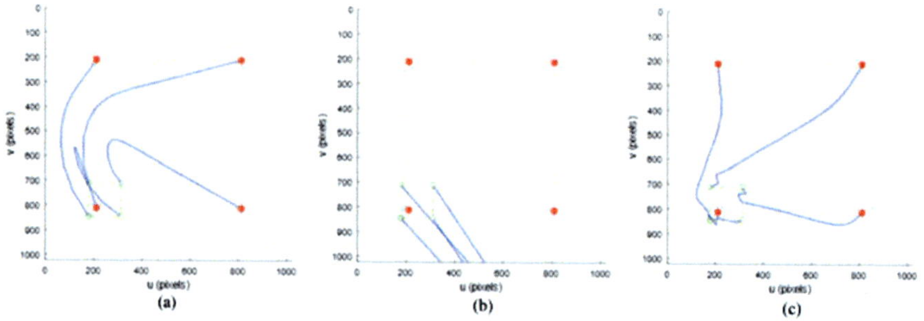

Fig. 3 The target image plane tracks the motion trajectory

seen from Fig. 4 that Fig. 4a and c run 84 steps and 92 steps respectively to reach the target, and the target tracking is successful. But Fig. 4b runs to step 61 and loses the target, the target tracking fails and the system diverges.

As shown in Fig. 5a and b, is the result that the convergence of systematic characteristic error when the depth value is unknown and manually set to 0.9 and 17 respectively. Figure 5c shows the result that the depth estimator is used. It can be known from Fig. 5 that Fig. 5a and c run 84 steps and 92 steps respectively to reach the target, and the target tracking is successful. But Fig. 5b runs to step 61 and loses the target, the target tracking fails and the system diverges.

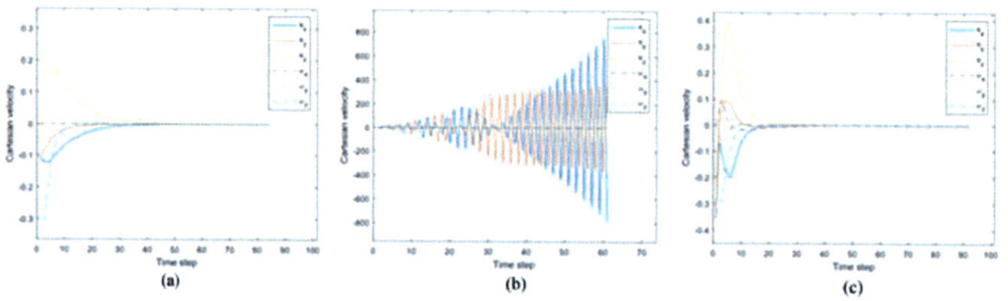

Fig. 4 Convergence of camera space velocity components

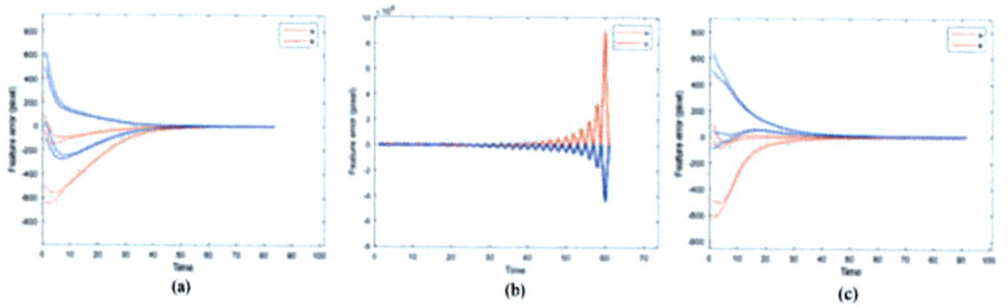

Fig. 5 Convergence of systematic characteristic error

Through the above comparison experiments, it can be found that under the same simulation conditions, when the depth value is 0.9, the amplitude of the tracking motion trajectory of the system image plane changes greatly, the system converges to the zero point, and the target tracking is successful, indicating that the system has good convergence. When the depth value is 17, the camera rolls back to diverge, and the target tracking fails, indicating that different depth values will have different effects on the system. For the sake of studying the convergence of the system when the depth value in this control system is unknown, under the same simulation conditions, an iterative least-squares method is used to design a depth estimator. The results show that the tracking is successful, and the trajectories are smoother and smaller in amplitude. Therefore, by using the iterative least-squares method, the system can achieve the convergence effect while ignoring the depth, indicating that the iterative least-squares algorithm can effectively improve the stability and robustness of the IBVS system.

5 Conclusion

In this essay, the simulation experiment is carried out on the MATLAB platform, and the simulation results obtained by the IBVS system at different depths are compared through simulation analysis. When the depth is underestimated, the target feature image plane tracks the trajectory curve with a larger amplitude, successfully tracks the target, and the system converges well. When the depth is overestimated, the system diverges, the target tracking fails, and the camera rolls back. The Image Jacobian matrix depth information is estimated online by designing the depth estimator using iterative least-squares. It can effectively improve the system output response. The results show that the system converges and the target tracking is successful, which is better than the cases when the depth values are 0.9 and 17. Therefore, it is verified by experiments that iterative least-squares can effectively enhance the stability and robustness of the system.

Acknowledgements This project was funded by Special Fund for Bagui Scholars of the Guangxi Zhuang Autonomous Region No.2019A08.

References

1. Hill, J., Park, W.T.: Real time control of a robot with a mobile camera. In: The 9th International symposium on Industrial Robots, pp.233–246 (1979)
2. Zhong, X.G., Zhong, X.Y., Peng, X.F.: Robust Kalman filter Elman neural network learning for vision-sensing-based robotic manipulation with global stability. Sensors 13, 13464–13486 (2013)
3. Conticelli, F., Allotta, B.: Nonlinear controllability and stability analysis of adaptive image-based systems. IEEE Trans Robotics Automation 17(2), 208–214 (2001)

4. Yang, J.C.: Research on online estimation method of image Jacobian matrix in uncalibrated visual servo. Hunan University of technology (2021)
5. Luo, Y. C., Li, S. P., Li, D.: Research on visual servo of uncalibrated robot based on image. Manuf. Autom. 42(11), 148–151 (2020)
6. Tsai, R. Y., Lenz, R. K.: A new technique for fully autonomous and efficient 3D robotics hand/ eye calibration. IEEE Trans Robotics Automation 5(3), 345–358 (1989)
7. Strobl, K. H., Hirzinger, G.: Optimal hand-eye calibration. IEEE international conference on intelligent robots and systems 2006(3), 4647–4653 (2006)
8. Zhao, Z.: Hand-eye calibration using convex optimization. Proceedi-IEEE international conference on robotics and automation (3), 2947–2952 (2011)
9. Hosoda, K., Asada, M., Versatile visual servoing without knowledge of true jacobian. In: Proceedings of IEEE/RSJ International Conference on Intelligent Robots and Systems (IROS'94).IEEE 1, 186–193 (1994)
10. Qian, J., Su, J.B.: Online Kalman filter estimation of image Jacobian matrix. Control and decision making 18(1), 77–80 (2003)
11. Zhong, X.G., Zhong, X.Y., Peng, X.F.: Robots visual servo control with features constraint employing Kalman-neural-network filtering scheme. Neurocomputing 151, 268–277 (2015)
12. Miljkovié, Z., Mitic, M., Lazarevié, M., Babi, B.: Neural network reinforcement learning for visual control of robot manipulators. Expert Systems Appl 40, 1721–1736 (2013)
13. Robotics; Investigators at Dalian Minzu University Describe Findings in Robotics (Unscented Particle Filter for Online Total Image Jacobian Matrix Estimation In Robot Visual Servoing).Jounal of Engineering (2019)
14. Xie, X. M., Ding, F.: Adoptive control system[M]. Beijing: Press of Tsinghua University 2002:6278
15. Corke P.: Robotics, vision and control: fundamental algorithms servoing of robots using a depth-independent interaction matrix. IEEE Trans Robotics 22(4), 80–817 (2006)

Research and Implementation of 5G Base Station Location Optimization Problem Based on Genetic Algorithm

Guoqing Chen, Xin Wang, and Guo Yang

Abstract The application requirements of 5G have reached a new height, and the location of base stations is an important factor affecting the signal. Based on factors such as base station construction cost, signal coverage, and Euclidean distance between base stations, this paper constructs a multi-objective planning and location model combined with genetic algorithm, and conducts algorithm simulation. Finally, the simulation experiment results are analyzed and it is concluded that the multi-objective 5G base station planning model combined with genetic algorithm has high coverage and feasibility in real life, and then provides a new direction for base station location selection.

1 Introduction and Literature Review

With the rapid development of 5G, communication bandwidth has become a key national development object, among which information and communication infrastructure is a key content for enhancing national strength, safeguarding national security, and enriching people's lives. Especially with the development and promotion of national 5G technology, the construction of 5G base stations is an important part of the future communication infrastructure. Therefore, base station site selection will become an important work of base station construction. At the same time, the types of base stations and antennas are gradually rich, which makes the planning and selection of communication network sites become more complex. In order to make the base station location of both complete information coverage and optimal cost. While realizing the sharing rate and planning accuracy, the choice of quantity and density becomes the focus of this planning. This paper uses the genetic

G. Chen (✉)
Department of Mathematics, ChengDu JinCheng College, ChengDu, China
e-mail: chenguoqing@scujcc.edu.cn

X. Wang · G. Yang
Department of Electronic Information, Chengdu Jincheng College, Chengdu, China

K. Nakamatsu et al. (eds.), *Advanced Intelligent Technologies for Information and Communication*, Smart Innovation, Systems and Technologies 365,
https://doi.org/10.1007/978-981-99-5203-8_33

algorithm to visualize the base station site selection and gives the optimal combination scheme under multiple objectives, proposing new ideas for the base station site selection. Sachan Ruchi applied the genetic algorithm to the optimal layout planning of 5G base stations based on traditional technology and differential evolution technology. The actual application results show that the application effect of this method in 5G network can reach 29%, which is in the same industry leading position [1]. The selection of base stations should comprehensively consider various indicators, such as sharing rate, planning accuracy rate, and planning depth. This is a multi-objective planning problem. Based on this, Dua Amit used the GA algorithm to set up dual-target coverage requirements in solving the wireless sensor network optimization problem, and used heuristic optimization differential evolution and particle swarm optimization to achieve both WSN lifetime and spatial coverage. Finally, the scheduling problem of wireless sensor networks with heterogeneous sensor nodes is solved [2]. Haifeng Ren simulated and studied the logistics distribution and vehicle routing modes by using the genetic algorithm of the VRP mode, and found that the final vehicle under the VRP mode planning greatly improved the efficiency, full load rate, and timeliness for different outlets and transportation areas, to achieve the expected effect [3]. Yu Pengfei established a communication network topology optimization model on the basis of genetic algorithm. Through simulation analysis and verification, it was found that the communication network with natural connectivity as the objective function is superior to the ordinary genetic algorithm model in the construction process, and has extremely strong feasibility and flexibility [4]. Ray Paromik optimized the project time cost problem based on the genetic algorithm, used the genetic algorithm to solve the single-objective TCTO problem, and proposed an algorithm formula for calculating the cost optimization result. Time and financial costs are reduced, and the project value is maximized to a certain extent [5].

To sum up, some scholars in the existing research have used the genetic algorithm for the application of planning problems, and can achieve good experimental results in different fields. Some scholars have combined the genetic algorithm and other algorithms to verify the effectiveness and ductility of the optimized genetic algorithm. Based on the strong and extensive genetic algorithm, an optimization genetic algorithm was proposed, using the Euclidean distance between planning base stations as a constraint to establish multiple objective function, the 5G base station planning solution, reduce the blind search algorithm efficiency, improve the local optimal accuracy of the genetic algorithm, and meet the experimental results of the base station coverage engineering standard, which provides a new method for the future 5G base station planning.

2 Mathematical Model of 5G Base Station Site Selection

2.1 Problem Description

With the promotion of 5G network to all walks of life, 5G network presents a huge demand, 5G base station planning problem has attracted the attention of experts and researchers, the traditional planning algorithm is prone to local optimal, cost, and coverage effect, this paper is using genetic algorithm for global optimal site selection for 5G base station planning simulation research, the base station planning overall description is as follows: according to demand, using constraint target function optimal solution, output planning points coordinate location and coverage. The specific solutions are as follows: (1) objective function: the base station needs to maximize the needs of the business volume, the base station must have a high standard after planning, and the cost of establishing the base station should be the lowest. (2) Constraints: Euclidean distance between base stations, whether the base station covers the test point, whether the coverage reaches the standard, etc. (3) Output: the specific planning point of the base station, the simulation diagram of the base station coverage test point. Overall, this 5G base station location is a problem of finding the optimal solution, using the genetic algorithm and constraints to obtain the final planning point data.

2.2 Constrained Condition

For the vast areas with low coverage and no coverage, how to reasonably configure base stations on the basis of high coverage and high service requirements to meet the needs of high coverage and high services, but there are huge costs in the construction process. The problem is, therefore, how to minimize the cost, this paper proposes a multi-objective programming solution model. In order to make the model simulation results more accurate, the following assumptions are made: (1) The test point can be directly represented by x and y, and the coverage of the base station can be simulated by a small circle. (2) The simulation only covers the horizontal direction and does not consider the vertical coverage. (3) During the MATLAB simulation, if the test point is inside the simulated small circle, it means that it is covered.

Decision variables: The coordinates x, y of the base station.

Objective function:
The base station planning goal is to achieve the signal enhancement effect, so the coverage rate and meet the business volume are the main factors to consider. When establishing the base station, the first consideration is the high coverage rate and the lowest cost of large meet the business volume, so the corresponding planning target function is

Business volume function:

$$W_{max} = \sum\nolimits_{i=1}^{M} W_M q_i \tag{1}$$

Base station construction cost:

$$S_M = \sum\nolimits_{i=1}^{M} q_i \tag{2}$$

Base station coverage:

$$p = \frac{n}{N} \tag{3}$$

where N is the total number of coverage points, n is the actual coverage number, p is the base station coverage, q is the newly built base station, M is the number of base stations to be selected, W is the traffic volume, and S is the base station cost.

Constraint condition:

When the base station construction considers the highest coverage, each test point is only uncovered, and the base station is only covered and the test point is only uncovered. In order to reduce the construction cost and limit the repeated coverage of base stations, the Euclidean distance is used between new base stations, and the Euclidean distance threshold between base stations is not less than 10, and the Euclidean distance between the new base station and the test point is less than 10, so the constraint function can be obtained as follows:

The status of the base station is planned or unplanned:

$$\mathrm{q_i} = 0, 1 \tag{4}$$

The purpose of the base station planning is to meet the signal coverage requirements, so the test point needs to be covered as full as possible, and the status of the test point is expressed as follows:

$$\mathrm{b_i} = 0 \text{ or } 1 \tag{5}$$

To verify the feasible line of this algorithm, engineering standards are introduced to restrict:

$$\mathrm{p} = \frac{\mathrm{n}}{\mathrm{N}} \geq 90\% \tag{6}$$

Base station planning considers cost factors to avoid repeated construction between base stations, so the Euclidean distance constraint between base stations is as follows:

$$\sqrt{(x - x_0)^2 + (y - y_0)^2} \geq 10 \tag{7}$$

Avoid the test point which is not covered by the built base station, and increase the construction cost. The constraints between the test point and the base station are as follows:

$$\sqrt{(x - x_0)^2 + (y - y_0)^2} \leq 10 \tag{8}$$

2.3 Genetic Algorithm

Genetic algorithm is originated from Darwin's theory of biological evolution and Mendel's theory of genetic evolution. It searches for optimal paths and shortest distances by simulating natural evolution in biological evolution theory and genetic variation in genetic evolution theory. Its essence is to place the population in an unknown location, let the population reproduce naturally and follow the natural law of survival of the fittest, and then control the genetics, mutation, and crossover of the population according to the set fitness function, and decode the new population obtained to obtain the optimal solution. Approximate solution. The genetic algorithm performs a global search for the whole, which has strong feasibility and high-level algorithm. In the process of coding, individual fitness calculation, genetic variation, etc., it is only necessary to deny the sub-populations that perform poorly in the evolution process according to the constraints. The optimal solution of the target can be obtained, and it is widely used in single workshop scheduling problem, traveling salesman problem, facility location problem, and so on.

Design of code:

Since site selection data belongs to spatial data and cannot be solved directly by genetic algorithm, binary operation is necessary to encode the data to transform the spatial data to the "chromosome" solved by genetic algorithm, that is, each chromosome represents a candidate site point. The 0–1 variables are introduced as follows:

$$A_b = \{b_1, b_2, b_3, \ldots, b_M\} \tag{9}$$

Base station status is

$$b_i = \begin{cases} 1, iSelected \\ 0, iNot\ selected \end{cases} \tag{10}$$

where b_i represents the case where the i th base station is selected.

Set the individual fitness:

Fitness size is an important measure of the quality of the results. Since the location of the base station is fixed, the sum of the optimal assigned European distance is considered as the fitness of the chromosome.

The European distance formula is

$$d(p_0, p) = \sqrt{(x_0 - x)^2 + (y_0 - y)^2} \tag{11}$$

The fitness expression formula is

$$Fit(p) = \frac{1}{\sum_{i=1}^{M}\sum_{j=1}^{M} d(p_0, p)b_i} \tag{12}$$

Cross-variant manipulation:

In order to establish a better base station site selection model, based on the optimization principle of the genetic algorithm, and to prevent the premature maturity of the genetic algorithm, the number of M base stations to be selected is ranked according to the fitness value, and the site selection strategy with a small fitness calculation value is evolved in the next step, that is, to become an individual in the next-generation population. When individuals with high fitness values are abandoned, the remaining number of individuals is randomly selected according to a certain probability. The following crossover and variation are schematic in Fig. 1.

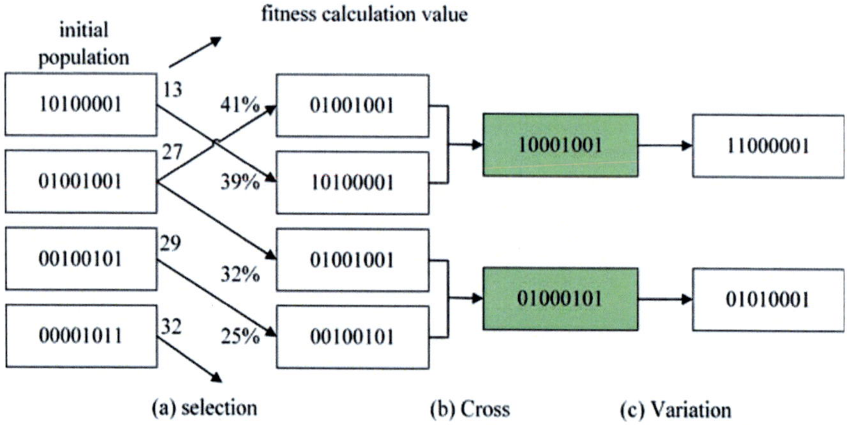

Fig. 1 Schematic representation of the selection, crossover, and variation

2.4 Overall Design of the Algorithm

Genetic algorithm simulates the evolution of evolution in nature, the algorithm using the initial population random evolution get better fitness population offspring, offspring population to the fitness function reproduction, get fitness stronger offspring, after countless iterations, genetic algorithm can through the initial population get the strongest global fitness population, namely the final global optimal solution. Such algorithms are often applied to robot and UAV formation and solving multi-objective function target point delivery problems. Such algorithms are also widely used in site selection. The global random search of genetic algorithm is used to continuously iteratively strengthen the characteristics of fitness, mapping the site search process into the optimization process of genetics. According to the optimization principle, the optimization of selection, crossover, compilation, and other processes are used to obtain the optimal solution. The flowchart of the genetic algorithm is shown in Fig. 2.

In this paper, 5G base station site selection uses genetic algorithm to global search and solve the optimal base station planning point. Combined with the constraints, the optimization in the test area can be approximated as a regional delivery class problem. The specific solution steps of the model are as follows:

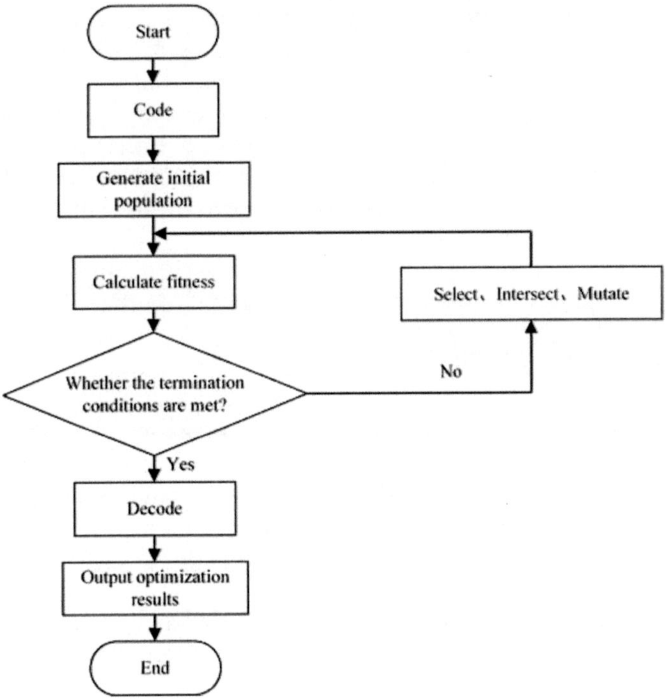

Fig. 2 Overall design diagram of the genetic algorithm

Step 1: Enter the test point data.

Step 2: Initialize the test point data and encode the test point.

Step 3: Randomly generate the initial population and randomly search the world.

Step 4: Initial population selection, variation, and genetic population with high fitness.

Step 5: The last population with the strongest viability is obtained from the fitness function.

Step 6: The optimal solution is obtained by decoding the last surviving population.

3 Simulation and Results Analysis

3.1 Algorithm Parameter Configuration

This paper selected 132 actual data points for base station site selection simulation using MATLAB in 500 km * 500 km square draw test points, test point position as shown in Fig. 3, each point in the square is the need for base station signal coverage test point, and different area test point network demand is uneven, test point centralized area network demand is large, and needs multiple base stations for repeated coverage. The basic parameters of the genetic algorithm are configured as follows: the population size is 150, the number of iterations is 200, the memory size is X (X = 7, 8, 9, 10), the probability of chromosome crossover is 60%, and the probability of variation is 10%. Using the above-configured parameters in the genetic algorithm simulation, we can obtain the specific experimental simulation results and base station planning data.

3.2 Algorithm Simulation

In this paper, 5G base station site planning will include base station construction cost, signal coverage, coverage and Euclidean distance between new base stations as a constraint, using genetic algorithm for simulation, site planning simulation diagram as shown in Fig. 4, purple point test point, blue circle for new base station and coverage area, and dotted line represents the Euclidean distance constraints between bases.

When the memory scale is 8, the number of base station plans is 64, as shown in Fig. 4. Meanwhile, this experiment can obtain different base station planning point simulation results by adjusting different memory scales.

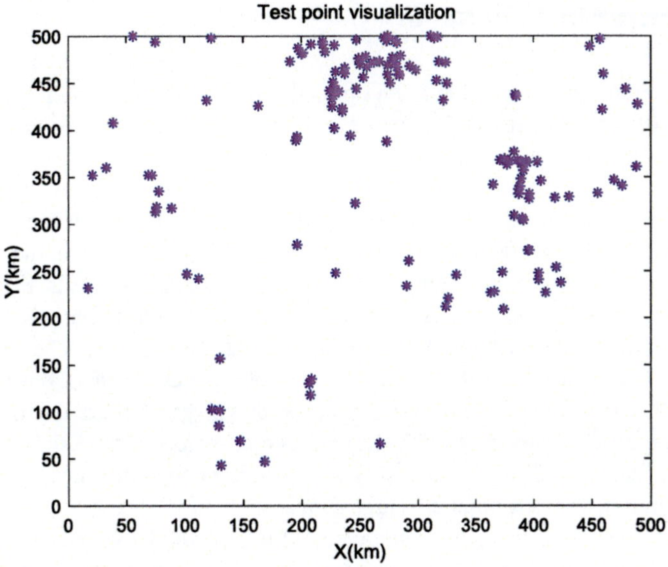

Fig. 3 Test point visualization

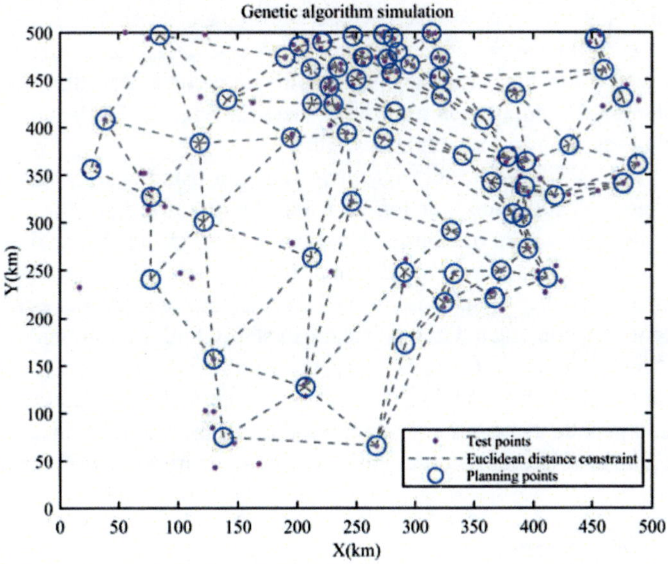

Fig. 4 Planning 64 base station simulation diagram

3.3 Interpretation of Result

The genetic algorithm can obtain the optimal base station selection point after the evolution of the population. This paper not only considers the cost and coverage rate when planning the site, but also introduces the Euclidean distance between base stations as a constraint in order to avoid the coverage effect of the base station not meeting the traffic demand. Due to the introduction of Euclidean distance, regional base stations with high demand cover more base stations. Therefore, the algorithm can reasonably adjust the planning points according to business needs. By adjusting the memory scale, the planned base stations are guaranteed to cover every test point in the diagram as much as possible, and allocate them reasonably. The base station plans resource points, so that there will not be multiple coverage of one point, and collaborative coverage will be provided at the point with greater demand to ensure the best coverage effect. However, for the test points that are too discrete, the introduction of the Euclidean distance will generate corresponding points that are not covered by the base station. The test point coverage data is shown in Table 1.

With 200 iterations using the genetic algorithm, the model simulation curves are shown in Fig. 5. We obtain the new base station planning data and the planning coverage rate, and the specific simulation results are shown in Table 1. Genetic algorithm through memory scale $X = 10$, the number of base station is set to 100, no coverage of 9, coverage reached $93.8\% > 90\%$, the experimental results show that this paper will have Euclidean distance as a constraint of the optimized genetic algorithm which has high accuracy, the optimized genetic algorithm for 5G base station planning, the simulation results can meet the engineering requirements, can be applied in real life, and has high feasibility.

In this paper, we build a multi-objective function using cost, business volume, and coverage, and solve the function by using the optimized genetic algorithm to obtain 200 iterations as in the figure above. According to Fig. 5, the algorithm can converge in the shortest time. For different hidden cells, different coverage rates reach a stable convergence state at the end of the experiment. It can be seen that the genetic algorithm constructed in this paper can stably find the solution of the target function. Moreover, due to the constraints, the algorithm obtains the global optimal planning point of the base station. To sum up, combined with Euclidean distance optimization genetic algorithm for base station planning is that the time complexity is low, and can be different according to adjust the memory-scale different solutions,

Table 1 Test point coverage

Total number of base station settings	No number of points covered	Fraction of coverage
49	28	80.60%
64	21	85.10%
81	17	88.20%
100	9	93.80%

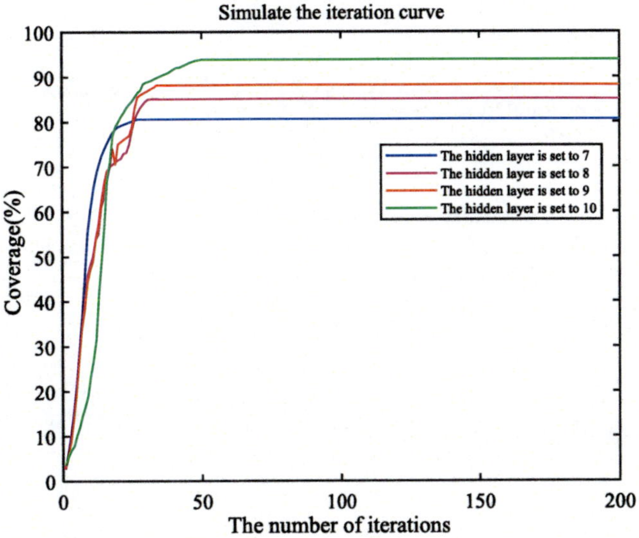

Fig. 5 Genetic algorithm simulation of iterative curves

but ultimately can use the constraints to obtain the global optimal solution of the target function, so the genetic algorithm provides a good reference scheme, can be applied to base station planning problem, and has a good accuracy and feasibility.

4 Conclusion

In order to meet the development trend of the fast pace of 5G, improve users' 5G use experience, reduce insufficient signal coverage, and other problems, more base stations need to be established to cope with the high requirements of 5G on the network. For the traditional planning model time-consuming, high cost, and local optimal problems, this paper proposes the genetic algorithm based on Euclidean distance optimization for 5G base station site planning. This method takes the three factors of Euclidean distance between base stations, the highest coverage and construction cost as the objective function, using the specific simulation point of the 5G base station, and achieve the overall coverage of 93.8%, which shows that the algorithm accuracy and result rationality are high, and the running time complexity is low. Compared with the traditional planning model, this algorithm runs the operation results with high accuracy, high operation feasibility, fast convergence speed, and good optimal planning solution effect. The planned base station can realize the basic full coverage of test points, which can provide a reference for the current 5G base station planning problems.

References

1. Sachan Ruchi, Dash Shatarupa, Sahu Bharat, J.R., et al.: Energy efficient base station location optimization for green B5G networks. 441–448 (2022)
2. Dua Amit, Krömer Pavel, Czech Zbigniew, J., et al.: A bi-objective genetic algorithm for wireless sensor network optimization.147–159 (2022)
3. Ren, H.: Vehicle routing optimization of logistics distribution based on genetic algorithm (VRPTW)[J]. Front. Econ. Manag. **3**(1), 294–299 (2022)
4. Yu Pengfei, Shi Yonghong, Wang Lei, et al.: A method for optimizing communication network topology based on genetic algorithm. 30–40 (2022)
5. Ray Paromik, Bera Dillip Kumar, Rath Ashoke Kumar, et al.: Time cost optimization using genetic algorithm of a construction project. 909–927 (2020)

Vector Control Strategy of Permanent Magnet Synchronous Motor Based on Fuzzy PID Control

Xiang Ji, Lin Zou, Yahong Li, and Mingming Dong

Abstract In recent years, the market demand for new energy vehicles is increasing, so more and more scholars are devoted to the research of this topic. Permanent magnet synchronous motors are widely used in the drive of electric vehicles because of their high efficiency and high power density. Therefore, the study of its control strategy is particularly important. The commonly used control strategy is vector control, but due to the complex working conditions faced by the vehicle, the traditional PID closed-loop control cannot meet the needs of real-time control. Based on the relevant theory of motor vector control and fuzzy PID control, this paper first builds a permanent magnet synchronous motor vector control model in Simulink, and validates the model by observing the electromagnetic torque changes during the simulation process and performing forward and reverse speed regulation. Then the fuzzy controller is designed to replace the traditional PID controller, and through simulation comparison, it is concluded that the model based on fuzzy PID has superior control performance and can solve the problem of real-time control extremely well.

1 Introduction

The development of new energy vehicles is an important strategy for social development in recent years. Therefore, research on pure electric vehicles has attracted much attention. The research of pure electric vehicle mainly focuses on three parts: battery, electric control, and electric drive [1]. Among them, the role of electric drive is particularly important, because it is closely related to the driving control of the vehicle. Whether an electric vehicle adopts a distributed drive or a centralized drive [2], the control of the motor is the key and difficult point. Some traditional control strategies are quite mature for solving the control problems of linear systems, but there are

X. Ji (✉) · L. Zou · Y. Li · M. Dong
Beijing Institute of Technology, School of Mechanical Engineering, Institute of Vibration and Noise Control, Beijing, China
e-mail: 527362190@qq.com

K. Nakamatsu et al. (eds.), *Advanced Intelligent Technologies for Information and Communication*, Smart Innovation, Systems and Technologies 365,
https://doi.org/10.1007/978-981-99-5203-8_34

many problems in solving the control problems of nonlinear systems. Therefore, it is difficult to satisfy various changing needs of the control problem of permanent magnet synchronous motor with strong nonlinearity that is widely used in electric vehicles.

In the research of scholars for many years, vector control has been widely used in the market as well as direct torque control [3]. Because direct torque control wound reduce the reliability of the system to a certain extent and might limit the speed range of the motor to a certain extent, vector control is more popular in the field of motor control [4]. Traditional vector control uses PI controller, which can achieve better speed regulation performance. However, PI controller is difficult to adapt to the changing environment, and the environmental noise has a great influence on its control effect, so it does not have good dynamic performance. The root cause is that its three control parameters Kp, Ki, and Kd are fixed and cannot change with the change of dynamic characteristics, so it is very important to study the self-tuning of PID controller control parameters.

Many scholars have carried out research on this topic. Reference [5] uses the related technology of adaptive fuzzy PID control to improve the control effect of traditional PID control. The results show that this method can not only improve the control effect, but also optimize the control performance. Reference [6] has the real-time speed regulation of the motor which adopts fuzzy PID control and neural network PID control, respectively. Experiments show that neural network has more advantages in anti-interference, and fuzzy PID is better in real-time performance. Reference [7] designed a variable universe adaptive fuzzy control method. The controller is composed of dual fuzzy controllers, which are, respectively, responsible for coarse adjustment and fine adjustment. The fuzzy PID control has better anti-interference ability and robustness than the traditional PID control.

This paper builds the vector control model of permanent magnet synchronous motor, and debugs the model based on the vector control theory. And fuzzy PID controller model is established on the base of the fuzzy control theory. In this model, the three control parameters Kp, Ki, and Kd are self-tuning. Finally, a specific simulation case is set, and the observation results which indicated that fuzzy PID control is more efficient than the PID control in terms of dynamic characteristics which are obtained.

2 Introduction to Vector Control

2.1 Introduction to the Motor Model

In the study of the PMSM model, the permanent magnet flux linkage is generally oriented on the d-axis, as shown in Fig. 1.

Therefore, the motor model is expressed in mathematical expression base on d-q-coordinate as follows:

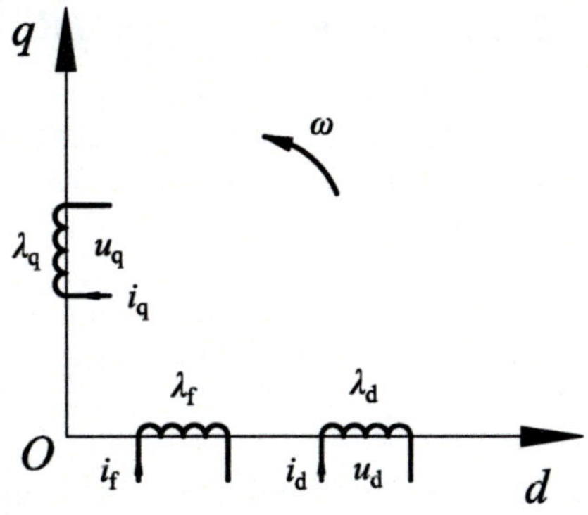

Fig. 1 Simplified model of permanent magnet synchronous motor

2.1.1 Stator Voltage Equation

$$\begin{cases} u_d = R_s i_d + \dfrac{d}{dt}\lambda_d - \omega\lambda_q \\ u_q = R_s i_q + \dfrac{d}{dt}\lambda_q + \omega\lambda_d \end{cases} \tag{1}$$

In the formula, λ_d and λ_q are the flux linkages, u_d and u_q are the voltages, and i_d and i_q are the currents of the stator in *d-q*-coordinate, respectively.

2.1.2 Stator Flux Linkage Equation

$$\begin{cases} \lambda_d = L_d i_d + \lambda_f \\ \lambda_q = L_q i_q \\ \lambda_f = L_{md} i_f \end{cases} \tag{2}$$

In the formula, L_d and L_q are the inductances in d-q-coordinate, L_{md} is the d shaft rotor equivalent excitation inductance, i_f is the equivalent excitation current of the permanent magnet rotor, and λ_f is the *d* shaft rotor flux linkage.

2.1.3 Torque Equation

$$T_e = \frac{3p}{2}\lambda_f i_q + \frac{3p}{2}(L_d - L_q) i_d i_q \tag{3}$$

In formula (3), the first term is called permanent magnet torque and the second term is called reluctance torque. Substituting Eq. (2) into Eq. (3), we can get relationship (4):

$$T_e = \frac{3p}{2}[L_{md} i_f i_q + (L_d - L_q) i_d i_q] \tag{4}$$

2.2 *Coordinate Transformation*

The mathematical models of permanent magnet synchronous motors are very complex, and their voltage equations, flux linkage equations, and torque equations have high nonlinearity and coupling. Therefore, in practical applications, the motor model must be simplified. To simplify the mathematical model, we need to start with the electromagnetic coupling relationship. Therefore, two coordinate transformations are introduced and they are Clarke transformation and Park transformation.

2.2.1 Clarke Coordinate Transformation

The formula for transforming the current vector from the three-phase stationary coordinate system to the two-phase stationary coordinate system is

$$\begin{bmatrix} i_\alpha \\ i_\beta \end{bmatrix} = \begin{bmatrix} \frac{\sqrt{3}}{2} & 0 \\ \sqrt{\frac{1}{2}} & \sqrt{2} \end{bmatrix} \begin{bmatrix} i_A \\ i_B \end{bmatrix} \tag{5}$$

In this formula,

$$C_{3/2} = \sqrt{\frac{2}{3}} \begin{bmatrix} 1 & -\frac{1}{2} & -\frac{1}{2} \\ 0 & \frac{\sqrt{3}}{2} & -\frac{\sqrt{3}}{2} \\ \sqrt{\frac{1}{2}} & \sqrt{\frac{1}{2}} & \sqrt{\frac{1}{2}} \end{bmatrix} \tag{6}$$

2.2.2 Park Coordinate Transformation

The formula for transforming the current vector from a two-phase stationary coordinate system to a two-phase rotating coordinate system is

$$\begin{bmatrix} i_d \\ i_q \end{bmatrix} = P_{2/2} \begin{bmatrix} i_\alpha \\ i_\beta \end{bmatrix} \tag{7}$$

In this formula,

$$P_{2/2} = \begin{bmatrix} \cos\theta & \sin\theta \\ -\sin\theta & \cos\theta \end{bmatrix} \tag{8}$$

2.3 Vector Control Theory

Equation (2) is the basic equation for the vector control of PMSM, but by analyzing Eq. (4), it can be seen that the reluctance torque in the electromagnetic torque will be affected by i_d, and the vector control strategy needs to meet the requirements of maximizing electromagnetic torque, so the $i_d = 0$ control strategy came into being. Let the stator d-axis current i_d be 0, Eqs. (2) and (3) will become Eqs. (9) and (10):

$$\begin{cases} \lambda_d = \lambda_r \\ \lambda_q = L_q i_q \end{cases} \tag{9}$$

$$T_e = \frac{3p}{2}\lambda_f i_q = \frac{3p}{2} L_{md} i_f i_q \tag{10}$$

It can be seen from the formula that the purpose of adjusting the electromagnetic torque can be changed only by adjusting i_q in the d-q-coordinate system. The relationship between the i_q and the electromagnetic torque is

$$i_q = \frac{2T_e}{3p\lambda_r} \tag{11}$$

According to the theory above, the control strategy is shown in the form of a flowchart in Fig. 2.

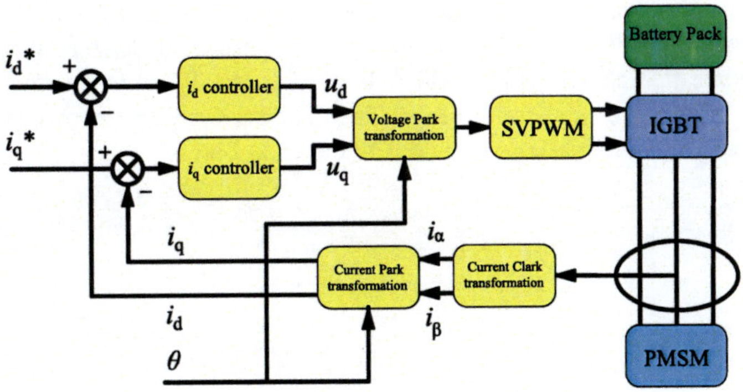

Fig. 2 Permanent magnet synchronous motor vector control block diagram

3 Construction of Motor Control Model

The control model of the motor base on the vector control theory consists of four parts: SVPWM generation module, coordinate transformation module, motor module, and PID control closed-loop module. In motor control, the control voltage signal is input to the SVPWM generation module through coordinate transformation, and then the generated SVPWM wave is input into the IGBT module to output the control voltage of the motor.

There are already packaged coordinate transformation and permanent magnet synchronous motor modules in the toolbox of Simulink. Select suitable output of the motor module as the direct-axis current, quadrature-axis current, and rotational speed, and compare them with the reference direct-axis current ($i_d = 0$) and the reference rotate speed, then input them to PID controller to form double PID closed-loop control.

3.1 Construction of the Motor Control Model

The SVPWM generation module is mainly composed of a sector judgment module and a switch state action time calculation module which is using center-symmetric PWM generation method to achieve efficient trigger control of power semiconductor switches, reduce power consumption, and improve system efficiency. The built SVPWM generation module is shown in Fig. 3.

On the basis of the SVPWM module, the Clarke and Park coordinate transformation modules, the PID controller module, and the permanent magnet synchronous motor module that comes with MATLAB/Simulink are added to set the output of the motor to form a closed control loop. Then build a complete permanent magnet synchronous motor vector control model as shown in Fig. 4.

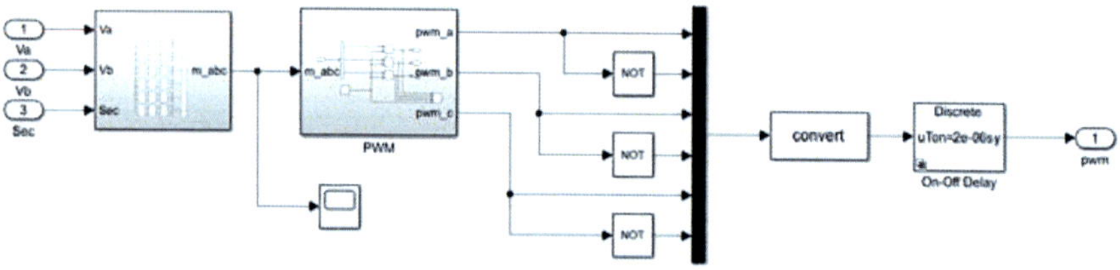

Fig. 3 SVPWM model in Simulink

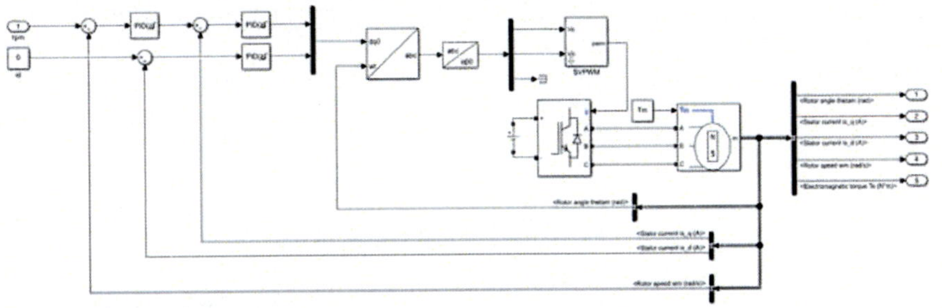

Fig. 4 Permanent magnet synchronous motor vector control model in Simulink

3.2 *Model Validation*

During simulation, set the DC bus voltage to 380 V, and set the given speed to 2000 rpm. In addition, the simulation time and sampling time are set to 0.02 s and 2e-6 s, respectively.

3.2.1 SVPWM Observation

At a certain moment, the six-way PWM waveform is shown in Fig. 5. The waveforms PWM1 ~ PWM6 are the six inputs of the inverter bridge, respectively. When the input signal of any channel is 1, it means that the switching device of this channel is in the ON state, otherwise it is in the OFF state.

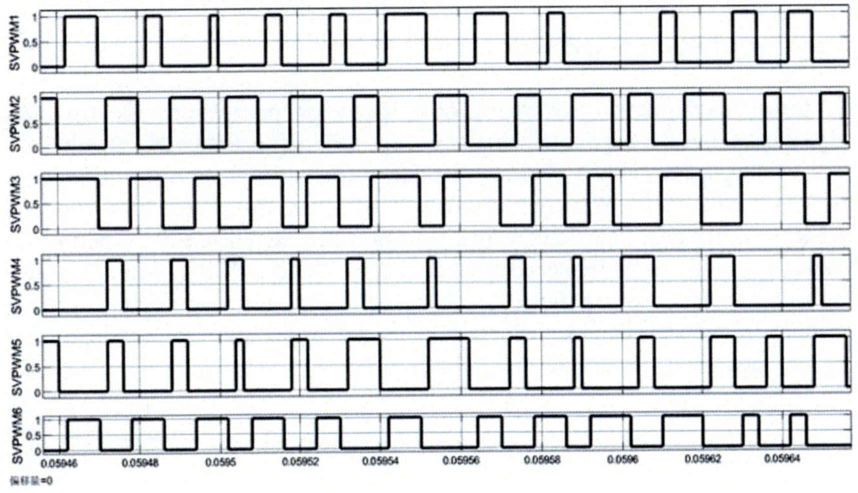

Fig. 5 SVPWM waveform at a certain moment

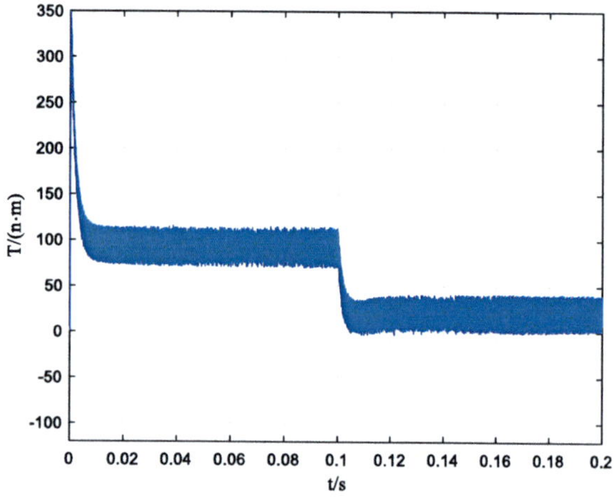

Fig. 6 Electromagnetic torque under sudden load torque condition

3.2.2 Electromagnetic Torque Observation

In the process of speed regulation, a step load torque is applied to the motor, and the load torque decreases at 0.1 s. It can be seen from the figure that at 0 ~ 0.02 s, the value of the electromagnetic torque is large, because the system needs to overcome the effect of the load torque during the starting process of the motor, so it shows a large amplitude. After that, the motor enters steady state. When the load torque decreases at 0.1 s, the electromagnetic torque of the motor quickly enters a steady state after a brief fluctuation. As shown in Fig. 6, the motor control model has good adaptability.

3.2.3 Speed Regulation Process

The forward and reverse speed regulation of the motor control model is carried out, respectively, and the results are shown in Fig. 7. When entering the steady state, the system speed is stable, indicating that the system has forward and reverse speed regulation capabilities. When entering the steady state, the system runs smoothly and achieves the expected effect, which verifies the characteristics of the permanent magnet synchronous motor speed control system. The simulation results indicated this motor control system can well reach the speed regulation of reference speed in constant. It is verified that the motor control model has fast start-up, strong overload capacity, and good speed regulation characteristics.

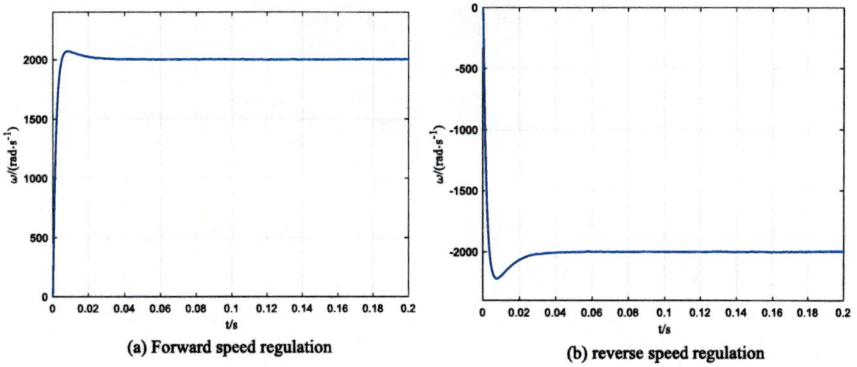

Fig. 7 Forward and reverse speed regulation. **a** Forward speed regulation **b** Reverse speed regulation

4 Fuzzy Control Theory and Construction of Fuzzy PID Controller Model

Fuzzy mathematics is the foundation of fuzzy control theory. The vast majority of engineering application problems in real life are nonlinear problems, and their mathematical models may have highly coupled characteristics or very complex nonlinear characteristics. Therefore, traditional control strategies which are good at dealing with simple linear problems are difficult to directly apply to the study of practical problems. If we try to linearize the actual complex system, the control effect will be unsatisfactory because the model is not accurate enough. So try to deal with these control problems with fuzzy mathematics [8].

4.1 Introduction to Fuzzy Control Theory

Fuzzy control uses the control rules summed up by experts based on engineering experience to control the system, so it has the advantages of simplicity, high efficiency, and strong reliability, and so it is popular in the field of intelligent control. Fuzzy PID control is actually based on fuzzy controller, according to expert experience, and according to the input system error and other information, the three control parameters of proportional, integral, and differential in traditional PID controller are carried out in real time with human thinking. The workflow of fuzzy controller is illuminated in Fig. 8.

According to the system shown in Fig. 5.1, the working scheme of the fuzzy PID controller can be summarized as follows:

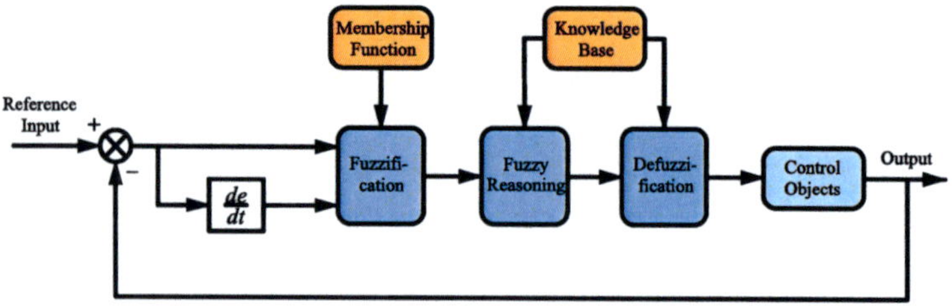

Fig. 8 Flowchart of fuzzy control

4.1.1 Fuzzification

Convert the error passed to the system and the derivative of the error into the corresponding fuzzy quantity.

4.1.2 Fuzzy Reasoning

Carry out fuzzy reasoning according to the summarized inference rules (fuzzy control planning table).

4.1.3 Fuzzy Decision

Clarify the inference result from the fuzzy quantity to the precise quantity that can be applied to the actual system [9].

4.2 Construction of Fuzzy PID Controller Model

4.2.1 Membership Function

The process of fuzzification is to judge which fuzzy subset the input belongs to through the membership function (as shown in Fig. 9), which is essentially to classify them into different intervals according to the value of the input error and its derivative, so as to carry out the analysis for different situations. Then through fuzzy reasoning and defuzzification, the control parameters of the PID controller are changed in real time.

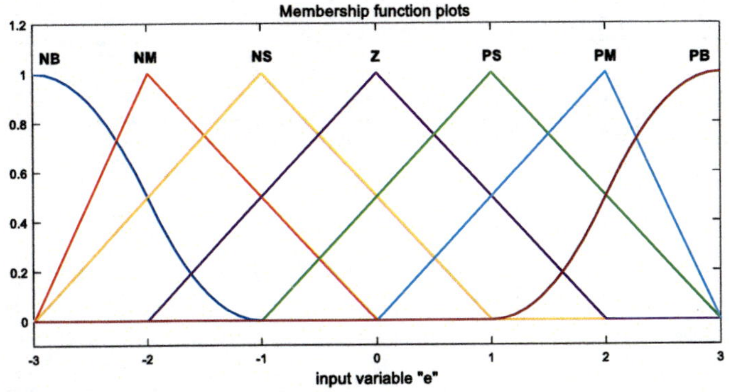

Fig. 9 Membership function

4.2.2 Fuzzy Inference Rules

Fuzzy inference rules are consisted of a battery of logical inference statements determined according to engineering experience. When the system is running, it will reasonably guide the system to output ideal values under certain conditions.

In PID control, when |e| (absolute value of error) is large, in most cases, we hope that the system can quickly reduce the error between the reference and the reference value, so a large proportional coefficient Kp is required. And for reducing the overshoot and vibration of this system, Ki and Kd are expected as small as possible. When |e| is moderate, it is not necessary to amplify the error too much, so the value of Kp ought to be smaller. For avoiding the occurrence of steady-state errors, Ki should be appropriately increased. Moreover, for ensuring the quality of quick response, Kd should also be appropriately increased. When |e| is small, a larger Kp and are required for the consideration of the steady-state characteristic quality of the system, and for avoiding vibration of the system, Kd should be smaller.

According to these basic rules, the fuzzy inference rules which are shown in Table 1 could be summarized, and such rules are called Mamdani-type fuzzy control rules.

4.2.3 Fuzzy Controller Model Construction

The fuzzy logic controller module in MATLAB/Simulink is a powerful tool for implementing fuzzy PID control. To use this module, create an m file which is used to generate fuzzy vector firstly. Then input the generated fuzzy vector name into the fuzzy logic controller module to use it. The fuzzy PID controller model built in Simulink is presented in Fig. 10.

Table 1 Fuzzy inference rules

e	ec						
	NB	NM	NS	ZO	PS	PM	PB
NB	PB/NB/PS	PB/NB/NS	PM/NM/NB	PM/NM/NB	PS/NS/NB	ZO/ZO/NM	ZO/ZO/PS
NM	PB/NB/PS	PB/NB/NS	PM/NM/NB	PS/NS/NM	PS/NS/NM	ZO/ZO/NS	NS/PS/ZO
NS	PM/NB/ZO	PM/NM/NS	PM/NS/NM	PS/NS/NM	ZO/ZO/NS	NS/PM/NS	NS/PM/ZO
ZO	PM/NM/ZO	PM/NM/NS	PS/NS/NS	ZO/ZO/NS	NS/PS/NS	NM/PM/NS	NM/PM/ZO
PS	PS/NM/ZO	PS/NS/ZO	ZO/ZO/ZO	NS/PS/ZO	NS/PS/ZO	NM/PM/ZO	NM/PB/ZO
PM	PS/ZO/PB	ZO/ZO/PM	NS/PS/PM	NM/PS/PM	NM/PM/PS	NM/PB/PS	NB/PB/PB
PB	ZO/ZO/PB	ZO/ZO/PM	NM/PS/PM	NM/PM/PM	NM/PM/PS	NB/PB/PS	NB/PB/PB

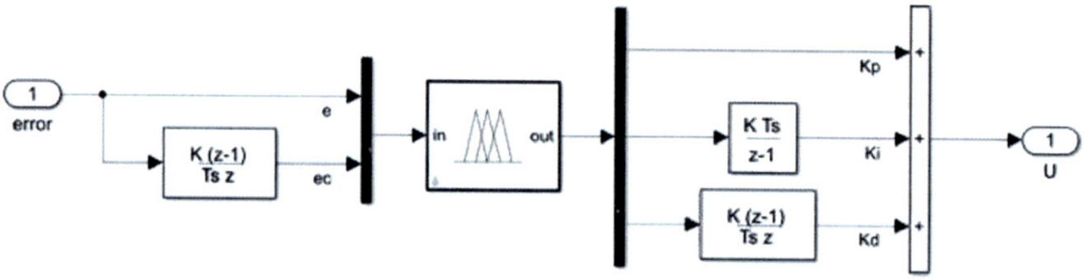

Fig. 10 Fuzzy controller model in Simulink

5 Simulation Comparison

The simulation conditions are set as the specific reference speed of 2000 rpm and the speed of sinusoidal change with an amplitude of 2000 rpm and a frequency of 50 Hz. The simulation duration and sampling time are still set to 0.2 s and 2e-6 s, respectively. Observe the static and dynamic speed regulation performance of the control model based on fuzzy PID control and the model based on PID control, and the control effect on the direct-axis current, then import the control results of the two models into the work area, draw comparisons, and the results in detail are shown in Figs. 11, 12, and 13.

Fig. 11 Comparison of static speed regulation

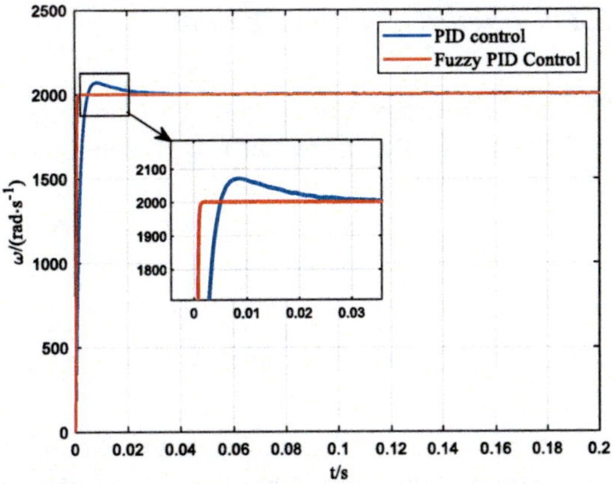

Fig. 12 Comparison of dynamic speed regulation

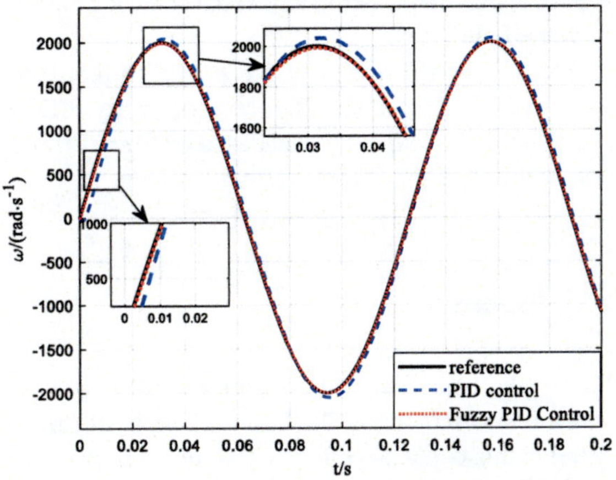

Fig. 13 Comparison of the control effect of ID

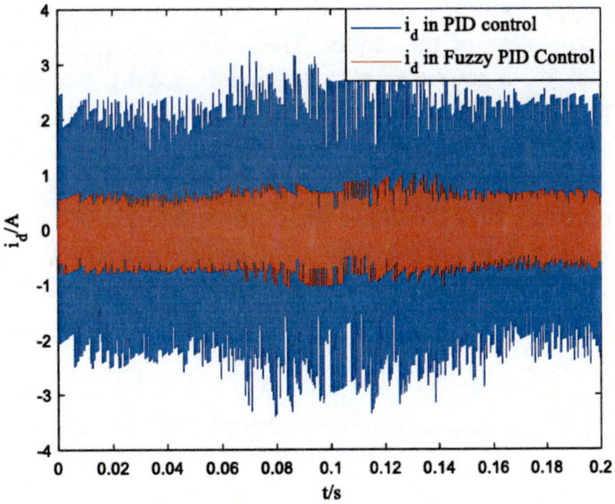

5.1 *Fixed Reference Torque*

The results are shown in Fig. 11.

The fuzzy PID-based model has a shorter rise time and significantly less overshoot, indicating that the system represented by this model can quickly reach the reference value and enter the steady state quickly. The results are shown in Fig. 11.

5.2 *Sinusoidally Varying Reference Torque*

Under the condition of sinusoidal speed change, the model based on fuzzy PID can follow the reference rotate speed with a small error, which reflects the advantage of the fuzzy PID controller to adjust the control parameters in real time. The results are presented in Fig. 12.

For the control results of the d-axis current (see Fig. 13), the amplitude variation range of the model ID based on fuzzy PID control is significantly smaller, which indicates that the model based on fuzzy PID can effectively reduce the energy consumption of the control system.

6 Summary

Firstly, the basic theory of vector control which is applied to permanent magnet synchronous motor and the basic principle of fuzzy PID control are introduced in detail in this paper. Secondly, based on these two theories, a vector control model is established in MATLAB/Simulink firstly, and then obtain the speed regulation results from Simulink and observe the change of electromagnetic torque, which proved the correctness of the model. Then use the fuzzy controller module in the Simulink toolbox to generate the fuzzy vector by writing an m file, and establish the model of the fuzzy PID controller. Finally, the two models are simulated, respectively, under the two working conditions of a specific reference rotate speed and a sinusoidal rotate speed. Through the comparison of the results, it is concluded that the control model based on fuzzy PID can well satisfy the demand of the nonlinearity of the electric vehicle to the motor. And it can greatly make up for the shortcomings of traditional PID control. It is proved that fuzzy PID is closer to the needs of market applications in terms of dynamic characteristics and real-time control.

References

1. Takeo, I., Yutaro, S., Nobuyuki, K.: Analysis for fault detection of vector-controlled permanent magnet synchronous motor with permanent magnet defect. IEEE Trans. Magn. **49**, 2331–2334 (2013). https://doi.org/10.1109/TMAG.2013.2243135
2. Rahman, M.F., Zhong, L., Haque, M.E., Rahman, M.A.: A direct torque-controlled interior permanent-magnet synchronous motor drive without a speed sensor. IEEE Trans. Energy Convers. **18**, 17–22 (2003). https://doi.org/10.1109/TEC.2002.805200
3. Osheba, S.M., Abdel-Kader, F.M.: Performance and analysis of permanent magnet synchronous motors part II: Operation from variable source and transient characteristics. IEEE Power Eng. Rev. **11**, 39–42 (1991). https://doi.org/10.1109/MPER.1991.88793
4. Rahman, M.A., Little, T.A.: "Dynamic performance analysis of permanent magnet synchronous motors." IEEE Power Eng. Rev. 4, 40–40 (1984). https://doi.org/10.1109/MPER.1984.5526098
5. Sun, Y., Su, B., Sun, X.: Optimal design and performance analysis for interior composite-rotor bearingless permanent magnet synchronous motors. IEEE Access **7**, 7456–7465 (2019). https://doi.org/10.1109/ACCESS.2018.2890020
6. Bo, X., Guoding, S., Wei, J., Feiyu, L., Shihong, D., Huangqiu, Z.: Design of an adaptive nonsingular terminal sliding model control method for a bearingless permanent magnet synchronous motor. Trans. Inst. Meas. Control. **39**, 1821–1828 (2017). https://doi.org/10.1177/0142331216649026
7. Pillay, P., Krishnan, R.: Modeling simulation and analysis of permanent magnet motor drivers, part I: The permanent-magnet synchronous motor drive. IEEE Trans. Ind. Appl. **25**, 265–273 (1989). https://doi.org/10.1109/28.25541
8. Zhou, Y., Shang, W., Liu, M., Li, X., Zeng, Y.: "Simulation of PMSM vector control based on a self-tuning fuzzy PI controller." 2015 8th International Conference on Biomedical Engineering and Informatics (BMEI), pp. 609–613 (2015) https://doi.org/10.1109/BMEI.2015.7401576
9. Zeghoudi, A., Chermitti, A.: A comparison between a fuzzy and PID controller for universal motor. Int. J. Comput. Appl. **104**, 32–36 (2014). https://doi.org/10.5120/18207-9347

Track Bolt Wrench Motor BP-Fuzzy Nerve PID Rotate Speed Control

Zhang Mou, Meng Jianjun, Li Decang, Xu Ruxun, and Chen Xiaoqiang

Abstract For the problem of inaccurate speed control and slow response time of the Track Bolt Operation Machine in different periods of bolt tightening, in the article, a BP fuzzy-neural speed control strategy for bolt wrench motor is proposed. The control strategy depends on a BP neural network architecture addressing the fuzzy control method, which optimizes the affiliation function in real time by training the center value and width, and finally outputs the three optimal bounds of the PID controller to control the speed of the motor at different periods of bolt tightening. Using MATLAB simulation, that's what the eventual outcomes shows: the control strategy comes forward in this paper which can significantly reduce the amount of overshoot, significantly reduce the adjustment time, and significantly improve the precision and efficiency of the bolt assembly. The suggested BP fuzzy-neural PID control in this research is compared to that of standard PID control and fuzzy PID control. This study reduces overshoot by 73% when compared to standard PID control and 58% when compared to fuzzy PID control, and the reaction speed is significantly improved.

1 Introduction

Due to the ever-changing improvement of China's railways, the gathering of affixing bolts on rail lines is likewise advancing new necessities [1]. At present, the bolt tightening wrenches used in domestic rail construction cannot meet the needs of the present period in terms of operation efficiency, operation precision, automation due

Z. Mou (✉) · M. Jianjun · L. Decang · X. Ruxun · C. Xiaoqiang
Mechatronics T&R Institute, Lanzhou Jiaotong University, Lanzhou, China
e-mail: 884075909@qq.com

M. Jianjun · L. Decang · C. Xiaoqiang
Gansu Provincial Engineering Technology Center for Informatization of Logistics & Transport Equipment, Lanzhou, China

Gansu Provincial Industry Technology Center of Logistics & Transport Equipment, Lanzhou, China

K. Nakamatsu et al. (eds.), *Advanced Intelligent Technologies for Information and Communication*, Smart Innovation, Systems and Technologies 365,
https://doi.org/10.1007/978-981-99-5203-8_35

to the low degree of intelligence and single control method, etc. While pursuing the operation efficiency, the operation precision is often reduced, so it is extremely important to flourish a speed control system that can further develop the system response speed, control precision, and control performance. Therefore, it is extremely important to flourish a speed control system that enhances the response speed, control accuracy, and security of the system without reducing the efficiency of the rail bolting machine.

For the precise control of bolt wrench, different scholars have adopted various research approaches. Jianliang [2] adopts fuzzy control method to adjust controller parameters in real time under the effect of membership function, although the control accuracy of this method is low. Chen [3] proposed a fuzzy self-adjusting controller which uses sensors to collect the parameter information of DC brushed motor rotation process using sliding weighted average filtering algorithm to eliminate noise interference and improve the control performance of DC brushed motor to some extent. Zhang [4] proposed the initial speed fluctuation of bolt wrench is reduced by adding harmonic current loop, and the speed fluctuation amplitude of bolt wrench is reduced by 50% in the start-up phase compared with the general control strategy. Englert [5] proposed a prescient force control plot for enlistment engines, which is using an upgraded Lagrangian method with altered inclination calculation to take care of the ideal control issue really. Klecker [6] adopts a hierarchical strategy to creating vigorous and versatile learning calculations for direction following, force control within the sight of vulnerabilities and exchanging imperatives. Ammar [7] proposes a strengthened predictive torque control program that replaces the PI in the external loop with a fuzzy controller for speed regulation, an improvement that guarantees precise following and powerful control for various vulnerabilities. The domestic and international research content is mainly focused on torque control accuracy, and although it is possible to control torque accurately, there are few speed control strategies for different periods of bolt tightening. And in the reality working process, for the sake of improvement of the assembly efficiency, the fast and then slow tightening speed is often used in the three periods of bolt assembly, so how to ensure the fast response and speed control accuracy when switching the speed between each two periods is crucial.

For the sake of improving the system performance without reducing the efficiency of the rail bolt working machine, the article proposes a control strategy which depends on Backward Propagation Fuzzy Neural PID (BPFN-PID) to control the motor speed at different periods of bolt tightening. Fuzzy control [8] is an intelligent control method with strong reasoning ability but deviating learning ability. Neural network [9] has a strong learning ability, but the inference ability is biased. The article joins the benefits of the two wise control techniques and the hindrances complete one another to propose the BPFN-PID control methodology, which can significantly improve the system stability and efficiency, has the advantages of small overshoot and fast response, and has good application prospects.

2 Motor Control System Model

2.1 *Workflow of Speed Control System of Rail Bolt Wrench*

The bolt wrench speed control system of orbital bolt working machine consists of motor, DSP, circuit module, sensor module, bolt wrench, etc. The control framework embraces three shut circle control design [10], the speed loop adopts BPFN-PID control, and the current loop and position loop adopt PID control. The orbital bolt wrench under the control system adopts the torque-turning method for tightening operation, and the workflow is divided into three periods [2]. The workflow is divided into three periods, the first and second periods are tightened by the torque method and the third period is tightened precisely by the corner method. The first period is the thread recognition period, where the orbital bolt wrench needs to overcome the frictional torque of the thread, and the stroke in this period is long, the resistance is minimal, the rotational speed is maximum, and the motor can run at full speed [11]. The torque value measured by the sensor exceeds the threshold value specified in the second period, it enters the fitting period, this period needs to overcome certain resistance, and the travel from the next period is short, the motor speed is reduced, the third period is the tightening period, when the torque measured by the torque sensor reaches the threshold value of the third period, it enters the tightening period, this period uses the corner method for tightening operation, it needs to control the motor to rotate at a certain angle, the motor is the lowest and the control is most accurate, after the torque measured by the sensor reaches the set torque value, the task is completed and the motor stops sharply.

2.2 *DC Brushless Motor Modeling*

Brushless Direct Current Motor (BLDCM) is the power source of the bolt wrench tightening system of orbital bolt working machine, which has outstanding speed regulation performance, low starting current, and smooth operation [12]. It is widely used in civil and industrial fields because of its excellent speed regulation performance, low starting current, and smooth operation. Figure 1 shows the structure of BLDCM.

Establish the mathematical model of BLDCM [13]. For the sake of simplifying the BLDCM model and facilitate the analysis, the following assumptions are made. The stator winding is connected in a star-shaped full bridge method.

(1) The three-phase winding is symmetrical, the rotor magnetic field and the stator current are completely symmetrical, and each opposite momentum is square wave.

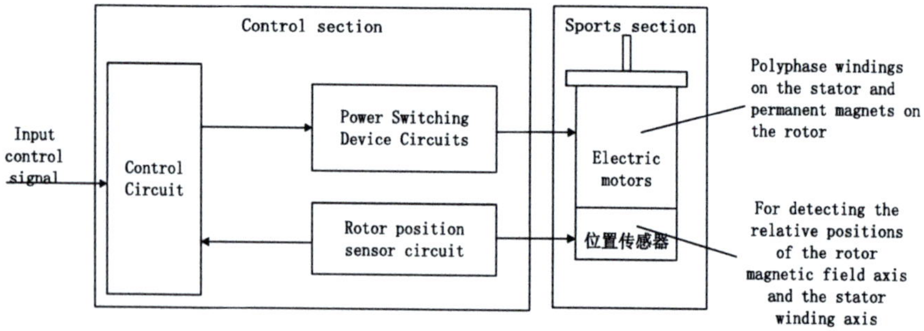

Fig. 1 BLDCM structure diagram

(2) Excluding the tube voltage drop in the circuit; the effect on the magnetic field produced by the tooth slot, armature reaction, and magnetic circuit saturation; and the hysteresis loss.
(3) When the rotor position changes, the reluctance of the magnetic circuit remains constant and the equivalent inductance of the phase winding is a constant.
(4) The change of motor speed during phase change is neglected.

The three-phase voltage of the BLDCM is represented in matrix form as follows:

$$\begin{bmatrix} u_1 \\ u_2 \\ u_3 \end{bmatrix} = \begin{bmatrix} R & 0 & 0 \\ 0 & R & 0 \\ 0 & 0 & R \end{bmatrix} \begin{bmatrix} i_1 \\ i_2 \\ i_3 \end{bmatrix} + \begin{bmatrix} L_a - M_a & 0 & 0 \\ 0 & L_a - M_a & 0 \\ 0 & 0 & L_a - M_a \end{bmatrix} \frac{d}{dt} \begin{bmatrix} i_1 \\ i_2 \\ i_3 \end{bmatrix} + \begin{bmatrix} e_1 \\ e_2 \\ e_3 \end{bmatrix} \tag{1}$$

This is the instantaneous voltage value of u_1, u_2, u_1; i_1, i_2, i_3 is the instantaneous present value; e_1, e_2, e_3 is the instantaneous counter-electromotive force value; R is the resistance value; L_a is the inductance value; and M_a is the mutual inductance value.

Select [14]: The phase current flowing in the direction of the center point of the winding is the forward current and the phase voltage takes the center point as the reference point. Because the reluctance of the magnetic circuit does not vary with the rotor position, L_a, M_a is a constant, let $L = 2(L_a - M_a)$, then the voltage equation of BLDCM can be expressed as

$$\begin{cases} u_1 = Ri_1 + L\dfrac{di_1}{dt} + e_1 \\ u_2 = Ri_2 + L\dfrac{di_2}{dt} + e_2 \\ u_3 = Ri_3 + L\dfrac{di_3}{dt} + e_3 \end{cases} \tag{2}$$

The total power of the motor is the sum of the power of the three-phase windings.

$$P_Z = e_1 i_1 + e_2 i_2 + e_3 i_3 \tag{3}$$

The equation for the electromagnetic torque of the motor is

$$\begin{aligned} T_w &= Fr = \frac{P_z}{V} r = \frac{P_z}{\omega} = \\ &\frac{e_1 i_1 + e_2 i_2 + e_3 i_3}{\omega} = \frac{2\pi(e_1 i_1 + e_2 i_2 + e_3 i_3)}{\Omega} \end{aligned} \tag{4}$$

where T_w is the electromagnetic torque; F is the generated torque; r is the radius of action; V is the linear velocity; ω is the angular velocity; and Ω is the motor speed.

For the electromagnetic torque T_w, the following expressions are commonly used:

$$T_w = K_w i \tag{5}$$

where K_w is the electromagnetic torque factor and i is the steady-state current.

The equation of mechanical motion of the motor under consideration of electrical load and viscous damping is

$$T_w = T_f + J\frac{dw}{dt} + B_w w \tag{6}$$

where T_f is the load torque, J is the rotational inertia, and B_w is the damping factor.

Neglecting the transient process of commutation, Eq. (2) yields

$$u_{12} = U_4 = 2Ri + L\frac{di}{dt} + 2e_1 = r_a i + L\frac{di}{dt} + k_e w \tag{7}$$

where U_4 is the bus voltage; r_a is the line resistance, $r_a = 2R$; L is the equivalent inductance, $L = 2(L_a - M_a)$; and k_e is the line counter potential factor.

From Eqs. (5), (6), and (7), it follows that

$$U_4 = \frac{LJ}{K_w}\frac{d^2\omega}{dt^2} + \frac{r_a J + LB_w}{K_w}\frac{d\omega}{dt} + \frac{r_a B_w + k_e K_w}{K_w}\omega \tag{8}$$

Applying the Laplace transform, Eq. (8) is converted to the three-phase transfer function of BLDCM.

$$G(s) = \frac{\omega(s)}{U_4(s)} = \frac{K_w}{LJs^2 + (r_a J + LB_w)s + (r_a B_w + k_e K_w)} \tag{9}$$

Considering that BLDCM operates with a load, the load can be regarded as the input of the system, then the structure of the tightening system of the bolt wrench of the rail bolt working machine is displayed in Fig. 2.

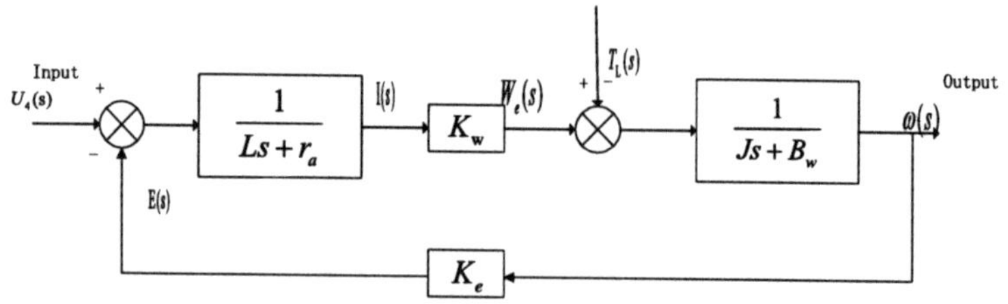

Fig. 2 Structure of tightening system

3 Motor Control Strategy

In the actual working process, the bolt wrench of the rail bolt working machine outputs the torque to the bolt after the motor input speed is increased by reducing the speed of the reducer connecting the sleeve; therefore, the controlled object of the article is the BLDCM model established in the first chapter.

3.1 Conventional PID Control

A complete PID control system should contain proportional, integral, and differential controllers and the target being controlled. The structure of a conventional PID system is displayed in Fig. 3.

The control deviation of the PID controller is determined by the difference between the input value $r_{in}(t)$ and the actual output value. $y_{out}(t)$

$$error(t) = r_{in}(t) - y_{out}(t) \tag{10}$$

The control law of the PID controller is

$$u(t) = k_p error(t) + k_i \int_{o}^{t} error(t)dt + k_d \frac{derror(t)}{dt} \tag{11}$$

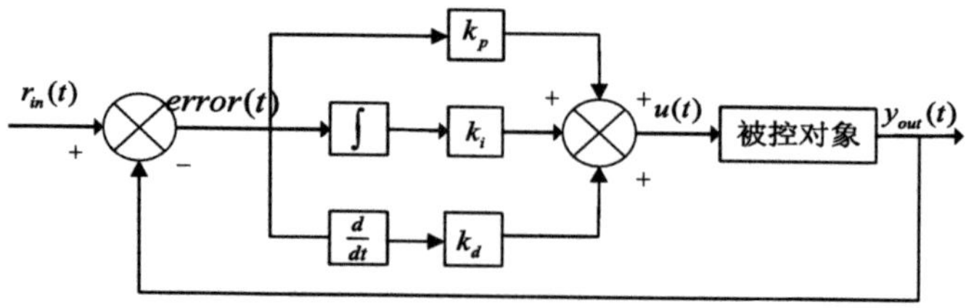

Fig. 3 Structure diagram of conventional PID system

where k_p is the scale factor; k_i is the integration factor; k_d is the differentiation factor; and $error(t)$ is the control deviation at the moment t.

PID controller [15]: The proportional link is used to decrease the deviation, the integral link is predominantly used to remove the steady-state error, and the differential link is used to suppress the unfavorable variation of the deviation.

3.2 *Fuzzy PID Control*

The so-called fuzzy control is to imitate the logic of human thinking, judging the input fuzzy concepts in an empirical way, using fuzzy logic to reason and get fuzzy results, and finally defuzzification to transform the fuzzy concepts in the results into control signals, a very critical step in fuzzy control is to determine the subordination function.

The commonly used affiliation functions are [16] Triangular, Gaussian, etc. For the sake of guarantee the running accuracy while reducing the running time, the article selects the triangular affiliation function.

Triangular affiliation function expressions:

$$f(x, d, e, f) = \max\left(\min\left(\frac{x-d}{e-d}, \frac{f-x}{f-e}\right), 0\right) \tag{12}$$

Figure 4 shows the fuzzy control simulation diagram.

The fuzzy rule tables for Δk_P, Δk_I, and Δk_D are displayed in Figs. 5, 6, and 7.

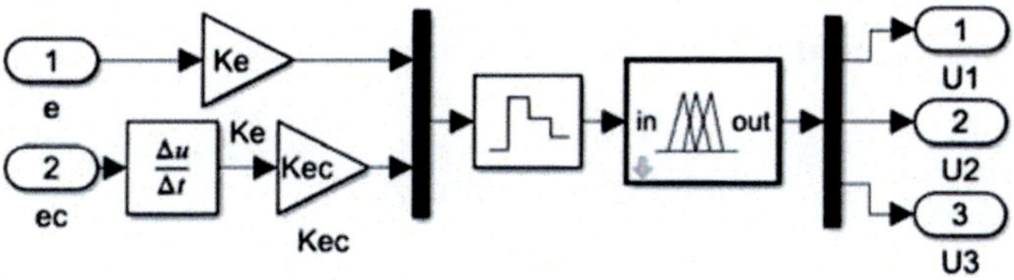

Fig. 4 Fuzzy control simulation diagram

Fig. 5 Δk_P fuzzy rule table

ec \ e	NB	NM	NS	ZO	PS	PM	PB
NB	PB	PB	PM	PM	PS	ZO	ZO
NM	PB	PB	PM	PS	PS	ZO	NS
NS	PM	PM	PM	PS	ZO	NS	NS
ZO	PM	PM	PS	ZO	NS	NM	NM
PS	PS	PS	ZO	NS	NS	NM	NM
PM	PS	ZO	NS	NM	NM	NM	NB
PB	ZO	ZO	NS	NM	NM	NB	NB

Fig. 6 Δk_I fuzzy rule table

ec \ e	NB	NM	NS	ZO	PS	PM	PB
NB	PS	PM	ZO	ZO	ZO	PB	PB
NM	NS	PM	NS	NS	ZO	NS	PM
NS	NM	PS	NM	NS	ZO	PS	PM
ZO	NM	ZO	NM	NS	ZO	PS	PM
PS	NS	NM	NS	NS	ZO	PS	PS
PM	NM	NS	NS	NS	ZO	PS	PS
PB	PS	ZO	ZO	ZO	ZO	PB	PB

Fig. 7 Δk_D fuzzy rule table

ec \ e	NB	NM	NS	ZO	PS	PM	PB
NB	NB	NB	NM	NM	NS	ZO	ZO
NM	NB	NB	NM	NS	NS	ZO	ZO
NS	NB	NM	NS	NS	ZO	PS	PS
ZO	NM	NM	NS	ZO	PS	PM	PM
PS	NM	NS	ZO	PS	PS	PM	PM
PM	ZO	ZO	PS	PS	PS	PB	PB
PB	ZO	ZO	PS	PM	PM	PB	PB

The article fuzzy PID controller used in the inference period is Mamdani fuzzy inference method [2], the fuzzy inference statement is "if A and B then C", A is the fuzzification matrix of the error signal, B is the fuzzification matrix of the rate of change of the error signal, C is the fuzzification matrix of the controller output, and the fuzzy relation matrix D is

$$D = (A \times B)^T \circ C \tag{13}$$

The output fuzzification matrices C' corresponding to the fuzzy input matrices A' and B' are

$$C' = (A' \times B')^T \circ D \tag{14}$$

After obtaining the output fuzzification matrix, it is necessary to defuzzify it.

$$g' = \frac{\int_a^b g\mu C'(g)dg}{\int_a^b \mu C'(g)dg} \tag{15}$$

where g' is the result of defuzzification;g is the element in the fuzzy domain U; $\mu C'(g)$ is the affiliation function of the output fuzzification matrix with respect to g; and a, b are the upper and lower bounds of the fuzzy domain U.

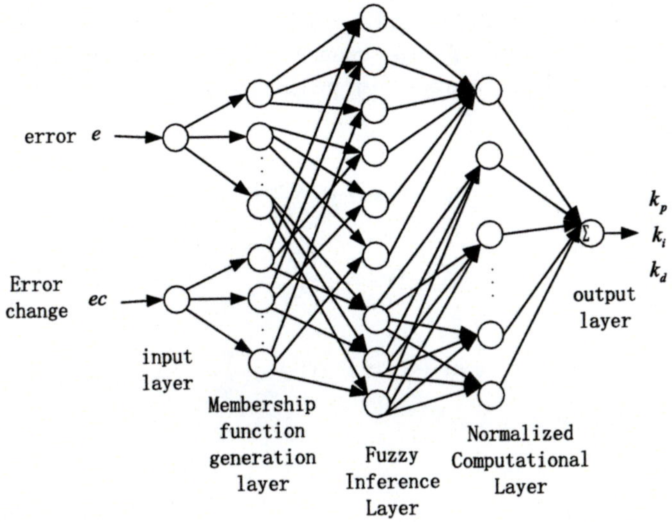

Fig. 8 BPFNN schematic diagram

3.3 BP Fuzzy Neural PID Control

Backward Propagation Fuzzy Neural Network [8]: Backward Propagation Fuzzy Neural Network (BPFNN) is a system with self-learning capability for fuzzy concepts created by fusing fuzzy and neural network techniques. Figure 5 shows the schematic diagram of BPFNN (Fig. 8).

In the orbital bolt wrench speed control system covered in the article, the analysis combined with Fig. 5 shows that the first layer has two inputs: corresponding to the deviation e and the deviation variation ec of the [17–19] control system, both inputs are fuzzily partitioned into seven linguistic variable values, corresponding to seven fuzzy subsets. The number of nodes in the second, third, and fourth layers are 14, 49, and 49; the third layer corresponds to the 49 fuzzy rules mentioned above; the fifth layer has 3 nodes, which corresponds to the three parameters of the PID control.

The BPFN-PID control schematic is displayed in Fig. 6.

The affiliation function of BPFNN $A_j^i(x_i)$ is

$$A_j^i(x_i) = \exp(-\frac{(x_{ij} - c_{ij})^2}{\sigma_{ij}^2}) \tag{16}$$

where $i = 1, 2; j = 1, 2 \cdots, 7$.

The PID parameters are modified, using the target cost function as

$$\varepsilon = \frac{1}{2} error^2(t) = \frac{1}{2}(r_{in}(t) - y_{out}(t))^2 \tag{17}$$

where ε is the target cost function and $error(t)$ is the control deviation.

The parameters to be trained are the connection rights of the output layer ω_{kl} (k = 1, 2 ,3; l = 1, 2..., 49), the centroid of the generating layer of the affiliation function C_{ij} and the width σ_{ij}.

$$\omega_{kl}(t+1) = \omega_{kl}(t) - \eta_1 \frac{\partial \varepsilon}{\partial \omega_{kl}} \tag{18}$$

$$c_{ij}(t+1) = c_{ij}(t) - \eta_2 \frac{\partial \varepsilon}{\partial c_{ij}} \tag{19}$$

$$\sigma_{ij}(t+1) = \sigma_{ij}(t) - \eta_3 \frac{\partial \varepsilon}{\partial \sigma_{ij}} \tag{20}$$

4 Simulation Results and Analysis

The specific parameters of BLDCM used for the bolt wrench speed control systems of the rail bolt working machine are displayed in Table 1.

In the bolt wrench control system of rail bolt handling machine, BLDCM needs to be combined with a reducer [20]. The BLDCM is used to reduce the speed and increase the torque.

Reduction gear ratio:

$$I = \frac{n}{n'} \tag{21}$$

where n is the input speed and n' is the output speed.

Reducer output torque:

$$T_{out} = 9549 \times \frac{P_{out}}{n} \times I \times fB \tag{22}$$

Table 1 Specific parameter table

Parameter name	Parameter value
Rated voltage (V)	220
Rated current (A)	2.04
Rated speed ($r/\min$)	1500
Stator winding line resistance (Ω)	2.88
Stator winding wire inductance (H)	0.008
Polar logarithm	5

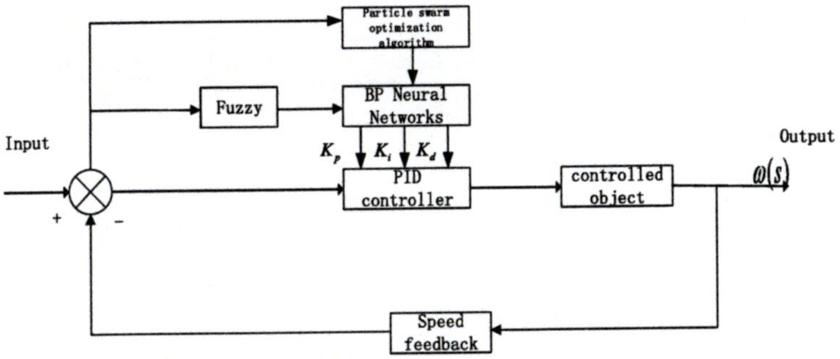

Fig. 9 BPFN-PID control schematic

where P_{out} is the motor power; *n* is the input speed; *I* is the transmission ratio; and *fB* is the usage factor.

The transmission ratio of the speed reducer used in the bolt wrench speed control system of the rail bolt working machine is 15 : 1, the usage factor is 1.2, and the transmission efficiency is 80% (Fig. 9).

According to the railway industry standard, the present parameters of the bolt wrench of the track bolt work machine are displayed in Table 2.

The speed control system of orbital bolt wrench designed according to the basic parameter table of bolt wrench is divided into three gears, which correspond to the recognition period, fitting period, and tightening period of bolt tightening process. The longest stroke is the recognition period, the speed is high speed, the speed in the fitting period is medium speed, and the speed in the tightening period requires tightening accuracy, and the speed is low speed. The three extremes correspond to specific rotational speeds as follows:

(1) The motor in the recognition period is in high-speed gear with a motor speed of 25 r/s.
(2) The motor in the laminating period is in medium-speed gear with a motor speed of 15 r/s.
(3) The motor is in low-speed gear during the tightening period, and the motor speed is 6.25 r/s.

Table 2 Table of basic parameters of bolt wrench

Projects	Technical parameters A	Technical parameters B
Sleeve no-load speed	≥ 70r/min	≥ 70 r/min
Tightening torque adjustment range	80N-m-150N-m	80N-m-150N-m
Spinning and loosening torque	300N-m	≥ 300N-m
Torque accuracy	± 10%	± 5%
Host quality (fuel not included)	≤ 120 kg (double sleeve) ≤ 95 kg (single sleeve)	≤ 120 kg (double sleeve) ≤ 95 kg (single sleeve)

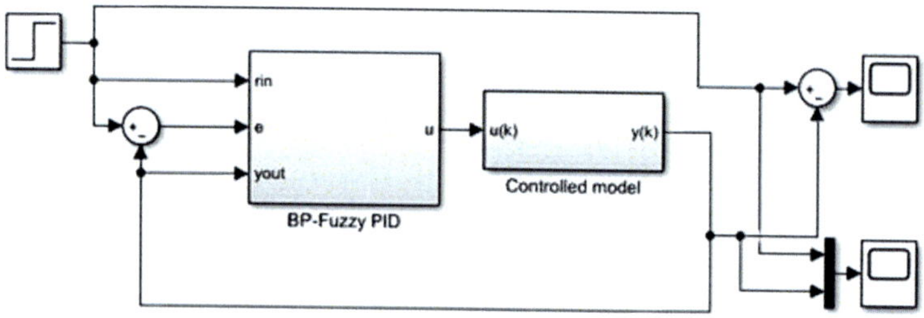

Fig. 10 BPFN-PID control simulation diagram

Simulation tests are performed in MATLAB, and Fig. 10 shows the BPFN-PID control simulation.

For the sake of simulating the actual control effect and ensure the validity of the simulation experiment, the simulation conditions are set as follows: the simulation time is 3 s, the initial speed is stepped from 0 to 25 r/s in the recognition phase, at 1.75 s, the speed is stepped from 25 r/s to 15 r/s in the fitting phase, and at 2.5 s, the speed is stepped from 15 r/s to 6.25 r/s. For the sake of test and verifying the response characteristics of the BPFN-PID control strategy, the three control strategies are compared. The simulation results are displayed in Fig. 11.

Table 3 shows the time taken by the three control strategies to reach steady state in the three periods of bolt tightening, and Table 4 shows the overshoot of the three control strategies in the three periods of bolt tightening.

In the recognition period, the conventional PID control strategy was stable at 0.2452 s with an overshoot of 2.875% overall; the fuzzy PID control strategy was stable at 0.2356 s with an overshoot of 1.552% overall; and the BPFN-PID control strategy was stable at 0.1413 s with an overshoot of 0.916% overall.

Table 3 Table of steady-state elapsed time

Assembly Control strategy	Silk recognition period	Lamination period	Tightening period
General PID	0.2452 s	0.2007s	0.2137 s
Fuzzy PID	0.2356 s	0.1894s	0.1735s
BPFN-PID	0.1413 s	0.1251 s	0.1097 s

Table 4 Overshoot volume table

Assembly Control strategy	Silk recognition period	Lamination period	Tightening period
General PID	2.875%	1.695%	1.703%
Fuzzy PID	1.552%	1.294%	1.263%
BPFN-PID	0.916%	0.427%	0.368%

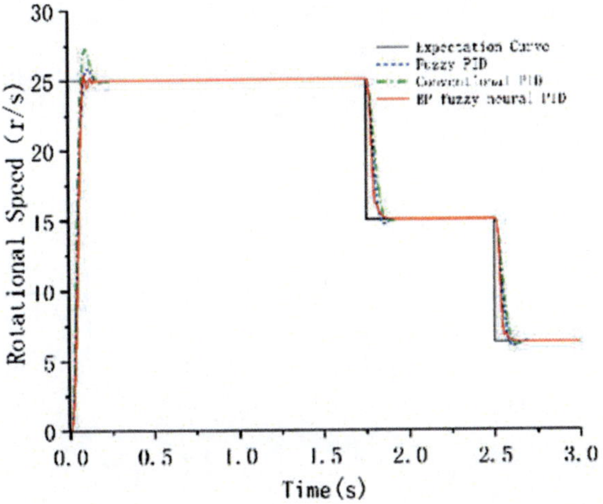

Fig. 11 Comparison of three control strategies

In the fitting phase, the speed input signal suddenly dropped from 25 r/s to 15 r/s, and the conventional PID control strategy stabilized after 0.2007s with an overall overshoot of 1.695%; the fuzzy PID control strategy stabilized after 0.1894s with an overall overshoot of 1.294%; and the BPFN-PID control strategy stabilized after 0.1251 s with an overall overshoot of 0.427%.

In the tightening period, the speed input signal suddenly dropped from 15 r/s to 6.25 r/s, and the conventional PID control strategy stabilized after 0.2137 s with an overshoot of 1.703% overall; the fuzzy PID control strategy stabilized after 0.1735s with an overshoot of 1.263% overall; and the BPFN-PID control strategy stabilized after 0.1097 s with an overshoot of 0.368%.

Figure 12 shows the speed error response curve during the adjustment of BPFN-PID control strategy. For the sake of improving the efficiency, the fast and then slow tightening speed is often used in the three periods of bolt assembly, which improves the assembly efficiency, but the response speed and control accuracy are often reduced. The BPFN-PID control strategy proposed in the paper can be seen from the curves in the figure that it can respond quickly and reach stability in a short time when the speed switches between each two assembly periods, from the start-up period, the wire recognition period to the lamination period, and from the lamination period to the tightening period (Fig. 12).

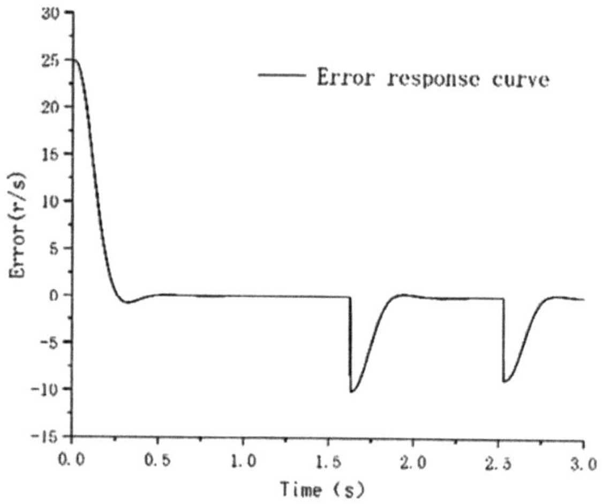

Fig. 12 Error response curve

5 Conclusion

For the sake of improving the performance of the system without reducing the efficiency of the rail bolt working machine, the article proposes to divide the bolt assembly into wire recognition period, fitting period, and tightening period, and the three periods adopt fast and then slow tightening speed, using BPFN-PID control strategy to represent the fuzzy control with the architecture of BP neural network and using 49 nodes of inference layer to formulate the 49 rules of fuzzy control can be optimized in real time by training the central value c_{ij} and width σ_{ij} to optimize the affiliation function, which makes up for the drawback that the affiliation function cannot be modified once it is determined in fuzzy control. Through the final simulation results, we know that the BPFN-PID control strategy has the fastest response, the smallest overshoot in each period, and the highest efficiency during the speed switching of the three assembly periods of the bolt wrench speed control system of the rail bolt work machine, which can well meet the accuracy and efficiency requirements of the rail bolt work machine.

References

1. Liu, J.D.: Research and design of measurement and control system for autonomous operational rail bolt work machine. Lanzhou Jiaotong University (2018)
2. Jianliang, L.: Research on the control technology of thread tightening torque. Zhejiang University (2019)
3. Chen, J.J., Xie, F., Wang, S.W., et al.: Design of bolt wrench torque control system based on fuzzy adaptive PID control algorithm. Mod. Manuf. Eng. **03**, 113–119 (2021)
4. Lei, Z., Deli, Z., Xuan, L., et al.: Study on control of screw screwing process based on permanent magnet synchronous motor drive . Mech. Electr. **37**(03), 62–67 (2019)

5. Englert, T., Graichen, K.: Nonlinear model predictive torque control and setpoint computation of induction machines for high performance applications . Control. Eng. Pract. **99**(5), 104415 (2020)
6. Klecker, S., Hichri, B., Plapper, P.: Robotic trajectory tracking: Bio-inspired position and torque control . Procedia CIRP **88**, 618–623 (2020)
7. Ammar, A., Talbi, B., Ameid, T., et al.: Predictive direct torque control with reduced ripples for induction motor drive based on T-S fuzzy speed controller . Asian J. Control **21**(4), 2155–2166 (2019)
8. Zhang, Z.-X.: Neural network control with MATLAB simulation. Harbin Institute of Technology Press, Harbin (2011)
9. Liu, J.: RBF neural network adaptive control MATLAB simulation. Tsinghua University Press, Beijing (2014)
10. Gu, D.Y., Wu, C.S., Hou, J.: BLDCM servo control based on compensated fuzzy neural network . J.Northeastern Univ. (Natural Science Edition) 34(01), 13–16 (2013)
11. Chen, C., Ma, T., Jin, H., et al.: Torque and rotational speed sensor based on resistance and capacitive grating for rotational shaft of mechanical systems. Mech. Syst. Signal Process. **142**, 106737 (2020)
12. Heng, X.: Flourishment of CVT electric fuel pump controller for hybrid vehicles. Hunan University (2017)
13. Jiancheng, T.: Permanent magnet brushless DC motor technology. Machinery Industry Press, Beijing (2011)
14. Li Dongwu, Xu Chao, Li Ruozhang et al. Contact parameters evolution of bolted joint interface under transversal random vibrations. Wear 500–501 (2022)
15. Jinkun, L.: Advanced PID control and its MATLAB simulation. Electronic Industry Press, Beijing (2011)
16. Liu, M., Song, H.: Application of adaptive fuzzy PID intelligent controller in brushless DC motor speed control system . Motor Control Appl. **39**(11), 22–25 (2012)
17. Liu, B., Guo, H.-X.: MATLAB neural network super learning handbook. People's Post and Telecommunications Publishing House, Beijing (2014)
18. Yu-Fang, S.: Design of column-mounted reactive power compensation device based on fuzzy neural logic. Zhongnan University (2014)
19. Xiaojing, W., Meizhen, L., Shuai, C., et al.: BP neural network predictive fuzzy control of hydraulic motor performance . Control. Eng. **27**(08), 1394–1400 (2020)
20. Xiaofeng, Z., Chunjiang, Z., Chuanqiang, Z., et al.: Flourishment of backlash detection device and analysis of measurement system for planetary reducer. Mech. Eng. 39(02), 238–243+275 (2022)

An MPC-Based Drifting Control System for Collision Avoidance in Autonomous Vehicles

Ran Duo, Cheng Hu, Xiaoling Zhou, Yu Qi, Lei Xie, and Honeye Su

Abstract For high-speed vehicles, sudden obstacles in front of them will put the vehicles in a dangerous situation. Drivers usually turn sharply and roll over because they cannot brake successfully in a short distance. In professional racing, racers often achieve the purpose of fast turning by actively controlling wheel slip named drifting. So it is feasible to use this agile maneuvering to avoid an emergency collision. This paper presents a control system using drifting equilibrium for collision avoidance based on model predictive control (MPC). The proposed control system can decide to drift when the car cannot avoid a collision at the minimum turning radius. Simultaneously, the control system can track the generated reference trajectory and drifting equilibrium. Then, the control system can make the vehicle quit the drifting condition and return to the original route. It acts successfully as a sudden obstacle that occurs when the vehicle speed is near 75 km/h in simulation.

1 Introduction

Ensuring the safe driving of autonomous vehicles is a critical factor for the sustainable development of automated driving. However, traditional methods cannot work well in making vehicles stable in high-speed situations. For this reason, the current vehicle anti-lock braking system (ABS) and the design principles of an automatic driving system were applied to autonomous vehicles in the past. The conventional techniques cannot adjust the situations like suddenly appearing obstacles. The obstacles can be animals or cargo dropped from forwarding vehicles, or even humans. So the main safety challenges for autonomous vehicles can be the vehicles running at high speed. Utilizing the drifting behavior in high-speed autonomous vehicles can successfully prevent the vehicles from a rollover in sharp turns or high-speed collision avoidance because of their low turning radius.

R. Duo (✉) · C. Hu · X. Zhou · Y. Qi · L. Xie · H. Su
State Key Laboratory of Industrial, Control Technology, ZheJiang University, Hangzhou, China
e-mail: 22060425@zju.edu.cn

K. Nakamatsu et al. (eds.), *Advanced Intelligent Technologies for Information and Communication*, Smart Innovation, Systems and Technologies 365,
https://doi.org/10.1007/978-981-99-5203-8_36

Most previous collision avoidance methods only concentrate on low-speed running and cannot solve the collision avoidance problem at high-speed running. We designed an MPC controller and verified its effectiveness through simulation to solve the emergency collision avoidance when sudden obstacles appeared.

The contributions are below:

(1) We designed a control system that can act as collision avoidance successfully when a sudden obstacle occurs.
(2) The control system includes decision-making part that meets the requirement of collision avoidance.
(3) The control system uses MPC controller to act collision avoidance when using the drifting maneuver to avoid collision and returns to the original path after collision avoidance.

Section 2 introduces previous research about collision avoidance, including control strategy and planning strategy. Section 3 presents the vehicle dynamics. Section 4 presents the whole control system. Section 5 reveals the simulation results and Sect. 6 discusses the conclusion.

2 Related Work

The work summary is mainly formed of three parts, trajectory planning for collision avoidance, the control strategy for collision avoidance, and drifting equilibrium tracking.

The early research mainly focused on the trajectory planning algorithm based on search, and the atlas method and interpolation method are the two usual methods in search algorithms. The main idea of the atlas method is to describe the traffic environment by the regional map composed of points or grids, and the non-passable area and passable area are divided. In [1], a method of performing an incremental search in feasible dynamic space is proposed, but it cannot work well in a complex environment because of extensive computation. So, the interpolation method is proposed to solve the problem of significant calculation. In [2], aiming at the issue of low production efficiency of industrial robots, the parameters of the artificial potential field are optimized by differential evolution algorithm to improve the collision avoidance performance of the potential synthetic field. In [3], the obstacle avoidance safety model is constructed through analyzing the behavior of human drivers. The neural network algorithm is presented in [4]. The outstanding advantages of the neural network are strong learning ability. It can plan through learning prediction samples, but the actual traffic environment is very complex, so the illustrations themselves have limitations.

At the tracking horizon, Levinson et al. [5] proposed a controller which can continuously select forward torque, brake, and steering wheel angle, maximize driving comfort and minimize trajectory tracking error to meet the requirements of obstacle avoidance in an emergency. Falcone et al. [6] proposed a model predictive control-based controller, using combined braking and steering. Brown et al. [7] proposed an alternative control framework combining local path planning and path tracking based on model predictive control. Ekim Yurtsever et al. [8] use the method of switching driving state and drift state to judge the obstacle avoidance of autonomous vehicles. Cheng et al. [9] proposed a longitudinal collision avoidance and lateral stability adaptive control system based on MPC.

During the cornering process, Gonzales et al. [10, 11] studied the professional drivers' behavior and proposed a mixed open-loop and closed-loop control strategy in order to achieve drifting maneuvering. Qi et al. [12] proposed an MPC controller to track both drifting and normal states. Gonzales et al. [13] achieved the drifting only through onboard sensors. Cutler [14] introduced a new strategy using reinforcement learning, tested it on a robotic car, and tested it successfully on autonomous drifting.

3 System Model

3.1 Dynamic Equations

We use the three-state vehicle model for steady-state turning and drifting and depict it in Fig. 1. Regardless of the model simpleness, the three-state model can efficiently express the vehicle's details. The states, V, β, and r, represent centroid velocity, centroid sideslip angle, and yaw rate. F_{yf} is the lateral force on the front wheel and F_{yr} is the lateral force on the rear wheel. The F_{xr} is the longitudinal force on the rear wheel and is also the driving force. a and b are the distances from the center of the vehicle to the front and rear wheels. δ is the steering angle. α_F and α_R explain the slip angle on the front and rear wheels.

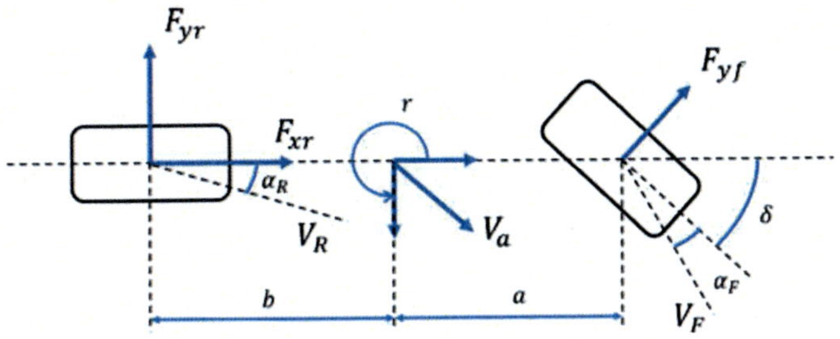

Fig. 1 Vehicle dynamic model when drifting

The basics of dynamic vehicle model are as follows:

$$\dot{V} = \frac{-F_{yf}\sin(\delta - \beta) + F_{yr}\sin\beta + F_{xr}\cos\beta}{m} \tag{1}$$

$$\dot{\beta} = \frac{F_{yf}\cos(\delta - \beta) + F_{yr}\cos\beta - F_{xr}\sin\beta}{mV} - r \tag{2}$$

$$\dot{r} = \frac{aF_{yf}\cos\delta - bF_{yr}}{I_z} \tag{3}$$

Additionally, m represents the mass of the vehicle and I_z represents the moment of inertia in the z-axis. The F_{yr} and F_{yf} can be calculated from tire magic model.

3.2 Tire Force Modeling

We can get F_{yf}, F_{yr} from the Pajecka model represented in [15]. The model is given in Eq. (4):

$$F_y = -\mu F_z \sin\left(C\tan^{-1}(B\alpha)\right) \tag{4}$$

The F_z is the normal load on the tire and the μ is the coefficient of friction. The front and rear tire slip angles are

$$\alpha_F = \tan^{-1}\left(\frac{V\sin\beta + ar}{V\cos\beta}\right) - \delta \tag{5}$$

$$\alpha_R = \tan^{-1}\left(\frac{V\sin\beta - br}{V\cos\beta}\right) \tag{6}$$

During the drifting, forces on the rear wheel reach saturation, so the lateral force of the rear wheel F_{yr} can be expressed by the friction circle in Eq. (7).

$$F_{yr} = \sqrt{(\mu F_{zr})^2 - (F_{xr})^2} \tag{7}$$

Table 1 Vehicle parameters in Carsim

Parameters	Value	Parameters	Value
m	1650 kg	μ	1
I_z	3234 kg m^2	B	11.3
a	1.4 m	C	1.53
b	1.65 m		

3.3 *Drifting Equilibrium*

The drifting equilibrium is presented in [16] by phase portrait of the specific equilibrium. The turning radius during the drifting maneuvering $R = V/r$. We can set the derivatives of vehicle states to zero and get the drifting equilibrium:

$$\dot{V} = 0 \quad \dot{\beta} = 0 \quad \dot{r} = 0 \tag{8}$$

The vehicle model parameters in the Carsim are given in Table 1.

3.4 *Linearization of Dynamic Model*

To achieve the goal of tracking reference trajectory, we additionally add two states, the lateral error e and the yaw error $\Delta\phi$, utilizing the curvature κ depicted in Fig. 2. And these two states can be expressed by Eqs. (9) and (10).

$$\dot{e} = V\sin(\psi + \beta) \tag{9}$$

$$\Delta\dot{\phi} = r - \kappa \cdot \frac{V\cos(\psi + \beta)}{1 - e \cdot \kappa} \tag{10}$$

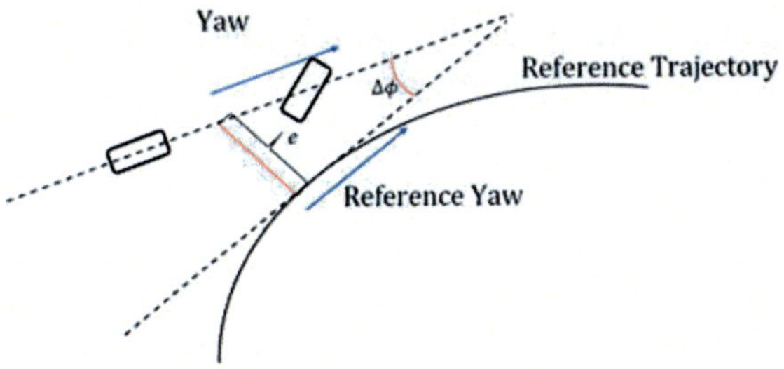

Fig. 2 The lateral error e and the yaw error $\Delta\phi$

The vehicle model is linearized to simplify the nonlinear problem and decrease the computing time. The original nonlinear vehicle model is linearized around an equilibrium point to establish the dynamic error model. Equations (1–3), (9, 10) are linearized as follows:

$$x(k+1) = Ax(k) + Bu(k) + D$$

$$D = x^{eq} - Ax^{eq} - Bu^{eq} \tag{11}$$

where

$$x(k) = \begin{bmatrix} e(k) \\ \Delta\phi(k) \\ V(k) \\ \beta(k) \\ r(k) \end{bmatrix}, u(k) = \begin{bmatrix} \delta(k) \\ F_{xr}(k) \end{bmatrix}$$

Then the discrete state-space function can be denoted by

$$\widehat{\xi}(k+1) = \widehat{A}\widehat{\xi}(k) + \widehat{B}\Delta u(k) + \widehat{d} \tag{12}$$

$$\widehat{\xi}(k+1) = \begin{bmatrix} x(k+1) \\ u(k) \end{bmatrix}, \widehat{A} = \begin{bmatrix} A & B \\ 0 & I \end{bmatrix}, \widehat{B} = \begin{bmatrix} B \\ I \end{bmatrix}, \widehat{d} = \begin{bmatrix} D \\ 0 \end{bmatrix}$$

where A and B, respectively, explain the Jacobian matrices computed at the equilibrium point.

4 Collision Avoidance Control System Design

The collision avoidance controller ensures the vehicle can track the drifting trajectory when a sudden obstacle occurs and return to its original path after avoiding the obstacle. Our control system is shown in Fig. 4.

4.1 Decision-Making Part

When a sudden obstacle occurs, the vehicle should make decision on collision avoidance timely. According to [17], the vehicle has an extensive minimum turning radius when the vehicle is in a high-speed condition. If the vehicle uses regular turning to avoid the collision, it will fail. So, the vehicle should utilize a drifting maneuver to avoid a collision, as shown in Fig. 3. At the rollover boundary condition, there is no force imposed on the wheels of one side. We assume that the left wheels have no force. We can analyze the torque balance of the vehicle on the x-axis. The equation can be expressed as Eq. (13). The L is the distance between the two rear wheels, the f is the friction, and the d is the distance between the centroid and the ground. The L' is the distance between front wheel and rear wheel (Fig. 4).

$$\mathrm{mg}\frac{L}{2} - fd = 0 \tag{13}$$

The friction f can be expressed as Eq. (14) and the turning angle can be expressed as Eq. (15). The above R and δ are the front wheel's minimum turning radius and maximum turning angle.

$$f = \frac{mv^2}{R} \tag{14}$$

$$\delta = \arctan\left(\frac{L\prime}{R}\right) \tag{15}$$

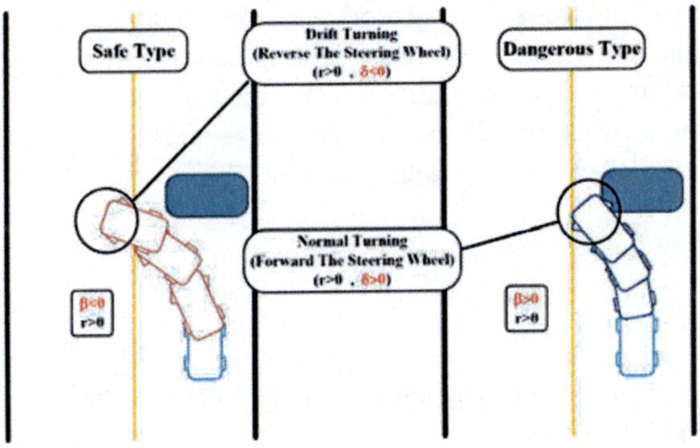

Fig. 3 Decision-making part

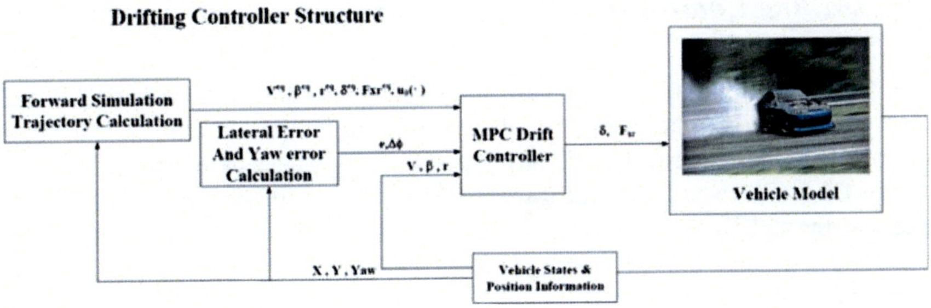

Fig. 4 Structure of controller

Based on the minimum turning radius and the maximum turning angle of the front wheel, the control system can analyze if the vehicle will avoid the collision successfully. If not, the control system will apply the drifting Maneuver to the collision avoidance.

4.2 Drifting Trajectory Generation

If the control system chooses drifting maneuver to avoid collision at the decision-making part, the drifting trajectory should be generated. Based on the dynamics of the vehicle discussed in Sect. 3, we can analyze the drifting equilibrium of the vehicle. From Eq. (1) to (3), there are five unknown factors and three equations, so we stabilize the current vehicle speed V as the speed of drifting condition and take the turning angle of the front wheel as searching factor. We set different turning angles as δ_i ($i = 1...N$), and get five factors from solving Eq. (8) referred to every δ_i. To achieve the drifting equilibrium, different factors can perform different trajectories. According to [10], the author presented LQR-based method to employ the vehicle to reach the drifting states based on the drifting equilibrium. We can apply different initial sequences (x_i^{eq}, u_i^{eq}) as the reference and calculate the input sequences $u_0(\cdot)$. For each δ_i, forward simulation trajectories can be calculated by $u_0(\cdot)$. Different trajectories can be calculated with the weighting factor based on Eq. (16). The H is the weighting factor, and it explains how far the trajectory is from the obstacle. The N' means the forward simulation trajectory is separated to N' points. The (X_{obs}, Y_{obs}) is the location of sudden obstacle.

$$H = \sum_{t=1}^{N\prime} (X_{obs} - X(t))^2 + (Y_{obs} - Y(t))^2 \tag{16}$$

After calculating the weighting factor, we will choose the sequences with the biggest weighting factor as the N horizon's reference of the control part.

4.3 Drifting Controller

MPC regards drifting states stabilization and path tracking as optimization goals. Then, the MPC controller takes vehicle dynamics as constraint, and we can make the optimization of the cost function a QP problem. The states of the dynamic model express that MPC must follow the drifting states V, β, r and keep tracking the reference path through the e and $\Delta\phi$. The e^{eq} and $\Delta\phi^{eq}$ are zero in order to track reference path. The sequences (V^{eq}, β^{eq}, r^{eq}, δ^{eq}, F_{xr}^{eq}) are calculated by forward simulation trajectory generation.

The cost function of the MPC controller occurred as

$$\min_{\Delta u(1|k),\ldots,\Delta u(N|k)} \sum_{i=2}^{N+1} ||x(i|k) - x^{\text{eq}}||_Q^2 + \sum_{i=1}^{N} ||\Delta u(i|k)||_R^2$$

st.

$$\hat{\xi}(k+1) = \hat{A}\hat{\xi}(k) + \hat{B}\Delta u(k) + \hat{d}$$

$$\Delta u_{\min} \leq \Delta u(k) \leq \Delta u_{\max}$$

The R and Q are the weighting matrices of each segment in the cost function. The primary goal of the first term is that the states should follow the drifting equilibrium point. The second term is to smooth the control inputs. At each time instant, Eq. (17) expresses the state-space function from k to $N + k - 1$.

$$\Xi_k = \Phi_k\hat{\xi}(k) + \Psi_{u,k}\Delta U(k) + \Psi_{d,k}d_k \tag{17}$$

The $\Xi_\mathbf{k}$, $\Delta U(k)$, d_k are given below:

$$\Xi_k = \left[\hat{\xi}(k)^T, \hat{\xi}(k+1)^T, \ldots, \hat{\xi}(k+N-1)^T\right]^T$$

$$\Delta U(k) = \left[\Delta u(k)^T, \Delta u(k+1)^T, \ldots, \Delta u(k+N-1)^T\right]^T$$

$$d_k = \left[\hat{d}(k)^T, \hat{d}(k+1)^T, \ldots, \hat{d}(k+N-1)^T\right]^T$$

The matrices Φ_k, $\Psi_{u,k}$, $\Psi_{d,k}$ are given below:

$$\Phi_k = \begin{bmatrix} \hat{A}(k) \\ \hat{A}(k+1) \\ \vdots \\ \prod_{i=k}^{k+N-1} \hat{A}(i) \end{bmatrix}$$

$$\Psi_{u,k} = \begin{bmatrix} \hat{B}(k) & \cdots & 0 \\ \hat{A}(k+1)\hat{B}(k) & \cdots & 0 \\ \vdots & \ddots & \vdots \\ \prod_{i=k}^{k+N-1} \hat{A}(i)\hat{B}(k) & \cdots & \hat{B}(k+N-1) \end{bmatrix}$$

$$\Psi_{d,k} = \begin{bmatrix} I & 0 & \cdots 0 \\ \widehat{A}(k+1) & I & \cdots 0 \\ \vdots & \vdots & \ddots \vdots \\ \prod_{i=k}^{k+N-1} \widehat{A}(i) & \prod_{i=k+1}^{k+N-1} \widehat{A}(i) & \cdots I \end{bmatrix}$$

The cost function can be transferred to

$$J = ||\Xi(k) - \boldsymbol{\Xi}_{\boldsymbol{k}}^{\boldsymbol{eq}}||_{Q'}^2 + ||\Delta U(k)||_{R'}^2$$

where

$$Q' = I_N \otimes Q$$

$$R' = I_N \otimes R$$

Then, the QP problem can be set as

$$J\left(\widehat{\xi}(k), \boldsymbol{\Xi}_{\boldsymbol{k}}^{\mathbf{eq}}, \Delta U(k)\right) =$$

$$||\Delta U(k)||_H^2 + 2f\left(\widehat{\xi}(k), \boldsymbol{\Xi}_{\boldsymbol{k}}^{\mathbf{eq}}\right)^T \Delta U(k) + g(\widehat{\xi}(k), \boldsymbol{\Xi}_{\boldsymbol{k}}^{\mathbf{eq}})$$

where

$$H = \Psi_{u,k}^T Q' \Psi_{u,k} + R'$$

$$f\left(\widehat{\xi}(k), \boldsymbol{\Xi}_{\boldsymbol{k}}^{\mathbf{eq}}\right) = \Psi_{u,k}^T Q'\left(\Phi_k \widehat{\xi}(k) + \Psi_{d,k}\boldsymbol{d}_{\boldsymbol{k}} - \boldsymbol{\Xi}_{\boldsymbol{k}}^{\mathbf{eq}}\right)$$

$$g\left(\widehat{\xi}(k), \boldsymbol{\Xi}_{\boldsymbol{k}}^{\mathbf{eq}}\right) = ||\Phi_k \widehat{\xi}(k) + \Psi_{d,k}\boldsymbol{d}_{\boldsymbol{k}} - \boldsymbol{\Xi}_{\boldsymbol{k}}^{\mathbf{eq}}||_{Q'}^2$$

Through the above transformation, the cost function of MPC can be transferred to a QP problem. Solving the QP problem can be faster than the normal optimal problem. In the simulation, we can calculate the QP problem through the Casadi toolbox [17] in MATLAB. The structure of drifting controller is above, and the reference of drifting controller can be calculated by the LQR-based forward simulation.

4.4 Exiting Part

After the vehicle successfully avoids the collision that a sudden obstacle will cause, the vehicle should exit the drifting state and return to its normal driving conditions. After drifting, the vehicle should quit the drifting condition, and the vehicle should reverse the steering wheel to stabilize the states away from floating out because of inertia. Utilizing the sideslip of the front wheel to the steering angle will make the sideslip of front wheel near zero and stabilize the vehicle quickly.

We form the open-loop control scheme as given below:

$$\delta \leftarrow \alpha_F, |r| \geq h_r \left| r^{\mathrm{eq}} \right|$$

The h_r is the weighting factor of the r^{eq}. It represents how loose the r should be away from r^{eq}. When the open-loop part works successfully, we use a simple lateral control based on MPC to make the vehicle return to original path.

5 Simulation Result

In Fig. 5, the comparison of collision avoidance using normal turning and drifting is revealed. The normal turning collision avoidance using minimum turning radius depicted in Sect. 4 fails with the minimum turning radius equal to 5.73°, and the drifting turning collision avoidance succeeds. When drifting, the β and r have different signs.

We use the vehicle dynamic system simulation software to actuate our control system, verified in Carsim software. Parameters of the controller are as follows: $N = 20$, $T = 0.05\mathrm{s}$, $Q = \mathrm{diag}\,(1,1,1,1,1)$, $R = \mathrm{diag}\,(1,1)$, and $h_r = 0.5$. N is the prediction time domain and T is the sample time. Q is the weighting factor of the first item and R is the weighting factor of the second item in the cost function. h_r is the weighting factors of open-loop controller. The drifting equilibrium of the drifting part is given as $V^{\mathrm{eq}} = 20.8123\ \mathrm{m/s}$, $\beta^{\mathrm{eq}} = -0.2532$ rad, $r^{\mathrm{eq}} = 0.4888\ \mathrm{rad/s}$, and $\delta^{\mathrm{eq}} = -10°$, $F_{\mathrm{xr}}^{\mathrm{eq}} = 4130\ \mathrm{N}$.

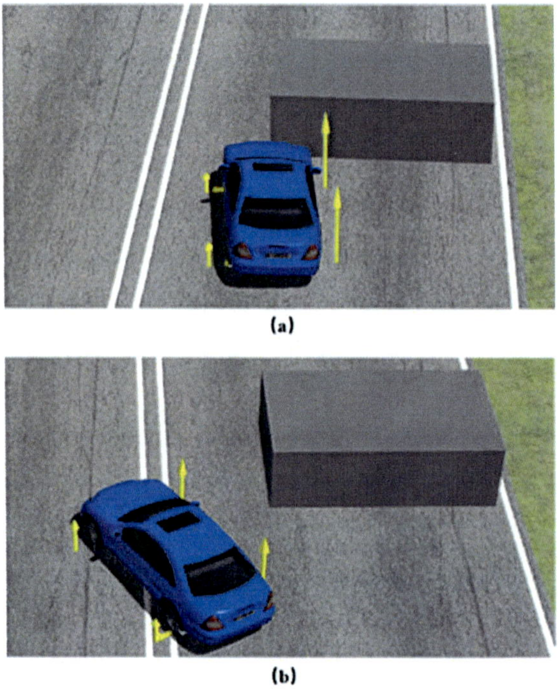

Fig. 5 Comparison of normal turning and drifting

5.1 Objective and Test Procedure

The collision avoidance control system should track the reference trajectory and return to the original path after collision avoidance. Figure 6 is the real X–Y-coordinates of the vehicle. Figure 7 depicts the lateral error of collision avoidance. The centroid speed V is near 75 km/h and met the obstacle at 10 s. The vehicle acts as collision avoidance after 10 s and returns to the original path after 11.3 s.

5.2 Results

In Fig. 6, the vehicle acts well in front of the suddenly occurred obstacle. The β and r have different signs, and it means the vehicle has entered the drifting condition. From the distribution of lateral error in Fig. 7, we can see that the largest deviation of the lateral error reaches -6 m, and it is within the limited deviation from a reference trajectory during drifting. During the time from 10 to 11.3 s, the reference path of the vehicle is the forward simulation drifting trajectory and is the original path at

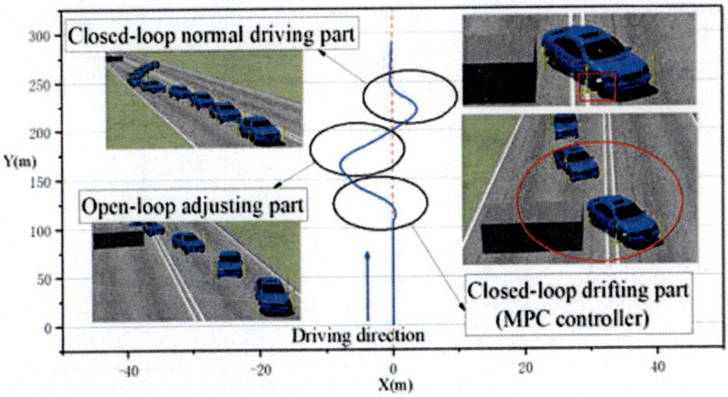

Fig. 6 Measured path in Carsim simulation

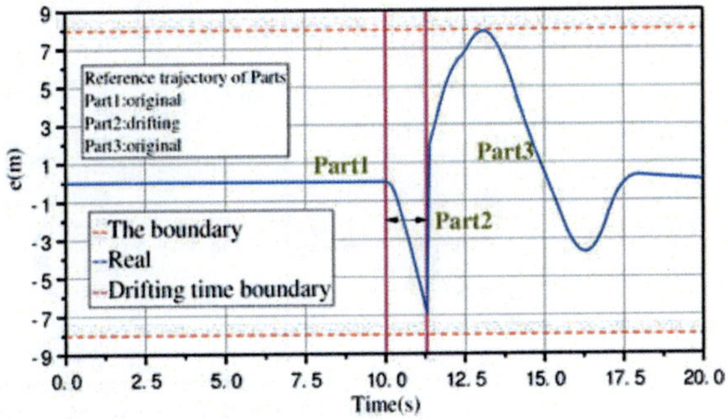

Fig. 7 Lateral error in Carsim simulation

other time. Although the lateral error grows during the drifting time, we only focus if the controller can avoid collision at a short time and the lateral error is within a reasonable range.

Figure 8 shows the vehicle's states and inputs during the collision avoidance time. Although the vehicle drifts for a short time, the controller can successfully track the reference drifting equilibrium.

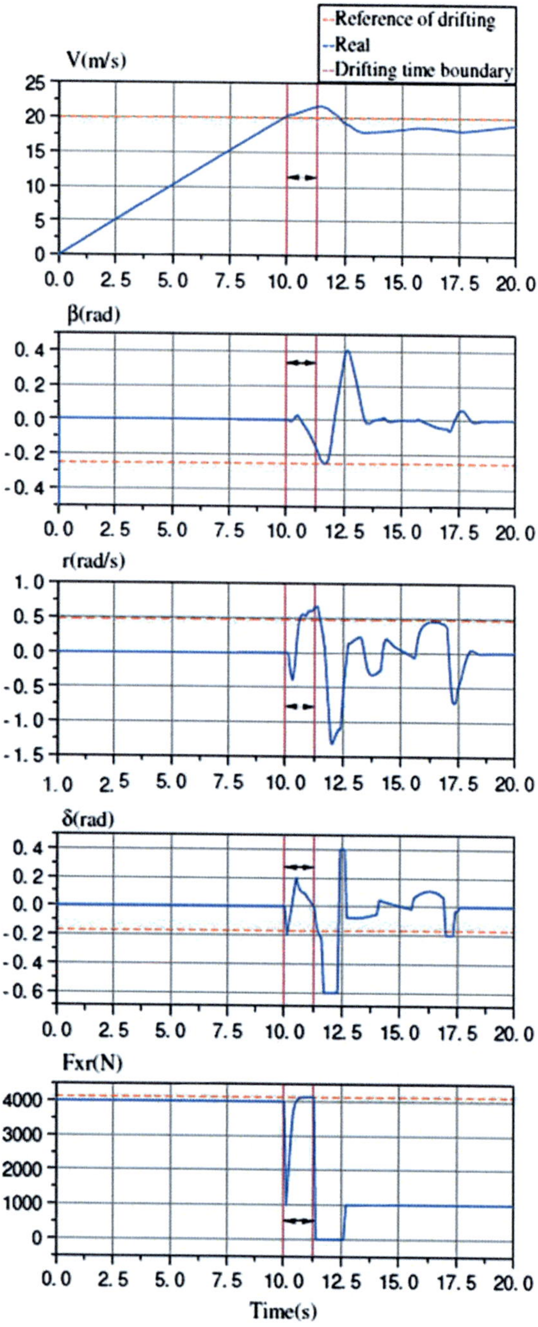

Fig. 8 States and inputs in Carsim simulation

6 Conclusion

This paper presents a control system that applies drifting to high-speed collision avoidance and tracks the reference trajectory. When a sudden obstacle occurs, the decision-making part makes the decision of collision avoidance using drifting. Then, we use forward simulation trajectory and MPC controller to make the vehicle achieve drifting. After drifting, the control system can stabilize the vehicle and make the vehicle return to its original path. The simulation result reveals the effectiveness of our control system proposed. The vehicle can successfully avoid the collision and reach the drifting states and return to its original path.

We will employ our control system in real cars and develop real-time drifting trajectory planning in the future.

References

1. Likhachev, M., Ferguson, D.: Planning long dynamically-feasible maneuvers for autonomous vehicles. Int. J. Robot. Res. (2009)
2. Lazarowska, A.: A discrete artificial potential field for ship trajectory planning. J. Navig. **73**(1), 233–251 (2020)
3. Wang, P., et al.: Obstacle avoidance path planning design for autonomous driving vehicles based on an improved artificial potential field algorithm. Energies **12**(12), 2342 (2019)
4. Gong, W., et al.: Palmprint recognition based on convolutional neural network-AlexNet. In: 2019 In: Federated Conference on Computer Science and Information Systems (FedCSIS). IEEE (2019)
5. Levinson, J., et al.: Towards fully autonomous driving: systems and algorithms. In: 2011 IEEE Intelligent Vehicles Symposium (IV). IEEE (2011)
6. Falcone, P., et al.: A model predictive control approach for combined braking and steering in autonomous vehicles. In: 2007 Mediterranean Conference on Control & Automation. IEEE (2007)
7. Brown, M., et al.: Safe driving envelopes for path tracking in autonomous vehicles. Control Eng. Pract. **61**, 307–316 (2017)
8. Zhao, T., et al.: Collision avoidance with transitional drift control. In: 2021 IEEE International Intelligent Transportation Systems Conference (ITSC). IEEE (2021)
9. Cheng, S., et al.: Longitudinal collision avoidance and lateral stability adaptive control system based on MPC of autonomous vehicles. IEEE Trans. Intell. Transp. Syst. **21**(6), 2376–2385 (2019)
10. Zhang, F., et al.: Drift control for cornering maneuver of autonomous vehicles. Mechatronics **54**, 167–174 (2018)
11. Zhang, F., et al.: Autonomous drift cornering with mixed open-loop and closed-loop control. IFAC-PapersOnLine **50**(1), 1916–1922 (2017)
12. Qi, Y., et al.: An MPC-based controller framework for agile maneuvering of autonomous vehicles. In: 2021 IEEE Intelligent Vehicles Symposium (IV). IEEE (2021)
13. Gonzales, J., et al. Autonomous drifting with onboard sensors. In: Advanced Vehicle Control AVEC'16. CRC Press, pp. 133–138 (2016)
14. Cutler, M., How, J.P.: Autonomous drifting using simulation-aided reinforcement learning. In: 2016 IEEE International Conference on Robotics and Automation (ICRA). IEEE (2016)
15. Bakker, E., Nyborg, L., Pacejka, H.B.: Tyre modelling for use in vehicle dynamics studies. SAE Trans. 190–204 (1987)

16. Hindiyeh, R.Y., Gerdes, J.C.: A controller framework for autonomous drifting: design, stability, and experimental validation. J. Dyn. Syst. Meas. Control **136**(5), 051015 (2014)
17. Qian, K.G.: Research on mechanical conditions of vehicle rollover at corners. Adv. Mater. Res. **1046** (Trans Tech Publications Ltd.,) (2014)

Research on Image Denoising Algorithm Based on Edge Enhancement Sparse Transform and Low Rank

Yicong Chen, Xiangwei Huang, Cuixiang Liu, and Qianqian Shan

Abstract Focused on the issue that detailed information of edge is easily lost in the process of image denoising, the image denoising algorithm based on edge enhancement sparse transform and low rank was proposed. Firstly, Canny algorithm was used to gain the image boundary to obtain the edge matrix, and secondly, the Non-subsampled contourlet transform (NSCT) was used to gain the multi-directional high-frequency sub-band map, and then the edge matrix was used to locate the edge position in the high-frequency sub-map. The edge sub-band coefficients were enlarged to enhance the edge to form the edge-enhanced noise image. Finally, the image was divided into image patches, then the local sparsity of image patches to sparse transform and non-local self-similarities of image patches to block match were used to achieve the purpose of image denoising. The experimental results show that, compared with K-Singular Value Decomposition (K-SVD), Group Low-Rank (GLR), Sparsifying Transform Learning and Low Rank (STROLLR), and DnCNN algorithms under different levels of noise, the proposed algorithm not only has a large improvement in the Peak Signal-to-Noise Ratio (PSNR), but also maintains clear edge details in visual effects while reducing noise.

1 Introduction

Noise of digital images is usually created during image acquisition or transmission. Noisy images will directly affect the performance of vision systems in image processing and understanding [1]. There are many types of noise, such as salt, pepper noise, and gaussian noise. The goal of image denoising is to gain a clean image X from a noisy image Y while preserving the details of the original image. The denoising problem is formulated as $Y = X + N$, where N is gaussian noise with mean zero and

Y. Chen (✉) · C. Liu · Q. Shan
School of Information Engineering, Hebei University of Technology, Tianjin, China
e-mail: cyc_1205_hebut@163.com

X. Huang
Glasgow College, University of Electronic Science and Technology of China, Chengdu, China

K. Nakamatsu et al. (eds.), *Advanced Intelligent Technologies for Information and Communication*, Smart Innovation, Systems and Technologies 365,
https://doi.org/10.1007/978-981-99-5203-8_37

standard deviation σ. The noise is additive noise, that is, the noise is added to the image indirectly. Therefore, if we want to get a clean image accurately, it is conceivable that it can be achieved by subtracting the noise part from the noise image. If we want to extract the noise, we need to know the source of the noise very accurately, but the reality is not easy to achieve [2].

Mallat and Zhang first proposed the concept of sparse representation based on overcomplete dictionary, entering the era of sparse representation for the first time in 1993. Elad et al. [3] proposed denoising algorithm of K-SVD, but it has 0-norm non-convexity and high computational complexity. In order to solve large-scale non-convex problems, Ravishankar et al. [4] proposed denoising algorithm for sparse transform learning, which uses sparse transform matrix to approximately sparse the signal, reducing the amount of computation. Non-local self-similarity has become a key feature of natural images. The Block-Matching and 3D filtering (BM3D) proposed by Dabov et al. [5] combined non-local characteristics and collaborative filtering. Wen et al. [6] proposed STROLLR, but ignored the edge detail characteristics of natural images. Zhang et al. [7] proposed non-local self-similarity low-rank sparse image denoising, using the block matching method of Mahalanobis distance to group natural image blocks. Zhang et al. [8] proposed quadratic joint sparse representation and low-rank estimation bas-relief optimization, which optimized the image twice and improved the quality of the image. Qian et al. [9] considered the spatial structure characteristics of the image by adding graph Laplacian regularization to constrain between pixels, and achieved better denoising performance.

In view of the above research, we consider the edge blur problem in the process of image denoising and the non-local self-similarity of image blocks, and propose sparse transform and low-rank image denoising algorithm based on edge enhancement. The algorithm mainly consists of edge enhancement part and sparse transform and low-rank denoising part. The edge enhancement part uses the Canny algorithm to detect the edge, and the high-frequency sub-band image obtained by NSCT transform is used for edge coefficient enhancement to obtain the edge-enhanced noise image; the sparse transform and low-rank denoising part divides the image into image blocks, and combines the sparsity and low rank of the image blocks to denoise the image blocks, and finally reconstruct the denoised image. Experiments show that the algorithm retains more abundant edge information and achieves better denoising effect in terms of peak signal-to-noise ratio.

2 Proposed Method

The sparse transform and low-rank image denoising algorithm based on edge enhancement is shown in Fig. 1.The red box is the edge enhancement part. Canny is used for edge detection, and the multi-directional sub-band image of NSCT transform is used for edge enhancement to obtain an edge-enhanced noise image; the blue box is the image denoising part, which is performed by using the local sparsity

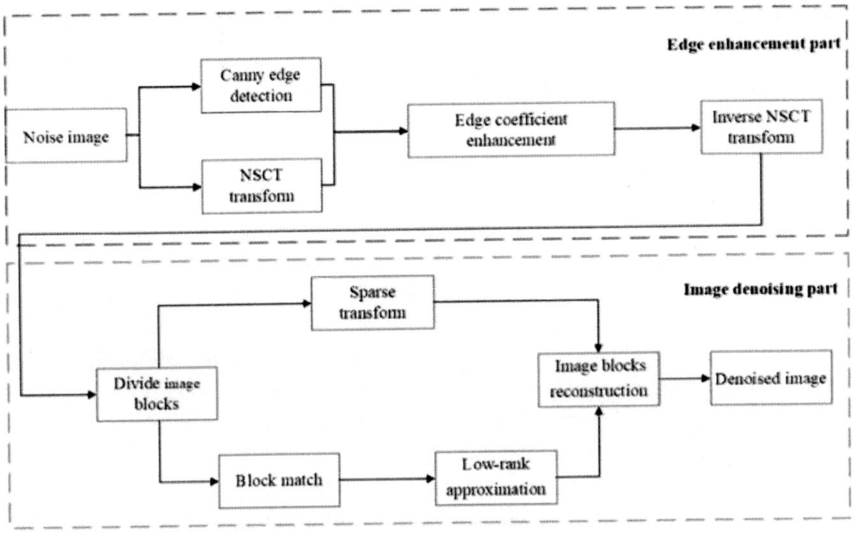

Fig. 1 Edge-enhancement sparse transform and low-rank image denoising model

of the image block. The sparse transform and non-local self-similarity are used for low-rank estimation to achieve the purpose of image denoising.

2.1 *Edge Enhancement Model*

Canny algorithm is an edge detection operator proposed by computer scientist John F. in 1986. It is now widely used in edge extraction of clean or noisy images. The processing steps of the algorithm are generally divided into five steps:

First step: Gaussian filter is used to smooth the image to eliminate noise interference. The size of the Gaussian kernel will affect the accuracy of edge extraction. Generally, a 3×3 or 5×5 Gaussian kernel is selected.
Second step: Calculate the gradient and direction of each pixel of the smoothed image. The direction of each pixel can be divided into horizontal and vertical components. By using Sobel or other operators to do convolution with the smoothed image, then we calculate the gradient of the pixel point in the horizontal direction and the vertical direction, thereby obtaining the gradient strength and gradient direction of the pixel point.
Third step: Perform non-maximum gradient suppression on each pixel of the image. By finding the local maximum point of the pixel gradient along the positive and negative directions, then we set the pixel as the edge point, and other non-polar values are removed. Large value points are suppressed to refine the edges.

Fourth step: Use dual thresholds to detect real and potential edges, set a maximum threshold and a minimum threshold. When the gradient intensity of the image pixel is higher than the maximum threshold, we set it as the strong edge of image, if it is less than the minimum threshold, it is discarded, and when it is between the maximum and maximum threshold, we set it as the weak edge of image.
Fifth step: Suppress weak edges to complete image edge extraction.

NSCT transform is a non-subsampled contourlet transform, which mainly consists of two parts: a Non-Subsampled Pyramid Filter Bank (NSPFB) for multi-scale decomposition and a non-subsampled directional filter for direction decomposition. The Non-subsampled Directional Filter Bank (NSDFB) has translation invariance, multi-scale, multi-direction, anisotropy, etc. The multi-scale and multi-directional sub-band images decomposed by NSCT transform have the same size as the original image, which can better represent the image edge and detailed texture information.

First, the low-frequency sub-band image and high-frequency sub-band image of the first layer of NSP scale decomposition are performed on the original image, and then the high-frequency sub-band image is multi-scale decomposed by NSDFB. After that, NSP is used to decompose the low-frequency sub-band images of the upper layer continuously, and then NSDFB is used to decompose the high-frequency sub-band images of this layer to gain direction sub-band images. The image is filtered through L layers of non-subsampling pyramid to obtain the low-pass sub-band image and $\sum_L 2^{m_i}$ direction sub-band images, where m_i is the number of directions for the decomposition of the i layer of band-pass sub-images, thus completing the NSCT transform decomposition of the image.

In order to solve blurred image edges in the process of image denoising, we first perform image edge enhancement before image denoising [10]. As shown in Fig. 1, the edge of the red box is enhanced. First, Canny edge detection is performed on the noise image. Secondly, the noise image is decomposed by NSCT to obtain multi-scale and multi-directional high-frequency sub-band coefficients, and finally the edge position of the high-frequency sub-band coefficients is obtained by using the edge coefficient matrix, and the corresponding edge sub-band coefficients are amplified to enhance edges. The specific steps of edge enhancement are as follows:

Step 1: Use the Canny algorithm to gain the image boundary, and obtain the edge coefficient matrix of the image, where the coordinate value is "0" position, indicating that the pixel is a non-edge, and the coordinate value is "1" position, indicating that the pixel is an edge.
Step 2: Use NSCT transform to decompose the noise image to gain the low-frequency sub-band image and the multi-directional sub-band image. The low-frequency sub-band image after NSCT transform includes the basic outline of the image, and there is almost no noise interference, while the corresponding high-frequency directional sub-band map includes edge, detail, noise, and other information. In order to enhance the edge of the image while suppressing the noise, the edge position of the high-frequency sub-band coefficients is obtained by using the edge coefficient matrix obtained in the first step, and the corresponding edge

sub-band coefficients are enlarged to enhance the edge, and the magnification in this paper is 1.2 times.

Step 3: After all multi-directional sub-band images are enhanced with edge coefficients, use NSCT inverse transform to reconstruct the low-frequency image and the sub-band images enhanced by edge coefficients, and finally obtain the edge-enhanced noise image, which is used as input for sparse transform and low-rank image denoising.

2.2 Sparse Transform and Low-Rank Denoising Model

For the noisy image $\mathbf{Y} \in R^{n \times N}$, the noise image is approximately sparsely decomposed $\mathbf{WY} = \mathbf{A} + e$ using sparse transform matrix $\mathbf{W} \in R^{m \times n}$, where $\mathbf{A} \in R^{m \times N}$ is the sparse coding matrix, $\mathbf{e}$ is the modeling error, and each column $y_i \in R^n$ in $\mathbf{Y} = [y_1, y_2, ..., y_N]$ represents a signal. Assuming that the denoised image is $\mathbf{X} \in R^{n \times N}$, that is, $\mathbf{X} = [x_1, x_2, ..., x_N]$, the denoised image can be calculated by $\|\mathbf{WX} - \mathbf{A}\|_F^2$.

In addition to the local sparsity of the image, the non-local similarity of the image is also exploited. Define block matching [11] operation V_i: a search box Q centered on the reference image block selects M vectors similar to the reference image blocks by calculating the Euclidean distance between the image blocks. The selected image blocks form the low-rank matrix $\mathbf{V_i Y}$ according to the size of the Euclidean distance from the reference image block.

Sparse transform and low-rank denoising model unites the adaptive sparse transform of image blocks and the low-rank nature of the data matrix composed of block matching, making good use of the local sparsity and non-local self-similarity between image blocks. As shown in Fig. 1, the blue box is the image denoising part. First, the noise image $\hat{\mathbf{Y}}$ with edge enhancement is divided into image blocks, and then sparse transform and block matching operations are performed, respectively, and finally the image block reconstruction is performed to obtain the denoised image, and the good denoising effect is obtained. The overall simplified model of the algorithm is as follows:

$$\begin{aligned} &\arg\min_{W,A,\{D_i,X\}} \|\mathbf{WX} - \mathbf{A}\|_F^2 + \gamma_s^2 \|\mathbf{A}\|_0 + \gamma_l \sum_{i=1}^{N} \{\|\mathbf{V_i X} - \mathbf{D_i}\|_F^2 + \theta^2 \mathrm{rank}(\mathbf{D_i})\} \\ &+ \gamma_f \sum_{i=1}^{N} \left\{ \left\| \hat{y}_i - x_i \right\|_2^2 \right\} \quad s.t. \mathbf{W^T W} = \mathbf{I_n} \end{aligned} \tag{1}$$

where γ_s, $\gamma_l, \gamma_f, \theta$ are regularization parameters; $\hat{\boldsymbol{Y}}$ is the noise image after edge enhancement; $\boldsymbol{X}$ is the denoised image; $\boldsymbol{D_i}$ is the low-rank estimate of block matching $V_i\hat{Y}$; rank(·) is the rank of the matrix; and $\boldsymbol{W}$ is the sparse transform matrix, which is identity matrix.

Our model is solved by an easy method that is the block coordinate descent to achieve image denoising. We solve this method by four procedures: (1) sparse transform, (2) renew transform, (3) low-rank estimation, and (4) image block reconstruction. When the step finishes, image denoising will be performed by performing (5) aggregation step. Specific procedures are as follows:

Sparse Transform. Given the image block and sparse transform matrix $\mathbf{W}$, to solve the problem of sparse coding,

$$\hat{\mathbf{A}} = \arg\min_{A} \|\mathbf{W}\mathbf{X} - \mathbf{A}\|_F^2 + \gamma_s^2 \|\mathbf{A}\|_0 \tag{2}$$

It is regarded as the sparse coding issue and the hard threshold $\mathbf{A} = H_{\gamma_s}(\mathbf{W}\mathbf{X})$ can be used to get the best sparse coding, where $H_{\gamma_s}()$ is written as $\left(H_{\gamma_s}(Q)\right)_{a,b} = \begin{cases} 0 & , \left|Q_{a,b}\right| < \gamma_s \\ Q_{a,b} & , \left|Q_{a,b}\right| \geq \gamma_s \end{cases}$.

Renew Transform. Optimize the sparse transform matrix $\hat{\mathbf{W}}$ with the fixed sparse coding A,

$$\hat{\mathbf{W}} = \underset{\mathbf{W}}{\operatorname{argmin}} \|\mathbf{W}\mathbf{X} - \mathbf{A}\|_F^2 \quad s.t. \, \mathbf{W}^{\mathbf{T}}\mathbf{W} = \mathbf{I_n} \tag{3}$$

The singular value decomposition (SVD) of $\mathbf{X}\mathbf{A}^T$ is denoted as $SVD\left(\mathbf{X}\mathbf{A}^T\right) = S\Sigma G^T$ under a single constraint and the final optimization result of the sparse transform matrix is $\hat{\mathbf{W}} = GS^T$.

Low-rank Estimation. Using the block matching operation, solve each low-rank matrix estimate $\mathbf{D_i}$ as

$$\hat{\mathbf{D}}_{\mathbf{i}} = \underset{D_i}{\operatorname{argmin}} \|\mathbf{V_i}\mathbf{X} - \mathbf{D_i}\|_F^2 + \theta^2 \operatorname{rank}(\mathbf{D_i}) \tag{4}$$

The similarity matrix is obtained using block matching in the search window centered on the reference image block, and the optimal solution is obtained as $\mathrm{SVD}(\mathbf{V_i}\mathbf{X}) = \Phi\Omega\Psi^T, \hat{\mathbf{D}}_{\mathbf{i}} = \Phi H_\theta(\Omega)\Psi^T$.

Image Blocks Reconstruction. Use fixed A, $\mathbf{W}$, and $\mathbf{D_i}$, and restore each image block $\hat{x}_i$:

$$\hat{x}_i = \underset{x_i}{\operatorname{argmin}} \|\mathbf{W}x_i - \alpha_i\|_2^2 + \gamma_f \left\|x_i - \hat{y}_i\right\|_2^2 + \gamma_l \sum_{j \in C_i} \left\|x_i - \mathbf{D_{j,i}}\right\|_2^2 \tag{5}$$

The first item is the constraint on the sparsity of the image block, the second item measures the degree of estimation between the denoised image block and the original image block, and the third item is the constraint on the low-rank estimation matrix of the image blocks. C_i is the column corresponding to the image block x_i in the low-rank estimation matrix. $\mathbf{D_{j,i}}$ is the columns corresponding to image block x_i in the low-rank estimation matrix $\mathbf{D_j}$.

Aggregation. Aggregate the reconstructed image patches to obtain the final denoised image.

$$\hat{x}_i = \left(\mathbf{W}^T\alpha_i + \gamma_f\hat{y}_i + \gamma_l\sum_{j\in C_i}\mathbf{D_{j,i}}\right)/(1 + \gamma_f + |C_i|\gamma_l) \quad (6)$$

3 Experimental Results and Analysis

3.1 *Experimental Parameter Settings and Datasets*

The image block size is $n = 8$ (the number of overlap between blocks is 1), the number of block matches is $M = 45$, the search window size is $Q = 40$, the edge enhancement coefficient is $\hat{e} = 1.2$, and the initial sparse transform matrix $\mathbf{W}_O$ is discrete cosine transform matrix. Regularization parameter settings $\gamma_f = 1/\sigma$, $\gamma_s = 2.5\sigma$, $\theta = 1.5\sigma$, and $\gamma_l = 1 \times 10^{-2}\sigma$, where σ is the noise variance and e is the edge coefficient of the high-frequency sub-band pattern.

In the experiment, we selected six images of Barbara, Lena, House, Cameraman, Coins, and Moon as test images for visual effect and quantitative analysis. In order to further verify the performance of the algorithm, quantitative analysis experiments were carried out using the datasets BSD68, which were commonly used in digital image processing. The BSD68 dataset consists of 68 grayscale images of different sizes.

3.2 *Result and Analysis*

The proposed method was combined with K-SVD [3], GLR [11], STROLLR [6], DnCNN [12] algorithms for image denoising comparison experiment. First of all, with the purpose of measuring the visual effect of image denoising, Gaussian noise $\sigma = 20$ was added to the image of Barbara. Figure 2 shows the image denoised result by K-SVD, GLR, STROLLR, DnCNN, and our algorithm. So as to further show the details of the image, we select some details in the red frame of the Barbara to zoom in and display it in the lower right corner of the image, compared with other algorithms, our method can obtain clear edge and detailed information. Specifically, from Fig. 2b–e, the texture on the pants of Barbara image is blurred, over-smoothed, shadows, and no obvious outline details, while Fig. 2f Barbara image preserves the texture on the pants.

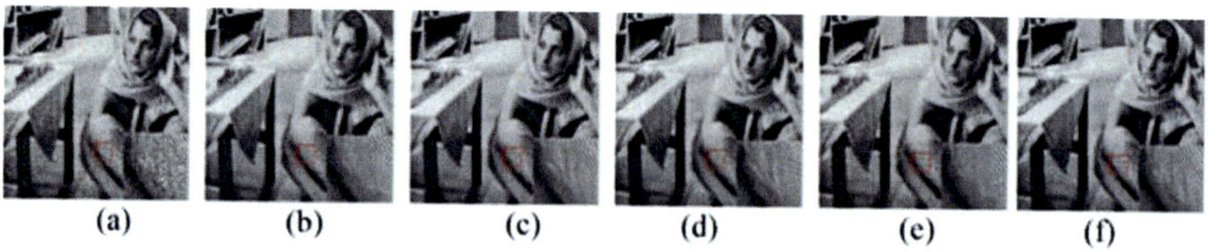

Fig. 2 Image Barbara denoising results in different algorithms under noise $\sigma = 20$, **a** noise image, **b** K-SVD, **c** GLR, **d** STROLLR, **e** DnCNN, **f** ours

With the purpose of further verifying the denoising performance of this algorithm, we added the six images of Barbara, Lena, House, Cameraman, Coins, and Moon, respectively, to the Gaussian white noise of and denoising in different algorithms for quantitative analysis. We use Peak Signal-to-Noise Ratio (PSNR) to determine image quality. Table 1 lists the noise images, the PSNR of the algorithm in this paper, and K-SVD, GLR, STROLLR, DnCNN algorithms after image denoising. The best results for denoising each image under the same noise are marked in bold. Under different algorithms, the algorithm in this paper has obtained good PSNR values. Our method has 0.664 dB better than K-SVD, 0.508 better than DnCNN, 0.297 dB better than GLR, and 0.275 dB better than STROLLR. It can be seen that the PSNR of our algorithm is the highest, with an average improvement of 0.2–0.6 dB compared with algorithms.

BSD68 dataset added different Gaussian white noise $\sigma = 15, 20, 35, 50$ which was used to evaluate the performance of the algorithms. Table 2 shows the average PSNR of BSD68 under different algorithms with different noise levels. Under the same noise, the best results of image denoising for BSD68 dataset are marked in bold. It can be known that on the BSD68 dataset, the peak signal-to-noise ratio of our method has been improved to a certain extent, and the maximum can reach 2.905 dB.

Table 1 PSNR of six images denoised by different algorithms

Image name	Noise image	K-SVD	GLR	STROLLR	DnCNN	Ours
Barbara	22.122	30.415	30.944	30.940	30.805	31.702
Lena	22.099	32.159	32.599	32.448	32.389	32.769
House	22.114	33.392	33.177	33.295	33.157	33.441
Camera	22.091	29.393	30.023	30.174	29.948	30.292
Coins	22.090	32.233	32.877	32.889	32.803	33.027
Moon	22.092	35.830	36.004	36.008	35.258	36.178
Average	22.101	32.237	32.604	32.626	32.393	32.901

Table 2 Average PSNR of different algorithms under the BSD68 dataset

σ	Noise images	K-SVD	GLR	STROLLR	DnCNN	Ours
15	24.603	30.738	31.920	32.084	31.205	32.317
20	22.111	29.590	30.100	30.202	29.855	32.495
35	17.257	28.863	29.061	29.253	28.961	30.321
50	14.149	27.725	28.131	28.447	28.335	28.602

4 Conclusion

Based on image denoising algorithm, this paper proposes sparse transform and low-rank denoising algorithm based on edge enhancement. The algorithm uses the Canny algorithm to obtain the edge of the image, and the NSCT transform is used to obtain the edge coefficient enhancement of the high-frequency sub-band image, thereby enhancing the edge of the image. Combined with the adaptive transform sparsity of image blocks and the low rank of the data matrix formed by block matching, image denoising is performed. Experiments show that under different noises and different datasets, the algorithm not only obtains good peak signal-to-noise ratio, but also retains clear edge details and better visual effects. There are still limitations in the research in this paper. Only Gaussian white noise is denoised. In the future, the realization of denoising algorithms such as real noise and mixed noise will be further studied.

References

1. Yan, M.: Restoration of images corrupted by impulse noise and mixed Gaussian impulse noise using blind inpainting. Siam J. Imaging Sci. **6**, 1227–1245 (2013)
2. Liu, L.P., Qiao, L.L., Jiang, L.C.: Overview of image denoising methods. J. Front. Comput. Sci. Technol. **15**, 1418–1431 (2021)
3. Aharon, M., Elad, M., Bruckstein, A.: K-SVD: an algorithm for designing overcomplete dictionaries for sparse representation. IEEE Trans. Signal Process. **11**(54), 4311–4322 (2006)
4. Ravishankar, S., Bresler, Y.: Learning sparsifying transforms. IEEE Trans. Signal Process. **61**, 1072–1086 (2013)
5. Dabov, K., Foi, A., Katkovnik, V., Egiazarian, K.: Image denoising by sparse 3-D transform-domain collaborative filtering. IEEE Trans. Image Process. **16**(8), 2080–2095 (2007)
6. Wen, B., Li, Y., Bresler, Y.: Image recovery via transform learning and low-rank modeling: the power of complementary regularizers. IEEE Trans. Image Process. **29**, 5310–5323 (2020)
7. Zhang, W.W., Han, Y.S.: Nonlocal self-similarity based low-rank spase image denoising. J. Comput. Appl. **9**(38), 2696–2700 (2018)
8. Zhang, Q.K., Ji, Z.P., Fu, X.F.: Bas-relief optimization based on twice-joint sparse representation and low-rank estimation. J. Image Graph. **25**, 1245–1259 (2020)
9. Qian, C., Chong, D.X.: Image denoising algorithm based on graph laplacian regularized sparse transform learning. Comput. Eng. Appl. **58**, 232–239 (2022)
10. Liu, J., Dong, Y., Zhang, X.: Image denoising based on edge enhancement and convolutional neural network. In: Thirteenth International Conference on Digital Image Processing (ICDIP 2021), vol. 11878, pp. 312–319 (2021)

11. Hu, H., Froment, J., Liu, Q.S.: Patch-based low-rank minimization for image denoising. Comput. Sci. (2015)
12. Zhang, K., Zuo, W., Chen, Y., Meng, D., Zhang, L.: Beyond a Gaussian denoiser: residual learning of deep CNN for image denoising. IEEE Trans. Image Process. **26**(7), 3142–3155 (2017)

An Intelligent Inter-Satellite Ranging System for High-Precision Satellite Constellation Configuration Control

Yifei Jiang, Zhong Chao, Wan Bei, Shufan Wu, Wang Wenyan, and Qiankun Mo

Abstract In this article, we propose an intelligent inter-satellite ranging system for constellation configuration. After the analysis of our mission, four different relative positional states are proposed. Four different technologies are applied to achieve the measurements. Microwave ranging technologies are used to correct the relative attitude and laser ranging technologies are used to achieve the ultimate high-precision ranging. After the detailed design is given, many experiments were used to verify the specific system.

1 Introduction

With the rapid developments of space technologies, especially in the huge satellite constellations, some intelligent systems become the focus of study and engineering. In many advanced system research, inter-satellite ranging systems are above all, because the ability of ranging instruments not only decide the orbit control strategy but also impact the performance of navigation system, tele-communication system, and formation flying control system. Especially for emerging satellite constellation, an accurate inter-satellite ranging technology is the key of high-precision configuration control.

In recent years, a large number of high-precision ranging techniques have been proposed, and some of them have been applied in practice. Optical position metrology for XEUS is reported as a coarse and fine sensor, which achieves accuracy of tens of microns, with a maximum relative distance of 120 m [1, 2]. The dual-wavelength interferometer (DWI) was chosen as a fine longitudinal sensor for the DARWIN mission, with a ranging accuracy of about 10 μm within 250 m distance [3, 4]. A compact high-precision frequency swept interferometer (FSI) has been used to

Y. Jiang (✉) · S. Wu · Q. Mo
School of Aeronautics and Astronautics, Shanghai Jiao Tong University, Shanghai, China
e-mail: jamesjiang@sjtu.edu.cn

Z. Chao · W. Bei · W. Wenyan
Institute of Shanghai Aerospace Control Technology, Shanghai, China

K. Nakamatsu et al. (eds.), *Advanced Intelligent Technologies for Information and Communication*, Smart Innovation, Systems and Technologies 365,
https://doi.org/10.1007/978-981-99-5203-8_38

autonomous ranging of satellite constellation control [5]. Especially, in the detection of space gravitational wave, a laser interferometry is used to measure gravitational-wave-induced motion between separated “free” masses in LISA project [6]. As can be seen from the results of these studies, a single high-precision ranging technique is already very advanced. Unfortunately, a single high-precision ranging instrument cannot be used in actual aerospace engineering without supporting facilities [7, 8].

Although high-precision ranging equipment can achieve high precision, it cannot achieve long-distance measurements. For inter-satellite ranging, the measurement requirements of tens of kilometers to hundreds of kilometers have exceeded the capabilities of most high-precision measurement equipment. More seriously, different ranging technologies also have different limitations on the angular range of the measured object.

Many studies have shown that ranging devices with large angular range and long-distance range suffer from low accuracy. To solve this problem, many scholars and engineers have proposed iterative measurement methods. In this method, multiple devices with different measuring ranges and measuring accuracies work together and form an intelligent ranging system. The design of this system is closely related to the characteristics of the ranging mission.

In this paper, we will propose an intelligent inter-satellite ranging system consisting of multiple parts with different ranging accuracies to complete our space detection mission.

2 Mission Analysis

For satellite constellation, multiple satellites will be carried by the same rocket to save costs. These satellites have to determine relative position to each other for configuration control after leaving the rocket immediately. Since the relative attitude and distance between satellites are random at the beginning, it is difficult to achieve high-precision ranging. A feasible ranging scheme should be compatible for any each relative position relationship, and ultimately achieve the highest precision measurement capability.

The main purpose of this article is to introduce an intelligent inter-satellite ranging system applied in satellite constellation configuration control. This system should operate with attitude controllers together, which can be momentum wheels or magnetic torques. From an initial random relative position state, while measuring the inter-satellite pointing and distance, adjust the attitude between satellites and repeat the above steps to gradually improve the ranging precision. Finally, after the two satellites are accurately aligned with each other, high-precision distance measurement is achieved. After our analysis, there will be four stages in the whole measurement process which will be discussed in detail below.

3 System Design

From the initial random state to the final precise state, our measurements need to go through multiple intermediate states. As shown in Fig. 1, we can divide the measurements process into four sections and apply different technologies to achieve measurement individually.

In Sect. 1, once multiple satellites are divided from rocket, the relative attitude and distance between them are random. In order to roughly determine the relative position between any two satellites, we propose a microwave coarse measurement method. The distance ranging from 0 to 10 km and the angle ranging from 0 to 360° can be measured by the specific method. Once the initial state is measured, the momentum wheel will be run to reduce pointing errors between satellites. The measurement and actuation are repeated alternately until the measurement results show that the two satellites are perfectly aligned. However, the angle accuracy is 90° and the distance accuracy is 10 m. Attitude control using microwave coarse measurement can only obtain measurement results with 90° angle accuracy and 1 m distance accuracy which is the boundary range for Sect. 2.

In Sect. 2, the distance ranges from 0 to 10 m and the angle range from 0 to 90° can be measured by our microwave fine measurement. Similar measurement and control steps will be performed to achieve an angular accuracy of 10° and a distance accuracy of 1 m.

In Sect. 3, a laser pointing measurement is used to further calibrate pointing errors between two satellites. Those steps make angle measurement accuracy better than 1 degree which is the condition for the ultimate high-precision distance measurement.

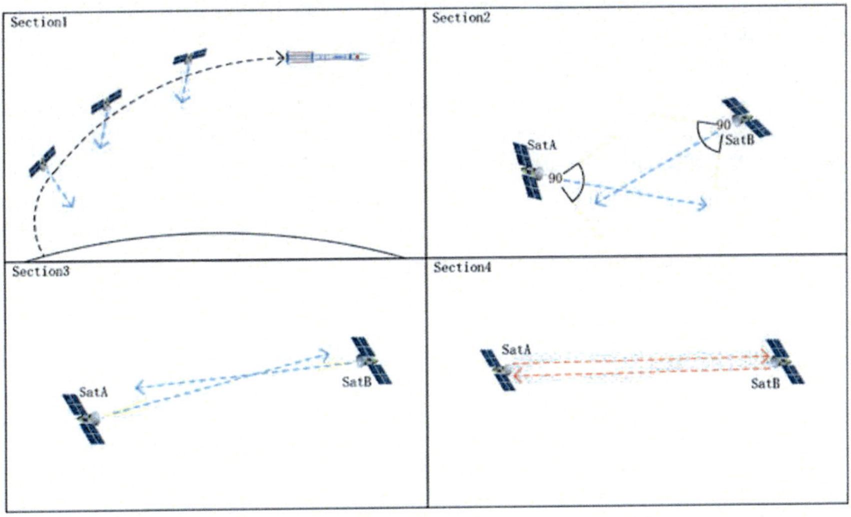

Fig. 1 Different states of measurements

Table 1 Different sections measurements accuracy

Item	Microwave coarse	Microwave fine	Laser pointing	Laser FSI
Distance range	0–10 km	0–10 m	0–1 m	0–1 m
Angle range (°)	0–360	0–90	0–10	0–1
Distance accuracy	10 m	1 m	/	100 μm
Angle accuracy	90°	10°	1°	/

In Sect. 4, a laser Frequency Swept Interferometer (FSI)-based high-precision distance measurement is proposed and a micron-level distance measurement accuracy is achieved. This technology is based on the previous measurements and controls.

As shown in Table 1, different parameters and ability of four measurement steps are compared. As the measurement progresses, the measurable angular range and distance range gradually decrease, on the contrary, the measurement accuracy gradually increases.

During the actual on-orbit operation, the system will automatically determine the current position status between two satellites, and select the corresponding measurement step. In addition to step selection, our system also needs to cooperate with the drive mechanism.

3.1 Microwave Measurement System

In the first and second steps, a microwave measurement system is designed as shown in Fig. 2. There are two different types of antennas installed on the surface of the satellites for coarse and fine measurements, respectively.

A omnidirectional quadrifilar Helix Antenna with low gain is responsible for coarse measurements. Once the exact pattern characteristics of the antenna are measured, the angular difference between the two satellites can be estimated from the magnitude of the received power. And a pulse signal is transmitted from sat1 to sat2, and forwarded back to sat1 by sat2. The distance can be measured by the time difference between sending and receiving signals.

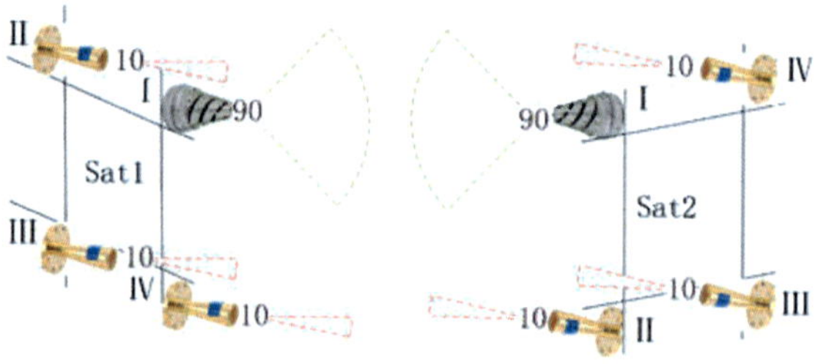

Fig. 2 Microwave measurement system

A high-gain horn antenna array is used to achieve microwave fine measurement. If one antenna in sat1 is selected to transmit and one antenna in sat2 is selected to receive at a time, there will be nine possible connections. By comparing the received signal power of nine different links, the angular difference can be reduced in 10°. And a carrier phase method can be used to achieve distance measurement through any horn antenna. Through the above two microwave technologies, 10° angle accuracy and 1 m distance accuracy can be achieved.

3.2 Laser Measurement System

The laser measurement system can be divided into two parts, namely, laser pointing measurement and laser frequency swept interferometer (FSI), which are responsible for angle and distance measurement individually.

As shown in Fig. 3, the laser pointing measurement consists of transmit laser probe in sate1 and a charge-coupled device (CCD) in sat2. A calibrated laser with a diameter of 1 mm is transmitted by sat1 and points on the CCD of sat2. According to the laser spot located on the CCD, the pointing difference between two satellites can be determined. And a 1° angular accuracy can be achieved by this method.

Within the precise pointing range, an FSI can achieve high-precision distance measurement with a 100um accuracy. The design detail of FSI will be discussed in the next chapter.

As shown in Fig. 4, a whole satellite configuration control should contain measurements and drive operations. This process can be divided into multiple individual steps. The final purpose is to maintain high-precision relative position between satellites.

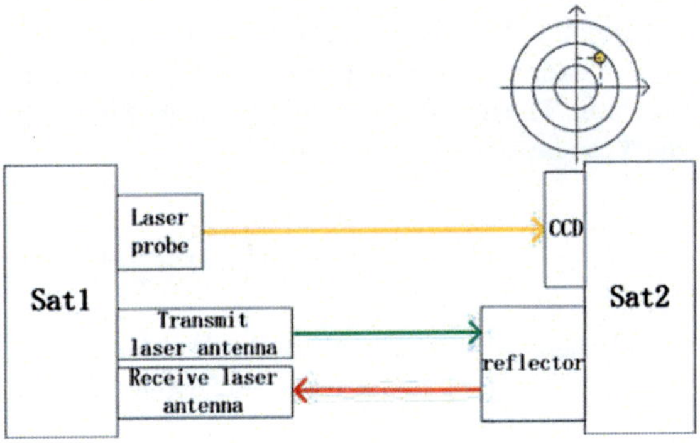

Fig. 3 Laser measurement system

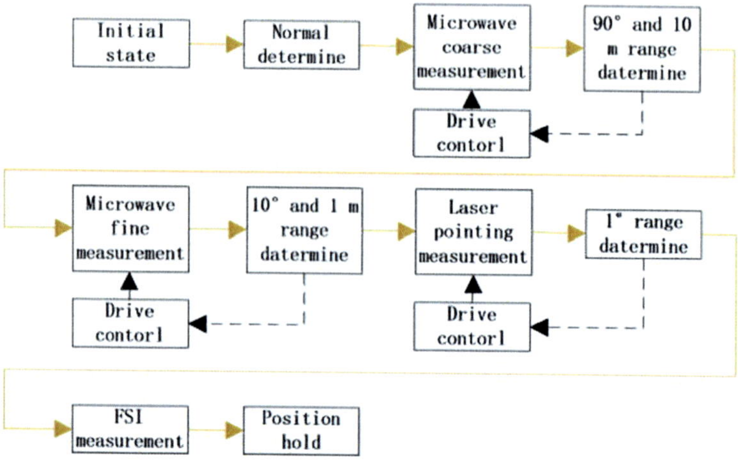

Fig. 4 System flow

4 The Design of FSI

The principles of the laser frequency swept interferometer are based on FMCW technologies. The linear frequency modulation takes the form of a triangle wave. The frequency of the laser can be expressed as follows:

$$f_L(\tau) = f_0 + \mu * t \tag{1}$$

where f_0 means the initial frequency, μ means the tuning rate, and t means the tuning time.

The tuning laser is split into two parts, one is local laser and another is test laser. The test laser reaches the device under test (DUT) and returns back to interfere with local laser. The results of interferometer are the phase information of the beat signal, which can be expressed as follows:

$$\varphi_b = 2\pi\left(\mu\tau t + f_0\tau - \frac{1}{2}\mu\tau^2\right) \tag{2}$$

In the above equation, τ stands the time delay between test path and the local path. The analytical resolution accuracy ΔR from the laser FSI, and can be expressed as

$$\Delta R = \frac{c}{2nS} \tag{3}$$

where c is light speed, n is refractive index of air, and S is tuning span.

To compensate for nonlinear errors caused by swept laser generator, a resampling method is introduced. As shown in Fig. 5, a modulated laser is divided into two

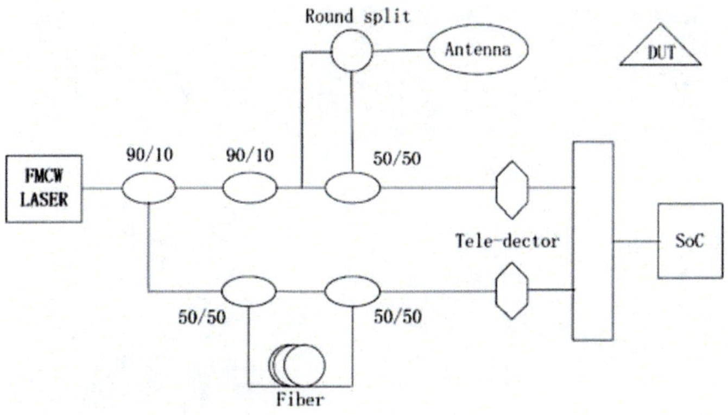

Fig. 5 The principle of FSI

Table 2 The detail of FSI design

Num	Subsystem	Items
1	Laser frontend	Telescope
2		Laser amplifier
3		Reflector
4		Splitter
5	Laser interferometer	Laser generator
6		Couple
7		Delay fiber
8		Solid-state storage

paths, namely, measure path and auxiliary path. The data detected from auxiliary path is applied as reference to resample the data detected from measure path. An FFT transform can be used to extract distance features from resampling signal.

The FSI contains two different subsystems as shown in Table 2, namely, laser frontend and laser interferometer.

The design of FSI is shown in Fig. 6, it satisfies the requirement of miniaturization.

5 Experiments

To verify the design of the specific system, a series of experiments are conducted. The verification experiment is divided into two types, one is a high-precision short-distance test in the laboratory environment and the other is a long-distance test in the field. The basic measurement principles of the two measurement methods are the same. All tests are measured by relative calibration.

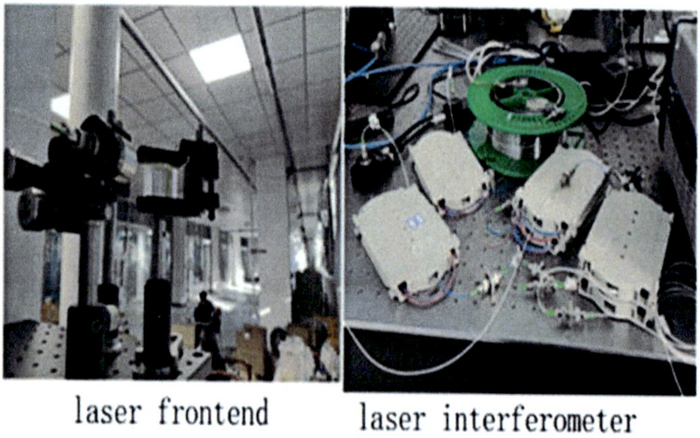

Fig.6 The design of FSI

Table 3 Test result

Num	Experiment		
	Standard distance (m)	Measure result (m)	Error (μm)
1	1	1.000053	53
2		0.999965	35
3		0.999955	45

As shown in Table 3, the close-up test results in the laboratory environment are shown in Table 3. In this state, the standard distance measured is 1 m. The maximum error is 83 μm. The relative accuracy is better than 0.13×10^{-5} and the power consumption is less than 0.5 W.

As shown in Table 4, the results show that our design system meets the accuracy requirements. The test status is a long-distance field test and the standard distance tested is 80 m. The relative accuracy is better than 0.13×10^{-5} and the power consumption is less than 5 W.

6 Conclusions

In this article, we propose an intelligent inter-satellite ranging system for constellation configuration. Four different relative positional relationships between satellites correspond to four different ranging technologies. On the basis of the large system, a laser ranging system based on FMCW principle is designed. The specific system can achieve a ranging accuracy of 100 μm by the synergy of four steps. The final precision measurement system has undergone a series of experiments to verify the effectiveness of the design. In order to fully meet the actual use requirements, two

Table 4 Experiment result

Num	Experiment		
	Standard distance (m)	Measure result (m)	Error (μm)
1	79.90	79.900083	83
2		79.900088	88
3		79.900075	75
4	80.00	80.000060	60
5		80.000103	103
6		80.000080	80
7	80.10	80.100078	78
8		80.100055	55
9		80.100100	100

measurement environments are resumed, namely, the laboratory state and the field state. The experimental results show that the modified design meets the requirements of the theoretical indicators.

Acknowledgements This work is supported by the SAST-SJTU Advanced Space Technology Joint research fund (Grant no: USCAST2019-22).

References

1. Bavdaz, M., Peacock, A.J., Parmar, A.N., et al.: XEUS mission and instruments. In: X-Ray and Gamma-Ray Instrumentation for Astronomy XII. International Society for Optics and Photonics, vol. 4497, pp. 31–40 (2002)
2. Hasinger, G.: The XEUS mission. Verhandlungen der Deutschen Physikalischen Gesellschaft **43**(2) (2008)
3. Defrère, D., Léger, A., Absil, O., et al.: Space-based infrared interferometry to study exoplanetary atmospheres. Exp. Astron. **46**(3), 543–560 (2018)
4. Shapshak, P.: Astrovirology, astrobiology, artificial intelligence: extra-solar system investigations. In: Global Virology III: virology in the 21st Century, pp. 541–573. Springer, Cham (2019)
5. Jiang, Y., Wu, S., Wang, X., et al.: A compact high-precision frequency swept interferometer (FSI) for autonomous ranging of satellite constellation. In: 2021 IEEE 6th Optoelectronics Global Conference (OGC), pp. 15–19. IEEE (2021)
6. Gong, Y., Luo, J., Wang, B.: Concepts and status of Chinese space gravitational wave detection projects. Nat. Astron. **5**(9), 881–889 (2021)
7. Li, Z., Zheng, J., Li, M.: Orbit insertion error analysis for a space-based gravitational wave observatory. Adv. Space Res. **67**(11), 3744–3754 (2021)
8. Li, Z., Zheng, J.: Orbit determination for a space-based gravitational wave observatory. Acta Astronaut. **185**, 170–178 (2021)

Author Index

K. Nakamatsu et al. (eds.), *Advanced Intelligent Technologies for Information and Communication*, Smart Innovation, Systems and Technologies 365,
https://doi.org/10.1007/978-981-99-5203-8